THE DISCOVERY OF INSULIN

Michael Bliss

THE DISCOVERY OF INSULIN

25TH ANNIVERSARY EDITION

with a new Preface

The University of Chicago Press

The University of Chicago Press, Chicago, 60637
 Published 1982
25th-anniversary edition, 2007
Printed in the United States of America
16 15 14 13 12 11 10 09 08 07 2 3 4 5

ISBN-10: 0-226-05899-9
ISBN-13: 978-0-226-05899-3

Library of Congress Cataloging-in-Publication Data
Bliss, Michael.
The discovery of insulin / Michael Bliss. — 25th anniversary ed.
p. ; cm.
Includes bibliographical references and index.
ISBN-13: 978-0-226-05899-3 (pbk. : alk. paper)
ISBN-10: 0-226-05899-9 (pbk. : alk. paper) 1. Insulin—History.
2. Diabetes—Research—History. I. Title.
[DNLM: 1. Insulin—history—Canada. 2. Diabetes Mellitus,
Type 1—drug therapy—Canada. 3. History, 20th Century—
Canada. WK 11 DC2 B649d 2007]
QP572.I5B58 2007
612.3'4—dc22

2006036298

⊗ The paper used in this publication meets the minimum
requirements of the American National Standard for
Information Sciences—Permanence of Paper for Printed
Library Materials, ANSI Z39.48-1992.

Contents

For my father, Jim, Mona,
Liz, Jamie, Laura, and Sally

PREFACE

The Discovery of Insulin was researched and written during a unique window of opportunity. The 1978 death of Charles Best, the last surviving member of the discovery team, happened to coincide with the release of the papers of Sir Frederick Banting. Suddenly it was possible to obtain access to complete documentation of the highly controversial events at the University of Toronto in 1921-23 that led to the isolation and emergence of insulin. At the same time, many individuals who had been witnesses to or participants in the discovery, and who were approaching the end of their lives, now felt free to speak frankly for the historical record. Working on this book, I not only uncovered many new collections of documents, but also found alive two of the original patients who had been treated with insulin in Toronto in 1922. Since publication, no significant new collections of documents have surfaced, but 66 of the 68 individuals I interviewed have died. No one can talk to them now, except through the notes of my interviews, which themselves are now part of the archival record.

My family has been untouched by diabetes, but deeply involved in medicine. As a professional historian I first became interested in the insulin story at the suggestion of a brother who had been exposed by one of J.B. Collip's colleagues to verbal accounts of the more dramatic incidents in the discovery saga. In 1978, fresh from publishing a biography of a major Canadian philanthropist/entrepreneur (Sir Joseph Flavelle), (and having been promoted to full professor at the University of Toronto, I decided that it should finally be possible to write the full story of the discovery of insulin.

There were serious obstacles, including much skepticism about a mere historian's qualification to write about advanced medical discoveries. Fully sharing that concern, I made a point of getting expert advice at every stage of the work, immersed myself in the history of endocrinology and diabetes, and worked and reworked my manuscript with more care and craft than I had ever applied to a project. The many people who helped me with the book have been thanked in earlier editions. Among those few who are still living, I feel particularly indebted to Richard Landon, who as head of the Thomas Fisher Rare Books Library at the University of Toronto, gave me crucial encouragement when it was most needed, and Dr. Anna Sirek, who, with her late husband Otto, gave me vital technical advice.

The archival and personal adventures generated by this project were remarkable, exciting, and life changing. They culminated in several of the most exhausting and rewarding days of my life in Cambridge, England, with the late Sir Frank Young, a grand old man of diabetes research and British science generally, as he challenged not only my conclusions, but my spelling and commas, insisting that a book that would be read around the world and for many years be as perfectly argued and polished as possible. "Bliss," he would say, "this book will be read by Fiji Islanders and Nobel laureates. You have to get it right."

The Discovery of Insulin received gratifyingly favourable reviews upon publication in 1982 and has remained in print since then and without need for significant revision or alteration. For this edition it has not been necessary to make any changes to the detailed narrative of the events of 1920-23. In addition to this new preface, it was, however, necessary to rewrite the final chapter, "A Continuing Epilogue," because so much has happened in the world of diabetes in our time.

Although I moved on to other work in the history of medicine and Canadian topics, I made a point of staying in touch with this subject, and expanded upon *The Discovery of Insulin* with a number of publications. The most important are *Banting: A Biography* (Toronto, 1984; 2nd ed., University of Toronto Press, 1992), and a scholarly article, "Rewriting Medical History: Charles Best and the Banting and Best Myth," *Journal of the History of Medicine and Allied Sciences*, 48 (July 1993): 253-74. Read singularly or together, these publications underline the foolishness of believing that insulin was discovered by Banting and Best. As I believe I make clear in *The Discovery of Insulin*, it was a collaborative process, drawing on the talents of at least four people as well as the comparatively great research capacity of the University of Toronto, where for many reasons a field of medical dreams had been built. I should have been more explicit in suggesting that J.B. Collip ought to have shared the Nobel Prize for insulin with Banting and Macleod, and in criticizing the sad attempts at historical falsification engineered by Charles Best, a troubled soul.

I also have published "Banting's, Best's, and Collip's Accounts of the Discovery of Insulin," *Bulletin of the History of Medicine*, 56 (Winter 1982-83): 554-68; and "J.J.R. Macleod and the Discovery of Insulin", *Quarterly Journal of Experimental Physiology*, 74 (1989): 87-96, along with several condensed summaries of this book. In "Growth, Progress, and the Quest for Salvation: Confessions of a Medical Historian," *Ars Medica*, I, 1 (2004): 4-14, I explain how *The Discovery of Insulin* relates to my 1991 study of smallpox, *Plague,* and to my 1999 biography, *William Osler: A Life in Medicine.* With the 2005 publication of *Harvey Cushing: A Life in Surgery,* I squared the circle, as it were, by writing about a surgeon who was both a true medical miracle worker and a great endocrinologist.

A substantial article literature, locatable through standard search engines, has developed since 1982 about the early development of insulin. Robert Tattersall's work on insulin in the United Kingdom is particularly noteworthy, as is the writing of Chris Feudtner in the United States, especially his book *Bittersweet: Diabetes, Insulin, and the Transformation of Illness* (2003). There now also exist biographies of the other members of the insulin team: *J.B. Collip and the Development of Medical Research in Canada,* by Alison Li (2003); *J.J.R. Macleod: The Co-discoverer of Insulin,* by Michael J. Williams (Royal College of Physicians of Edinburgh, Supplement to *Proceedings,* 1993); *Margaret and Charley: The Personal Story of Dr. Charles Best,* by Henry Best (2003). E.C. Noble, who lost the famous coin toss to Charles Best, finally receives attention in M. Jurdjevic and C. Tillman, "E.C. Noble in June 1921, and his account of the discovery of insulin," *Bulletin of the History of Medicine,* 78, 4 (2004): 864-875. With the development of the Internet the University of Toronto has been able to make more than 7,000 pages of the original documents available on its "Discovery and Early Development of Insulin" website, http://digital.library.utoronto.ca/insulin. An Oxford-based British team has done marvelous work creating an oral history of patients' experiences with diabetes, which may be accessed at www.diabetes-stories.com.

Publication of *The Discovery of Insulin* dismayed and offended some of Fred Banting's and Charles Best's less critical admirers. Getting this history right has practical importance in Canada, for ceremonies and plaques, and in steering historical preservation priorities. Gradually, sometimes grudgingly, most devotees of Banting's or Best's legacies have come to accept most of my conclusions. Not so the scientific chauvinists in Romania, who have continued their noisy advocacy of Nicolas Paulesco, a campaign that substitutes repetition and agitation for scholarship. To dispute my findings without addressing them, one basic trick is to cite authorities who pronounced on credit before the evidence cited in *The Discovery of Insulin* became available.

The central argument of this book has now been generally accepted. In several adaptations, the story told here has educated and, to my surprise and delight, inspired people whose lives have been touched by diabetes. *The Discovery of Insulin* has been read by diabetic teenagers, by their parents, by insulin sales representatives, by medical students, by research scientists, by historians of science, by Nobel laureates, and possibly by Fiji Islanders. It has inspired students to go into diabetes research and at least one researcher to revisit the potential of fish islet cells. In 1988 the story was made available to tens of millions of people around the world through Gordon Hinch/Gemstone Productions' beautifully done 1988 television adaptation, "Glory Enough for All."

Of the books I have written, *The Discovery of Insulin* is my favourite. I hope it will still be read long after a cure for diabetes been found and no one needs to take insulin. I look forward to rewriting this preface on the centennial of insulin's discovery in 2021-22, and to celebrating the occasion, once again, with my dear wife Elizabeth and our children and grandchildren.

Toronto, October 31, 2006 Michael Bliss

INTRODUCTION

What Happened at Toronto?

The discovery of insulin at the University of Toronto in 1921-22 was one of the most dramatic events in the history of the treatment of disease. Insulin's impact was so sensational because of the incredible effect it had on diabetic patients. Those who watched the first starved, sometimes comatose, diabetics receive insulin and return to life saw one of the genuine miracles of modern medicine. They were present at the closest approach to the resurrection of the body that our secular society can achieve, and at the discovery of what has become the elixir of life for millions of human beings around the world.

This book is an attempt to re-create the discovery of insulin as accurately and fully as can be done in a single volume. It draws on a vast body of primary source material never before available to researchers. It reflects no point of view other than a professional historian's obligation to be as objective and fair as possible. It is written to be read by anyone from a scientist to a high school student, and especially by those in between.

Many readers will begin this book believing they have a reasonably clear understanding of the discovery of insulin. It is a story told in several books, in textbook accounts, in films, tapes and television programs. In broad outline, the conventional history is something like this:

By the early years of the twentieth century it was understood that the disease named diabetes mellitus involves the body's inability to metabolize or utilize its food, especially carbohydrates. It was also understood that the pancreas holds the key to carbohydrate metabolism. When experimental animals had their pancreases removed, they immediately lost the ability to utilize carbohydrates, the amount of sugar in their blood and urine rose sharply, and they soon died from severe diabetes. Various researchers speculated that the pancreas, which secretes digestive enzymes into the gut (its external secretion), must also produce another kind of secretion, one enabling the body to utilize its fuel. The search for the internal secretion of the pancreas had occupied a number of physiologists throughout the world, but by 1920 it had not produced any practical results.

In the autumn of 1920 Frederick Banting, a young surgeon in London,

Ontario, happened to be reading an article about the pancreas. Banting began thinking about the problem of the internal secretion, and late that night jotted down an idea for an experimental procedure - ligating the pancreatic ducts - that might be a way of isolating an internal secretion. He took his idea to his alma mater, the University of Toronto, where the Professor of Physiology, J.J.R. Macleod, was an internationally known expert in carbohydrate metabolism. Macleod was at first skeptical of Banting's suggestion, but reluctantly agreed to give him a lab and some dogs for a few weeks during the coming summer. He assigned him a young science student, Charles Best, to do the chemical tests necessary for the work, and then went off to Scotland for his summer holidays.

Banting and Best experienced a number of problems with their work in that summer of 1921, the story goes, but soon found that their approach was yielding remarkable results. With the extract of pancreas they had made from duct-ligated dogs they were able time after time to lower the blood sugar and remove other symptoms from diabetic dogs. Prof Macleod came home to a pair of excited researchers who, by the autumn of 1921, were keeping a severely diabetic dog, Marjorie, alive with their extracts. Marjorie eventually lived for seventy days before being sacrificed; until then diabetic dogs had died within a week or two of their pancreas being removed.

By the winter of 1921-22 Banting and Best were giving their first papers on the internal secretion of the pancreas. They were also ready to test their extract on humans. In Toronto General Hospital a young boy, Leonard Thompson, became the first diabetic to receive insulin. His life was miraculously saved.

Professor Macleod put his whole laboratory to work on insulin. An American drug firm, Eli Lilly and Company, was brought in to help prepare it in commercial quantities. At the same time, however, the University of Toronto patented the process in order to control the quality of insulin sold to diabetics. By 1923 insulin was being produced in virtually unlimited quantities, and was the stuff of life itself for thousands of diabetics.

Late in 1923 the Nobel Prize was awarded for the discovery of insulin. It was awarded to Banting and J.J.R. Macleod. This raises what seems to be the single really controversial point about the discovery: why should Macleod have shared a Nobel Prize for work done in his lab while he was on holiday? It is fairly well known that Banting was dissatisfied with the Nobel Committee's decision. He immediately announced that he was sharing his half of the award with Best. Macleod announced that he would share his half with J.B. Collip, a biochemist who had joined the team late in 1921 and worked on the development of the extract.

There are several commonly held views about this problem of credit. Perhaps the Nobel Committee just made a mistake, possibly because

Macleod's name was on some of the early publications. Perhaps Macleod, a German-trained professor, held Teutonic-type notions about the head of a lab meriting credit for everything done in his fiefdom. Perhaps it was a case of human weakness - perhaps Macleod deliberately tried to steal credit from the inexperienced young men who had actually made the discovery. Whatever happened, the judgment of history, at least in North America, has been to remember Banting and Best as the discoverers of insulin. And, of course, it was a magnificent discovery, a medical fairy tale come true of the lone doctor and his partner overcoming all obstacles to realize an idea and save the lives of millions and millions of people. Surely this is truth stranger than fiction, or it is the truth that makes fiction plausible.

Even at first glance, however, we are left with some curiosities. For completeness' sake, it would be interesting to know exactly why Macleod got his half of the Nobel Prize. More curious, come to think of it, who was J.B. Collip? Why did he end up with the same share of the Nobel Prize money that Banting and Best each got? More generally, why was insulin discovered by two inexperienced researchers in a city and a country which had no particular stature in the world of medical research? Was existing research so poorly developed that a total outsider could confound all the experts with a brilliant, untried suggestion? Or was there somehow a large element of chance involved? Perhaps the Canadians were just lucky.

A few readers will know that some articles have already been written on the points raised by these and similar questions. Even most experts, though, will be surprised to know how early the controversy about the discovery of insulin actually began. The first attempt at serious historical assessment of the Toronto work was made almost immediately. In December of 1922 a physiological researcher in Cambridge, England, Dr. Ffrangcon Roberts, wrote a long letter to the *British Medical Journal* reviewing Banting and Best's first publications. It was a scathing critique of the Toronto investigation. "The production of insulin," Roberts concluded, "originated in a wrongly conceived, wrongly conducted, and wrongly interpreted series of experiments."[1]

Roberts was immediately rebuked for the intemperance of his letter, indeed for writing the letter at all, by Dr. Henry H. Dale, a leading figure in British research who had recently been in Toronto studying insulin at first hand. Roberts' review, Dale wrote, was "armchair criticism," the kind of destructive comment that "seldom leads to anything but verbal controversy." Whatever might or might not finally be decided about Banting and Best's experiments, nobody could deny that a first-rate discovery had been made. "It is a poor thing," Dale scolded, "to attempt belittlement of a great achievement by scornful exposure of errors in its inception."[2]

Dale's view that critical discussion of Banting and Best's work amounted to belittlement of a great achievement prevailed in medical and historical circles for the next three decades. There was no point to be served, it was believed, in discussing the issues raised by Roberts, who himself immediately lapsed into silence and was not heard from again; or, for that matter, in discussing the controversy about credit and the Nobel Prize. Most accounts of the discovery of insulin tended to slide over matters that might impair the dignity of science or the glory of the achievement. Yes, it could be revealed that Banting had been extremely bitter about Macleod and the Nobel Prize, and had spoken of the professor in the harshest terms. But why describe the nature of their quarrels in print, let alone print the vulgar terms Banting had actually used about Macleod? As for the technical but extremely important points raised by Roberts about the experiments, some of these could be mentioned in footnotes without appearing to belittle the great discoverers. The one serious biography of Banting, by Lloyd Stevenson, published in 1946, handles the issues this way. For the time Stevenson showed remarkable candour: Banting and Macleod were dead, but the other participants and their colleagues were alive and influential.

Well into the 1950s the oral history of the discovery of insulin was more interesting than the written history. There was a kind of underground of gossip, centring in Toronto medical circles and usually becoming more interesting after each round of drinks. Everybody who had been on the spot in 1921 and 1922 - professors in the university, medical students, residents and nurses in the hospital, friends of those involved - had stories to tell about what had *really* happened in those days, what the discoverers were really like, what their fights had been really about. The best stories were the ones the discoverers themselves told. Banting, who died in 1941, and Best, who lived until 1978, tended to be the most talkative. J.J.R. Macleod, who left Toronto in 1928 and died in 1935, had let slip an occasional bitter remark. J.B. Collip, who was never employed in Toronto after 1922, was a very discreet professor at McGill University and then dean of medicine at the University of Western Ontario before his death in 1965. But even Collip would sometimes get talking about the insulin days. For all of them, after all, it was the greatest event of their lives.

Outsiders and insiders alike assumed that the truth about the birth of insulin would eventually come out after all the principals had died. A few insiders knew, and a few others guessed, of the existence of important unpublished documents. A verbal reference to these documents made about 1967 by my older brother, at that time a professor of physiology at McGill, first interested me in the possibility of some day writing a book on the discovery of insulin. This would be a much better book if he had lived to help write it.

The critical silence had meanwhile been broken by an American doctor, Joseph H. Pratt, whose lifelong interest in the pancreas and diabetes went back well before the discovery. In his eighties in 1954, Pratt published in the *Journal of the History of Medicine* a thirty-five-hundred-word article entitled "A Reappraisal of Researches Leading to the Discovery of Insulin." It was actually a condensed version of a much longer article Pratt had been circulating for some years and agreed to tone down under repeated pressure not to reopen old wounds. Even so, Pratt's publication was a major critical review of the insulin work. He repeated and expanded upon Roberts' old criticisms, and made a special point of drawing attention to the contribution Macleod and Collip had made in refining both Banting and Best's flawed experiments and their crude pancreatic extract. "Credit for the discovery of a preparation of insulin that could be used in treatment," Pratt wrote, "belongs to the Toronto investigators Banting, Best, Collip, and Macleod working as a team. Each of these men made an important contribution."[3]

Pratt's attempt to rewrite the history of the discovery of insulin prompted a sharp reply from a medical historian in Toronto, Dr. W.R. Feasby, who was also an ardent admirer of C.H. Best. The burden of Feasby's 1958 article, "The Discovery of Insulin," in the *Journal of the History of Medicine,* was that the conventional history of the discovery was correct in all important particulars. "The published and unpublished records of Banting and Best's work establish the fact that convincing proof of the presence of insulin was available in the summer of 1921, when they were working alone..." Banting and Best discovered insulin, Feasby reiterated; the others helped somewhat in its development.[4]

Pratt had died. Feasby died before finishing the biography of Best on which he worked for several years. Frederick Banting's second wife, Henrietta, died before making any significant progress on the biography she planned to write of her husband. The Toronto doctor who took over her work, Ian Urquhart, also died. In the meantime medical historians in other countries were beginning to consider the discovery of insulin from the point of view of other people besides the Torontonians who had been working on pancreatic extracts.

Before his death (the death rate among those trying to write about the discovery of insulin sometimes seems higher than it is among diabetics), a Scots medical historian, Ian Murray, published several articles in the late 1960s and early 1970s on the search for insulin. His aim was to show how the Toronto work related to half a century of earlier investigation of the pancreas and diabetes. Insulin had not emerged out of a vacuum, but was the culmination of years of work by dozens of scientists in many countries. Murray was particularly interested in a Romanian scientist, Nicolas Paulesco, who in 1921, just as Banting and Best were starting to work, published very important papers describing successful experiments with pancreatic extracts. Unfortunately for Paulesco, the North Americans moved so quickly into the testing and production of insulin that he never got serious clinical tests of the material he called "pancréine" under way.

Paulesco and his work disappeared from history.

Now they were resurrected. "Banting and Best are commonly believed to have been the first to have succeeded in isolating insulin," Ian Murray wrote. "They have been hailed as its 'discoverers'. Their work, however, may more accurately be construed as confirmation of Paulesco's findings."[5]

Murray's work revived Romanian interest in a countryman who had apparently achieved so much and been so little honoured. Influenced by the impending fiftieth anniversary of the discovery, members of the Romanian School of Medicine in Bucharest launched a campaign to have Paulesco given his due. As a result of their agitation, the International Diabetes Federation decided to establish a special blue-ribbon committee to prepare a factual account of the various researches leading to the discovery of insulin. The report, published in 1971, was a careful, tightly written summary of historical knowledge about the discovery. Its conclusions, difficult to simplify because of the subtlety of the argument, were to the effect that Paulesco might indeed have discovered insulin as a therapy for diabetes had not the North Americans been able to move so swiftly and successfully to develop the results of Banting and Best's research. Pancréine probably contained insulin – so did the pancreatic extracts prepared by several earlier researchers, especially a German named Zuelzer – but it was the Canadians who made insulin suitable for the treatment of diabetes.[6]

The Romanians were not satisfied. Their continued complaints about the composition and work of that committee were secondary to their deep anger at an egregious error Banting and Best had made in their first paper, published in February 1922. In their only reference to Paulesco's work, published before theirs, Banting and Best imply quite wrongly that his results were negative. It is such an odd error, with apparently such devastating consequences for Paulesco's reputation. Was this why the Nobel people neglected him? the Romanians asked. The leading Romanians interested in Paulesco's rehabilitation decided that Banting and Best's misrepresentation of his work was too suspicious to explain away as a simple mistake. It was a deliberate distortion of Paulesco's work by Banting, wrote Dr. Constantin Bart in a 1976 article entitled "Paulesco Redivivus." Bart went on to deduce what he thought was the real truth behind Macleod getting half of the Nobel Prize: Macleod, well versed in the literature, must have found out about Banting's falsification and threatened him with public exposure unless Banting shared the credit and glory with him, Macleod. The history of the discovery of insulin seems to have included scientific blackmail and a vicious conspiracy to cheat Paulesco out of his rightful share of honour and prizes. Truth indeed stranger than fiction.[7]

Fanciful as their speculations were, the Romanians had a point in wondering why more had not been written about the events at Toronto. Their reviews of the literature on the history of the discovery alerted them to the quarrelling among the discoverers and to all the unresolved historical controversies about Banting and Best's research. With European authorities writing almost jeeringly about the "vrai panier de crabes" at Toronto in 1921-22,[8] it was surely time to find out what had really happened.

There was one more important publication in the late 1970s. J.J.R. Macleod had died in Scotland in 1935. Thirteen years later a copy of a document found among his papers reached North America. Dated September 1922, it was entitled "A History of the Researches Leading to the Discovery of Insulin," and was Macleod's personal account of the events at Toronto. From 1948 to 1978 the Macleod manuscript had had an underground circulation among a small circle of scholars. Fearful of reopening a controversy that might do no one any good, the president of the University of Toronto in the mid-1950s had quite improperly used his influence to prevent its publication.[9] Lloyd Stevenson, who had written Banting's biography many years earlier, finally published the Macleod document in the *Bulletin of the History of Medicine* in 1978.

As the research for my book developed, Macleod's account turned out to be only one of many new documents shedding light on the discovery. It was clear from a careful reading of Macleod that Banting and Best had prepared similar accounts at the same time in 1922. Manuscripts of these were found. In the Banting Papers was a second long account that Banting had written in 1940. So was the correspondence Banting and Best had had with Macleod in the summer of 1921. So were the original index cards on which Banting and Best had recorded the notes taken from their reading, including their note on Paulesco's prior article. Banting's original notebook, in which he recorded his first idea and the first series of experiments, was discovered. So were many other documents. Some of them were coming to light in the natural course of events, as when the University of Toronto made the Banting Papers available for scholarly research, and when the Nobel Committee of the Caroline Institute in Stockholm agreed to open its archives to qualified researchers. Others emerged because of this project. At the outset I decided to make a more determined and careful search for documents than anyone had previously undertaken, and that search was rewarding.

My aim was to carry out the historian's job of re-creating the discovery of insulin. As far as possible I wanted to work from contemporary sources. I wanted to ignore the judgments of later writers and put aside the partisan recollections of the discoverers themselves, at least until I had found out from the documents generated at the time - laboratory notebooks, correspondence, published articles, etc. - exactly what had happened. I wanted to reconstruct the insulin research dog by dog, day by day, experiment by experiment. After that it would be proper to reflect on the fallibility of the participants' memories and the validity of the scientists' claims and counter-claims.

More documents were found than I had expected. They include the voluminous records of the University of Toronto's Insulin Committee, complete with the droppings deposited by experimental mice; every scrap of paper relating to the introduction of insulin into Britain, preserved by the Medical Research Council in London; and the priceless letters a diabetic child wrote to her mother from Toronto in 1922, which were handed to me by their author. I have been able to reconstruct the events in far more detail than I first thought possible. Even so, it will be seen that aspects of the discovery of insulin are still unclear. In some places gaps in the record have to be acknowledged. Some of them are small or insignificant enough to permit a careful, explicit speculative leap. More often than a purist would wish, I have to fall back on memories, on the written accounts generated and distorted by the sensation of the discovery, and on the sixty-year-old recollections of the many people who talked to me out of their concern that the truth be recorded, but who well understood that the truth might not be exactly as they remembered it. These interviews, with everyone I could find who had something to say and was willing to say it (there were more of these people alive than I or anyone else had imagined), were invaluable in conveying colour, anecdotes, and the appreciation of personality absolutely crucial to understanding the men and the events.

Historians who want their books to be widely read often aim at producing seamless narratives. They weave the material spun from their sources so skilfully that readers are not distracted or bothered by signs of the documentary origins. For this book, however, I decided not to go out of my way to cover up the references to the sources. At times the controversial nature of both the subject and the sources absolutely requires that they be noticed and quoted from at length. At other times the documents are so eloquent that summarizing and paraphrasing destroyed their impact. Changing metaphors, the job is always to build as carefully and finally as one can, but this is the kind of architectural situation where it is best not to try to hide the pipes or disguise the cementwork.

The book is aimed at any intelligent reader. So all readers will have to put up with minor inconvenience. Laymen will have to learn a few not very difficult medical terms. Medical men must understand that I cannot use their shorthand and am sometimes forced to simplify or ignore some very complex issues – without ever, I hope, doing so to the point of distortion. The extensive documentation in the end notes and bibliography is designed to satisfy scholars without intruding upon the normal reader.

My aim has been to write a readable and definitive history of the discovery of insulin. I have tried to make the book definitive in two senses: first, that readers who disagree with my conclusions will agree that the book contains a full and fair presentation of all the evidence; second, that the research has been thorough enough to guarantee that any new documents found after publication will not significantly change the account.

The history of the discovery of insulin, as it emerges in the following pages, is a much more intricate, complex event than our conventional accounts have suggested. It is also a richly dramatic event, and I have tried to present every facet of that drama - the tension, interludes, crises, climaxes, ironies, and occasional absurdities - exactly as it happened. In offering this history, I reject the view that the truth will lead to a belittlement of the discovery of insulin or of the discoverers. This is a book about life, disease, death, salvation, and immortality. It is a wonderful thing to be a witness to the struggles of men, weighed down with all the burdens manhood bears, to find a way of enlarging the possibilities of our human condition.

CHAPTER ONE

A Long Prelude

A person becomes diabetic when the body starts losing the ability to utilize its fuel. The food taken in is no longer fully transformed, or metabolized, into energy. Instead, nutrients begin passing through the system; hence the origin of "diabetes," from the Greek word meaning siphon or pipe-like.

A seventeenth-century English surgeon called diabetes "the pissing evile." The frequent and voluminous urination by severe diabetics (as much as ten to fifteen quarts a day), accompanied by their unquenchable thirst, had caused the disease to be recognized thousands of years before by the Egyptians and Greeks. In the first century A.D., Arataeus described the disease as "a melting down of the flesh and limbs into urine." When early physicians found that the urine of diabetics was sweet to the taste, they realized it was heavy with sugar. Gradually the Latin word for honey/sweet, "mellitus," was added to distinguish the disease from diabetes insipidus, a pituitary disorder in which a large volume of sugar-free urine is passed (containing no sugar, the urine is insipid to the taste). Diabetes mellitus was also called the sugar disease or the sugar sickness. That was a fair description, for the most obvious problem in diabetes mellitus is the body's failure to be able to burn much of the simple glucose made from its food, especially from carbohydrates. Instead of being absorbed into the cells, the glucose remains in the bloodstream. The kidneys normally remove sugar from the body's waste water, but in a diabetic's system the sugar overload is too great. Glucose spills into the urine; the quantities of urine greatly increase; and as the body loses liquids a terrible thirst develops as the system craves renewal. Its craving for sugar leads to a terrible hunger, especially for carbohydrates.

Frequent urination (polyuria), constant thirst (polydipsia), and excessive hunger (polyphagia) are the classic symptoms of diabetes. They are often accompanied by fatigue or weakness, and then rapid weight loss as the body begins to fail from lack of nourishment.

Diabetes seems to be brought on by a variety of factors. It is most commonly found in people over forty whose metabolic system has had to work hard during their lives to cope with over-nourishment leading to obesity. That kind of "maturity-onset" (or type 2) diabetes often develops gradually over many years, and the early symptoms are hardly noticeable. In younger people, however, the system's failure is more commonly sudden and serious. "Juvenile-onset" (type 1) diabetes may in some instances have a viral cause. In both types, however, there is often an hereditary predisposition to diabetes, an inherited genetic or systemic weakness that worsens, either gradually or suddenly, under the influence of other factors.

Tasting the urine was doctors' original test for diabetes. Early in the nineteenth century chemical tests were developed to indicate and measure the presence of sugar in the urine, that is, the condition of glycosuria. A patient showing glycosuria was generally deemed to be diabetic (other disorders that could cause sugar in the urine were far less common than diabetes and were usually ignored), so diabetes was sometimes defined as a condition in which glycosuria exists.

Perhaps the continual thirst and the constant pissing would develop gradually, as it often did in adults. Perhaps a ten-year-old boy would suddenly want quarts and quarts of milk or water, or be eating to extremes ridiculous even in a ten-year-old. A severe illness might set off the symptoms, which could also include a constant itching in the genital areas, erratic skin sensations, sometimes blurred vision. The symptoms would mount until you visited your doctor. He tested the urine, found sugar, and pronounced you diabetic.* By the early twentieth century, urine tests were often being made routinely on hospital patients and as part of life insurance examinations; these disclosed a substantial number of fairly mild diabetics.

But there was no agreement on the exact definition of diabetes. Diagnostic methods were uncertain and changing. So were statistical methods. This all meant that it was impossible to know how many diabetics there were in any given country in, say, the year 1920. There tended to be more diabetics among peoples who were prosperous and well-nourished rather than among the poor and lean. In the early twentieth century the disease was particularly noticed among wealthy Jewish people, and seems to have been most visible in the richest countries, notably the United States and Germany. As nations became richer and peoples became better nourished, and as vaccines, anti-toxins, and sanitary measures began to reduce the death rate from infectious diseases, the prevalence of diabetes was increasing. By 1920 between 0.5 and 2.0 per cent of the population of industrialized countries had diabetes.

*Sometimes, if you were a careless male, the doctor could make a Holmesian diagnosis the moment you walked into his office. The dry white sugar spots on your shoes or pants gave it away.

I

It was easier to diagnose diabetes than it was to treat the disease. Without treatment the "progress" of diabetes was downwards. The effects of the disease were far more wide-ranging than weight loss and a general weakening of the system. The blood vessels of the eyes and lower extremities of an untreated diabetic are particularly liable to be damaged. Longstanding diabetics often suffered from cataracts, blindness, and severe foot and leg infections which were often accompanied by gangrene. They had lowered resistance to disease of all kinds, and were as likely to be destroyed by tuberculosis or pneumonia as by the deterioration caused by diabetes itself. Boils and carbuncles plagued diabetics, often fatally. Doctors often let gangrene and other operable conditions take their course because few diabetics survived the complications and trauma arising from surgery. All wounds healed badly. Severely diabetic people were often impotent or sterile; those women who could conceive were seldom able to carry the foetus to full term.

The infections and the other complications were often the cause of death in older diabetics whose condition developed slowly. In the young, and in the severely diabetic older patients, the diabetes itself destroyed the body, often very quickly. The life expectancy of juvenile diabetics was less than a year from diagnosis. The wasting away of the flesh from lack of nourishment could be dreadful in itself: "When he came to the hospital he was emaciated, weak and dejected; his thirst was unquenchable; and his skin dry, hard and harsh to the touch, like rough parchment."[1] But the breakdown was more general, for the body was unable to metabolize its fats and proteins properly either. As it struggled to assimilate fats in place of carbohydrates, the system became clogged with partially burned fatty acids, known as ketone bodies. When the doctors found an abundance of ketones in the urine (ketonuria), they knew the diabetes was entering its final stages. They could smell it, too, for some ketone bodies were also volatile and were breathed out. It was a sickish-sweet smell, like rotten apples, that sometimes pervaded whole rooms or hospital wards.

The diabetic suffering from acid-intoxication or acidosis (often used synonymously with ketosis) was losing the battle. Food and drink no longer mattered, often could not be taken. A restless drowsiness shaded into semi-consciousness. As the lungs heaved desperately to expel carbonic acid (as carbon dioxide), the dying diabetic took huge gasps of air to try to increase his capacity. "Air-hunger" the doctors called it, and the whole process was sometimes described as "internal suffocation." The gasping and sighing and sweet smell lingered on as the unconsciousness became a deep diabetic coma. At that point the family could make its arrangements with the undertaker, for within a few hours death would end the suffering.

II

Turn-of-the-century doctors tried to neutralize the fatty acids by giving comatose diabetics alkali solutions, most commonly sodium bicarbonate. The procedure was seldom effective in the early stages of diabetic coma, never effective in deep coma. If diabetes was to be treated at all, it had to be in the early stages. Perhaps something could be done about the sugar problem.

Like almost all other patients, diabetics before the mid-1800s were done more harm than good by doctors' bleeding and blistering and doping. The last vestige of these futile practices was the use of opium to treat diabetes; it was still being mentioned by William Osler in 1915, and in 1919 the leading American diabetologist, Frederick Allen, complained that the opium habit in diabetic treatment "is very difficult to break even at the present time."[2] Opium dulled the despair.

Another treatment lasting into the twentieth century was based on the notion that a diabetic needed extra nourishment to compensate for the nutritive material flowing out in his urine. Therefore the patient should eat as much as possible. A French doctor in the late 1850s, Piorry, refined the idea and advised diabetics to eat extra large quantities of sugar. A physiologist who became an advocate of his views had the misfortune to become diabetic himself, practised what he preached, and died very quickly. In the early 1900s there were still ignorant diabetics and ignorant doctors for whom diabetes therapy involved increasing the sugar consumption. Even sophisticated doctors were constantly tempted to try to help diabetics gain weight. Allen believed it was still vital to combat "the modern fallacy of replacing through the diet the calories lost in the urine."[3]

The first important advance came when doctors gradually came to espouse the reverse of the extra-feeding idea. If the system could not handle all its food, perhaps it should not be given so much food to try to handle. Perhaps the extra food diabetics took in because of the body's lust for nourishment actually increased the strain on the system, making things worse. Carbohydrates seemed particularly villainous. If the diabetic's body could not metabolize them, perhaps he should be given a diet low in carbohydrates.

Another French doctor, Bouchardat, more than made up for Piorry's disaster by beginning to work out individual diets for his diabetic patients. Already experimenting with the use of periodic fast days, on which no food would be taken, Bouchardat observed the actual disappearance of glycosuria in some of his patients during the rationing while Paris was besieged by the Germans in 1870. He also noticed that exercise seemed to increase a diabetic's tolerance for carbohydrates. "You shall earn your bread by the sweat of your brow," Bouchardat remarked to a patient pleading for more of what was then everyone's staple.[4]

The unwillingness of diabetics to follow diets was and still is the single most difficult problem physicians had to face as they tried to treat the disease. The important late nineteenth century Italian specialist, Cantoni, isolated his patients under lock and key. A disciple of his system, the German physician Bernard Naunyn, would lock patients in their rooms for up to five months when necessary to obtain "sugar-freedom."[5] Because diabetes was then thought to involve only a failure of carbohydrate metabolism, the diets contained a minimum of carbohydrates and a very high proportion of fat, sometimes extremely high if a doctor believed he should replace lost calories and build up a diabetic's weight and strength.

Any low carbohydrate diet, even if fats more than compensated for the calories lost, was unappetizing over a long period of time. So it seemed a great breakthrough in 1902 when the German, von Noorden, announced his "oat-cure" for diabetes. Suddenly a diabetic could increase his carbohydrate rations so long as they were in the form of foods made from oatmeal. An enormous research effort was begun by nutritionists to find out what it was that made oatmeal more assimilable than other carbohydrates (bananas, the von Noordenites found, seemed to be the next best). Actually, the oat-cure was only the most popular of a long line of carbohydrate "cures" offered from time to time - the milk diet, the rice cure, potato therapy, and others.[6] There may be a direct link between these early fads in diet therapy for diabetes and popular fad diets of the late twentieth century.

Low-carbohydrate diets did often reduce or eliminate glycosuria (leading almost as often to the conclusion that the diabetes was cured, followed by a resumption of normal diet, followed by more glycosuria). Milder diabetics, usually older ones, who kept to a diet reasonably well were sometimes able to live with their disease for years without too much discomfort. Severe diabetics, especially children, seemed seldom helped by high-calorie, low-carbohydrate diets. They deteriorated almost as quickly as before, and in fact it was later argued that the high fat content of the diets speeded the development of acidosis leading to coma. Like cancer, diabetes was not a satisfying disease to treat. (It could be financially rewarding to treat, of course, particularly if a doctor specialized in mild cases and thereby claimed a high success rate as measured by the long lives of his patients; it also helped if all patient deaths from infections, tuberculosis, or other complications were not counted as deaths from diabetes.) A British doctor made a famous flippant remark about a French diabetologist: "What sin has Pavy committed, or his fathers before him, that he should be condemned to spend his life seeking for the cure of an incurable disease?"[7]

III

The quip was actually a tribute to the dedication of medical scientists. Their basic strategy in the search for a cure for diabetes involved first finding the cause of the disease. The common-sense assumption that the problem was in the stomach gradually faded as physiologists came to understand the role of other organs in metabolism. Claude Bernard, for example, showed that it is the liver, transforming material assimilated in digestion, that dumps sugar into the bloodstream. So perhaps diabetes was a liver disease. Except that from the middle of the nineteenth century there was a gradually accumulating body of evidence from autopsies on diabetics that the disease was sometimes accompanied by damage to a patient's pancreas - and, more important, that patients with extensively damaged pancreases almost always had diabetes.[8]

The pancreas is a jelly-like gland, attached to the back of the abdomen behind and below the stomach. It is long and narrow and thin, irregular in size, but in humans usually measuring about 20 x 6 x 1 centimetres and weighing about 95 grams. To the layman the pancreas appears to be a not very interesting cluster of blobs of fleshy material. Animal pancreases, along with thymus glands and sometimes testes, have long been considered delicacies; their gourmet name, sweetbreads, appears to have nothing directly to do with sugar or diabetes.

The main function of the pancreas appeared to be to produce digestive enzymes. These are secreted through the pancreatic ducts into the duodenum (or small intestine), where they become the important constituents of the juices working to break down foodstuffs passing down the alimentary canal. Surely a straightforward enough job for an organ.

Close studies of the pancreas under the microscope revealed a situation not quite so straightforward. In 1869 a German medical student, Paul Langerhans, announced in his dissertation that the pancreas contains not one, but two systems of cells. There are the acini, or clusters of cells, which secrete the normal pancreatic juice. But scattered through the organ and penetrating the acini in such a way that they often seem to be floating in a sea of acinar cells, Langerhans found other cells, apparently unconnected to the acini. He declared himself completely ignorant of their function. Several years later the French expert, Laguesse, named these mysterious cells the islands or islets of Langerhans (*îles de Langerhans*). He suggested that if the pancreas has some other function in the system besides secreting digestive juice, the islet cells are probably involved.

Evidence connecting the pancreas and diabetes was still tenuous in 1889 when an astonishing discovery was made in the medical clinic of the University of Strasbourg. Oskar Minkowski and Joseph von Mering had disagreed on whether or not the pancreatic enzymes were vital to the digestion of fat in the gut. To settle the issue they decided to try the very difficult experiment of removing the pancreas from a dog, and then observing the result. What would happen to digestion without pancreatic juice?

In an account written many years later,[9] Minkowski described how he had kept the depancreatized dog tied up in his lab while waiting for von Mering to return from a trip. Even though the animal was housebroken and regularly taken out, it kept urinating on the laboratory floor. Minkowski had been taught by his supervisor, Naunyn, to test for the presence of sugar in urine whenever he noticed polyuria. His tests revealed 12 per cent sugar in the dog's urine, the realization that it was suffering from something indistinguishable from diabetes mellitus, and the hypothesis, subsequently demonstrated in case after case, that without its pancreas a dog becomes severely diabetic. Somehow the absence of the pancreas *caused* diabetes. This was a great experimental breakthrough, due not just to good luck and close observation, but also to the skill of researchers who apparently were performing some of the first successful total pancreatectomies. (Much of the fair amount of skepticism with which their finding was greeted related to doubts that they had actually excised the whole pancreas, for parts of it could be easily missed.)*

The next problem was to discover how the pancreas regulated sugar metabolism. Was it the absence of pancreatic juice, for example, that brought on the diabetes in a depancreatized dog? Apparently not, for Minkowski confirmed the observations of other experimenters who had ligated and/or cut the ducts leading from the pancreas to the duodenum. Stopping the flow of pancreatic juice in this way caused minor digestive problems, but it did not cause diabetes. Only total pancreatectomy did. When critics pointed out that duct ligation often failed to work, for tied ducts were by-passed and new ducts often formed to replace cut ones, the French researcher, Hédon, in 1893 devised a compelling proof. In the first stage of his operation he would take out almost all of the pancreas, completely and irrevocably cutting off the supply of pancreatic juice. He would leave only a small remnant of pancreas, still nourished by its blood supply, which he pulled out through the wound and grafted under the dog's skin. Although the dog had lost most of its pancreas, and was getting no pancreatic juice at all, it did not become diabetic. But when Hédon completed the pancreatectomy by cutting off the remnant of the graft (without having to open the abdomen again), diabetes immediately deve-

*The more common story of this discovery is that Minkowski's attention was directed to the dog's urine because a lot of flies were being attracted to it because of the sugar. Minkowski explicitly denied this version.

loped. Minkowski and von Mering did similar experiments.

It was hard to dispute the conclusion that the pancreas must have two functions. The digestive juices, poured into another organ, were the pancreas's *external* secretion. Its other function must be to produce some other substance, an *internal* secretion, which fed directly into the bloodstream and regulated carbohydrate metabolism. In 1901 an American at Johns Hopkins University in Baltimore, Eugene Opie, supplied a missing link in the argument by showing a pathological connection between diabetes and damage to the mysterious cells Langerhans had discovered. From then on it was widely believed that the islets of Langerhans produced an internal secretion of the pancreas. That hypothetical internal secretion was the key. If it could ever be discovered, actually isolated, it would unlock the mystery of diabetes.

The new ideas about the pancreas fitted with exciting new concepts and empirical findings about organs and their secretions. There were several ductless glands - such as the suprarenals, thymus, thyroid, ovaries, and pituitary - whose chief function appeared to be to produce powerful internal or endocrine secretions. In the 1890s a great deal of excitement was generated by the discovery that several diseases - endemic goitre, cretinism, and myxoedema - could be succesfully treated by feeding patients extracts of thyroid. Evidently the gland produced a secretion whose deficiency could be supplied artificially. The discovery of a secretion from the suprarenal, or adrenal medulla, named adrenalin, was another exciting milestone at the turn of the century. Adrenalin was a bit of a disappointment in that it could not keep animals who had lost their suprarenals alive, but it was obviously a powerful secretion of the greatest physiological importance. There were more to come: after Bayliss and Starling discovered secretin (a secretion from the duodenal epithelium that triggers the flow of pancreatic juice) in 1902, Starling coined the term "hormone" to describe these chemical messengers. The body's endocrine system seemed to be as important as or more important than the nervous system in regulating its vital functions.

How many more hormones were there? How did they work? Within a few years thousands of articles were being published on research in this new field of endocrinology. It was a young field in terms of solid achievements, and a highly speculative one (leading to wild quackery) when people thought about the ultimate discoveries that might be made regarding the secretions of the sex organs. Back in the 1880s one of the eccentric pioneers of endocrinology, Brown-Séquard, had received much attention with his announcement that extracts of tissue of the testicle were the secret of his own rejuvenation. If nothing else this somewhat premature revelation helped spread the idea that these hormones, the "vital juices" of popular lore, could be very potent. In the less exotic field of diabetes research, it certainly seemed that both theory and experimental observation pointed towards a potent hormone being produced in the pancreas to regulate metabolism.[10]

IV

As soon as it was realized that the pancreas controls diabetes, attempts began to treat the disease, literally, with pancreas – just as diseases of the thyroid were being treated with thyroid. Minkowski was the first of many researchers to try to restore the pancreatic function to diabetic animals (others experimented on human diabetics) by preparing and administering extracts of pancreas. The extracts could be made in a variety of ways; they could also be administered in a variety of ways, although the most obvious were orally and by injection. The important observation would be of sugar in the urine. If an extract reduced glycosuria it might be potent. It might contain the internal secretion; indeed, it might supply the proof that there actually *was* an internal secretion, for until its effect could be practically demonstrated, the internal secretion of the pancreas was merely a good-looking hypothesis.

The results of the early experiments with pancreatic extracts were mixed, tending towards the negative. Some extracts had no effect; some had decidedly harmful effects, throwing the animals into shock or worse. Others had temporary sugar-reducing effects that were more than cancelled out by harmful side-effects – so much so that it was impossible to tell whether it was the extract or its toxic effect on the system that was the true cause of the reduction in glycosuria. If an extract caused kidney failure, for example, it might be changing the contents and quantity of the urine without affecting the diabetic condition at all. A few researchers did report encouraging results with extracts, but others who tried to repeat their work got discouraging results. It will never be known precisely how many researchers tried giving pancreatic extracts to diabetic animals and humans. Estimates run to more than four hundred. It was an easy experiment for even a country doctor to try, but if the results were not encouraging many would decide there was no point publishing. As it was, there was no shortage of publication, on every conceivable aspect of the problem of diabetes and the pancreas, it seemed. In 1910 Opie complained that the literature on diabetes was voluminous. A few years earlier Lydia Dewitt estimated that more thought and investigation was going into the islets of Langerhans than any other organ or tissue of the body.[11]

Despite the discouragement, the search for a workable pancreatic extract continued. Perhaps the problem with extracts was that somehow the pancreas's external secretion, or the tissues producing it, destroyed the internal secretion in the extirpated organ. Laguesse suggested using extracts made from foetal pancreases, because it seemed that the islet cells develop well before the acinar cells in gestation. If the experiment was tried, it failed. So did a number of other experiments involving fish. In certain species of fish the islet tissue had been found to be anatomically distinct from the acinar tissue, making it possible, it seemed, to get an extract which was more purely an extract of the islets of Langerhans. Between 1902 and 1904 two Scots researchers in Aberdeen, John Rennie and Thomas Fraser, fed an extract of boiled fish islets to four diabetic patients. After inconclusive results, including a toxic reaction when they tried to inject the extract into a fifth patient, they gave up.[12]

The most persistent and important of the early extractors was Georg Ludwig Zuelzer, a young internist in Berlin who in the early 1900s became interested in the theory that diabetes was actually caused by adrenalin. Experimental evidence that large doses of adrenalin could produce glycosuria convinced Zuelzer that the function of the internal secretion of the pancreas was simply to neutralize adrenalin in the system. He decided to try to prove this by injecting an extract of pancreas into rabbits along with adrenalin. When no glycosuria developed, Zuelzer was encouraged to go on and see if his extract could reduce diabetic symptoms in depancreatized dogs. When it appeared to reduce the sugar excreted in the urine of two diabetic dogs, Zuelzer was encouraged to go further.

Dying diabetics were hopeless cases, so it must have seemed that nothing could be lost in experimenting on them. On June 21, 1906, Zuelzer injected eight cubic centimetres of his pancreatic extract under the skin of a comatose fifty-year-old diabetic in a private clinic in Berlin. The next day he injected another ten cc. Whatever effect the extract was having on the patient's glycosuria could not be measured, for the man had lost control of his bladder and was wetting his bed. What was clear was that the patient seemed to be coming back from the edge of the grave. His overall condition improved, his appetite returned, and his severe dizziness disappeared.

But there was no more extract. The patient sank into deep coma on June 30 and died on July 2. What Zuelzer had seen was tremendously encouraging, a diabetic momentarily pulled out of coma. "Whoever has seen how a patient lying in agony soon recovers from certain death and is restored to actual health will never forget it," he wrote years later just after insulin had been discovered in Toronto. He was almost certainly referring to his first experience with his own pancreatic extract, which he named "acomatol."[13]

Zuelzer had immense practical difficulties carrying out his experiments. It was hard to get a supply of pancreases, for example. Workers at local slaughterhouses thought the doctor who wanted them to give him fresh sweetbreads for medical research must be a little crazy. The extract was not at all easy to make, and had a frustrating tendency to lose its potency (Zuelzer tested his batches on rabbits, measuring the potency by the amount of extract needed to neutralize the sugar-creating effects of a unit of adrenalin). But there were those early results, and it was obvious that a workable pancreatic extract would be a wonderful thing. When Zuelzer approached the Schering drug company with his idea they offered him financial support and technical help and applied for patents on his methods. By the summer of 1907 he was ready to try again on humans.

The extract produced the amazing effect of completely suppressing for a few days glycosuria and acidosis in a twenty-seven-year-old man. Other diabetics – a six-year-old, a thirty-five-year-old, and two sixty-five-year-olds – had their symptoms dramatically relieved by acomatol. (Some others, it appears, did not; Zuelzer reported only the most interesting cases.) On the other hand, in every case after the first two there were serious reactions to the injections: vomiting, high fevers, sometimes convulsions. Knowing that his preparation was not yet a practical therapy, Zuelzer was still confident enough to publish his results in 1908. He came to the triumphant conclusion "that it is possible through the injection of a pancreatic extract to eliminate the excretion of sugar, acetone, and acetoacetic acid by a diabetic without making any changes in the patient's diet."[14]

These exciting findings caught the attention of a worker in the clinic directed by Minkowski in Breslau. J. Forschbach obtained samples of Zuelzer's extract and tested it on three dogs and three humans. His verdict was negative. Yes, Zuelzer's was the first pancreatic extract to suppress glycosuria in both the short and the long run. But it did so at the cost of severe toxic side-effects, especially fever, so severe that Forschbach stopped his human experiments for fear of doing permanent damage to achieve only temporary relief. "It will be difficult to convince a patient who has been made severely ill by a single injection," he wrote, "that this result was connected to a significant beneficial effect upon his diabetes." Forschbach was fairly convinced, especially after some impotent extract caused no ill effects in one case, that the cause of the potency and the cause of the side effects were the same. So there was no future in it. Forschbach's 1909 paper on his tests of Zuelzer's extract was decidedly discouraging, and must have been more so because of Forschbach's association with the great Minkowski himself. The giants in the field had passed judgment.[15]

At about the same time the Schering company decided that the results did not justify the cost of the work and withdrew their support. Zuelzer's application for a grant of 500 marks (about $125 at that time) to spend six weeks at a zoological station seeing if he could make an extract from those interesting fish pancreases was rejected by the University of Berlin. Zuelzer was evidently neither wealthy nor well-connected, an outsider in Berlin medical circles now left on his own with his erratic extract. He published nothing more about it.

In fact he carried on, a big, shambling doctor hawking his idea and his method from one drug company to another, bribing slaughterhouse workers to give him pancreases. In 1911 the big Hoffman–La Roche chemical firm put him back in business, funding a small lab and some co-workers. The next year Zuelzer took out an American patent on his "Pancreas Preparation Suitable for the Treatment of Diabetes." The patent was wishful thinking, though, for there were still problems with the extract. When the first big batch was made in the new lab, from 100 kilograms of pancreas, the animals on which it was tested went into severe convulsions. Zuelzer had never seen anything like this before. He decided it was the old story of toxic side-effects, perhaps caused this time by the use of copper containers, and threw out the batch. After more problems with ineffective extracts from horse pancreas, Zuelzer was ready for another round of experiments with what looked like promising material in the summer of 1914. When the war began, the hospital he was working in was turned over to the military. Georg Zuelzer was called to the front.[16]

The main effect of Zuelzer's work was probably to set back the search for an effective pancreatic extract. His published findings, plus Forschbach's report, seem to have convinced researchers of the impossibility of the enterprise: even if you did get an extract with anti-diabetic effects, whatever good effects it might have would be more than cancelled out by its bad effects. Experienced scientists had learned to be cautious, and it became something of a mark of professional prudence to qualify any findings about pancreatic extracts.*

A classic example of this learned cautiousness was the treatment of a student's work at the University of Chicago in 1911-12. The student, E.L. Scott, who had been deeply affected by the death of a friend from diabetes, took up the search for the internal secretion as the research project for his master's degree. He reasoned that previous failures had been caused by the external secretion, the powerful proteolytic (protein-destroying) enzymes, destroying the internal secretion. Perhaps the answer lay in getting rid of all traces of these enzymes. One way might be to ligate the ducts of the pancreas; this apparently would cause the tissues producing the external secretions to atrophy; from the remaining tissue an extract could be prepared and then tested. (Lydia Dewitt had already tried this method in 1906, but had tested her extract only on test-tube solutions, not living animals.) Scott abandoned the idea as impractical when, working under primitive conditions in a very hot summer, he found that it was almost impossible to get a pancreas to atrophy after ligation. Instead he turned to alcohol, a fairly common solvent and one that Zuelzer had also used in the preparation of his extract, to do the same job. Using extracts which had gone through various stages of development through mixing the pancreas with alcohol, filtering it, treating the residue, and other chemical procedures, Scott found one formula that gave encouraging results on three of

*I return to Zuelzer's methods and his problems on p. 177 below.

the four diabetic dogs he treated with it. Not only did their sugar excretion diminish, but "if one dared to say it," Scott wrote, the dogs "seemed even brighter for a time after the injection than before it."

Like Zuelzer before him and others afterwards who observed the subjective signs of improvement in diabetic animals and patients, Scott was convinced that he had been successful. The first two conclusions of his master's degree thesis were:

> 1st. There is an internal secretion from the pancreas controlling the sugar metabolism.
> 2nd. By proper methods this secretion may be extracted and still retain its activity.

Scott's thesis adviser, the noted physiologist Anton Carlson, did not share his student's confidence. Having just read of recent work by Hédon questioning the effectiveness of pancreatic extracts, Carlson worried that Scott had not sufficiently controlled his experiments. He urged that the conclusions be rewritten, probably supplying the new wording himself:

> It does not follow that these [good] effects are due to the internal secretion of the pancreas in the extract. The injections are usually followed by a slight temporary rise in the body temperature, and this may be a factor in the lowered sugar output. Physiologists are not agreed as to whether the internal secretion acts by diminishing or retarding the passage of sugar from the tissues into the blood, or by increasing the oxidation of the sugar in the tissues. The pancreas extract may decrease the output of sugar from the tissues by a toxic or *depressor* action, rather than by a specific regulatory action of the pancreas secretion.... The work is being continued in the hope of clearing up these points.[17]

Despite his conservatism, Carlson urged Scott to continue the research and work out his "salvation or damnation along the pancreas extract line.... There is something ahead in that line - possibly both shoals and open water. Puzzle: find the channel." Scott tried half-heartedly, attempting to buttress his urinary sugar results with studies of his extract's effect on the blood sugar of cats. He reported the "very surprising" result that it caused an increase in their blood sugar.[18] Having struck a shoal, Scott veered away from pancreatic extracts to study problems relating to blood sugar.

Before giving up the work, Scott chatted about it with some of the other experts in the field. One of these was a professor at Western Reserve University in Cleveland, Ohio, John James Rickard Macleod. Macleod was a Scotsman, trained in Aberdeen, Germany, and London, who had emigrated in 1903 to take his American appointment at the age of twenty-seven. He had been working for several years in the area of carbohydrate metabolism. A competent researcher and a prolific writer and synthesizer of current knowledge in physiology, Macleod was particularly knowledgeable about the literature in his field. The only knowledge we have of his discussion with Scott is that it began with a consideration of how to cure Scott's child's diarrhoea. When the talk turned to pancreatic extracts, Macleod may have discouraged the younger man, for about this time he was working on his own main research contribution to the search for the internal secretion of the pancreas. Macleod was able to show that the findings of two leading Britishers, Knowlton and Starling, who thought they had a pancreatic extract which assisted the heart of a diabetic dog to utilize sugar in the blood, were not repeatable.[19]

Knowlton and Starling's joined Scott's and Zuelzer's in the list of apparently ineffective pancreatic extracts. Two young Americans, John R. Murlin and Benjamin Kramer, continued to fiddle with pancreatic extracts similar to Knowlton and Starling's, but their work led them off into examinations of the influence of alkaline solutions on metabolism.[20]

Macleod summarized the state of the search for an internal secretion in his 1913 book, *Diabetes: Its Pathological Physiology*. After due deliberation he concluded that there was an internal secretion of the pancreas, but suggested several reasons why it might never be captured in a pancreatic extract. The powerful pancreatic juice might destroy it; there might be no reserves of it in the pancreas to be captured by extraction; or it might exist in the pancreas only in latent form and not be activated until secreted into the blood. Macleod's own interest and his work tended to be on the behaviour of blood sugar rather than pancreatic extracts. He thought the most convincing proof of the existence of an internal secretion came in Hédon's early work (now questioned by Hédon himself) using grafts of pancreatic remnants to show that a small, isolated portion of the pancreas could stave off diabetes.[21]

In 1913 Dr. Frederick Allen pronounced what seemed to be the epitaph of a generation's attempts to treat diabetes with pancreatic extracts: "All authorities are agreed upon the failure of pancreatic opotherapy in diabetes.... injections of pancreatic preparations have proved both useless and harmful. The failure began with Minkowski and has continued to the present without an interruption.... The negative reports have been numerous and trustworthy."[22]

V

Frederick Madison Allen wrote with particular authority. Born in Iowa in 1876, trained in medicine in California, Allen had come east to do medical research, drifted into a poorly paying fellowship at the Harvard Medical School, and found himself working on problems of sugar consumption. The study turned into three years of intensive research concentrating on diabetes. Most research is reported upon in journal articles. Allen's was not. His first publication, subsidized by his father, was a remarkable 1913 volume entitled *Studies Concerning Glycosuria and Diabetes.* "Its spirit is that of an enlarged journal article," Allen wrote in the introduction, claiming it was a book in which he hoped to give "simplicity and order" to the study of diabetes.[23] The reader who waded through the following 1,179 pages – in which Allen set his own research, involving experiments on more than two hundred dogs, the same number of cats, and assorted guinea-pigs, rabbits, and rats, in the context of everyone else's research (his bibliography contained approximately twelve hundred listings) – knew that the subject was anything but simple and orderly, except possibly in the head of Dr. Allen. Both the research and the book were prodigious achievements in themselves, and even more significant for the revolution in diabetes therapy that flowed from them.

Most researchers created experimental diabetes in animals by taking out the whole pancreas. Allen's approach was to remove a large part of the pancreas, about 90 per cent in dogs, but leave the rest. He thus created a state of mild diabetes in animals which was probably much closer to the diabetes most humans experienced than was the severe, quickly fatal diabetes arising from total pancreatectomy. Although he tried some experiments involving pancreatic extracts, Allen's chief interest was in the effect of diet on the diabetic animals. What kinds of diet would enable an animal with a partial pancreas to keep metabolizing his food without becoming more diabetic? What kinds of diet were harmful to animals with these crippled pancreases, making the diabetes worse?

Allen's work undercut the view that diabetes was mostly a problem of carbohydrate metabolism. It was not just the carbohydrates, but the proteins and fats as well, that the diabetic's body was having trouble with, Allen argued. All kinds of food tended to over-burden the system. Diets which involved cutting back sharply on carbohydrates and then increasing the proteins or fats to compensate, achieved nothing – or, worse, caused a higher rate of acidosis and death in coma because of their fat content. The answer was to continue to cut back on carbohydrates, but to cut back on everything else, too, so that the diabetic's total calorie intake was reduced. If over-nourishment or normal nourishment produced diabetic symptoms, notably glycosuria, then the trick was to find the degree of *under-nourishment* that would enable a diabetic to live sugar- and symptom-free. Any previous diabetic diets that had actually been effective, Allen claimed – high-fat, oatmeal cure, or whatever – had been characterized by a low total calorie count. There was no way a diabetic could save his carbohydrates and eat his calories too.

An outsider with no advanced degrees, Allen had trouble getting a job until the Rockefeller Institute in New York, impressed by his book, offered him a junior position in 1914. The appointment gave him access to a small ward of diabetic patients, and turned out to be a marvellous opportunity to begin applying his theories to humans.

After four years' clinical work, Allen and his associates published their results in 1919 in a second massive volume, *Total Dietary Regulation in the Treatment of Diabetes,* which ran to 646 pages plus charts. Almost half the book consists of exhaustive case records of seventy-six of the one hundred patients Allen had treated.

His methods were tried on all sorts of diabetics, mild and severe, recently diagnosed and terminally comatose, old and young, educated and ignorant, well-to-do and desperately poor. The therapy was almost always the same: When a diabetic was admitted to hospital, he or she was put on a fast (liquids only) until the glycosuria and, in the severe cases, the acidosis disappeared. Then there would be a gradual building up of diet, measuring by carbohydrate tolerance, but with strict weighing of all foods, to see how much the patient could take before becoming glycosuric. When sugar appeared in the urine, the limit had been reached. A fast day would clear the urine again and the diet would be fixed at a total calorie intake just under this tested tolerance.

This quick description of the Allen method might go unremarked by readers unfamiliar with serious diabetes and in an age when most of us have to diet occasionally. At the time he introduced what came to be called the "starvation treatment" of diabetes, Allen was advocating serious dieting in a country where being well-fed was still a sign of good health. More ironically, he was advocating serious dieting to patients two of whose complaints were their terrific hunger and their rapid weight loss. They came to the doctor to be treated for these symptoms and the doctor seemed to be telling them that they had to be hungry more often, that they had to lose even more weight.

The ironies, the Hobson's choices, the catch-22's of the treatment were staggering. An adult diabetic, weak, emaciated, wasted to perhaps ninety pounds, would be brought into hospital and ordered to fast. If the patient or the patient's family complained that he or she was too weak to fast, Dr. Allen replied that fasting would help the patient build up strength. If the patient complained about being hungry, Allen said that the fasting would help ease the hunger. Suppose the method didn't seem to work and the symptoms seemed to get worse. The answer, Allen insisted, was more rigorous under-nourishment: longer fasting, a maintenance diet even lower in calories. To top it off, Allen and others were also urging diabetics to take as much physical exercise as possible, claiming it would help them burn more food and increase in strength.

Where was the limit to the dieting? Where would you stop? In fact there was no limit. In the most severe cases the choice came to this: death by diabetes or death by what was often called "inanition."

"The plain meaning of this term," Allen wrote, "is that the diabetes was so severe that death resulted...from starvation due to inability to acquire tolerance for any living diet." "The best safeguard against inanition," he added, "consists in sufficiently thorough undernutrition at the outset." In those situations where the awful choice between death from diabetes and death from starvation could not be avoided, "comparative observations of patients dying under extreme inanition and those dying with active diabetic symptoms produced by lax diets or by violations of diet have convinced us that suffering is distinctly less under the former program."[24]

To illustrate, consider Rockefeller case 60, a forty-three-year-old housewife who came into the hospital on New Year's Day, 1916, having lost 60 pounds in the few months since the onset of her diabetes. She weighed 36 kilograms or 79 pounds on admission, and was so weak that even Allen hesitated to go ahead with severe fasting:

> The experiment was tried of feeding more liberally for a short time in the attempt to restore some strength, so as to get a fresh start for further fasting....the attempt caused only harm instead of benefit, as always in genuinely severe cases. The question thereafter was whether the glycosuria could be controlled without starving the patient to death....Though the food was thus pushed to the utmost limit of tolerance, it was not possible to prevent gradual loss of weight.

She was utterly faithful in following her diet, which during hospital stays averaged 750 calories a day and about 1,000 calories when she was at home. When Allen last saw her, in April 1917, her weight was down to 60 pounds and falling. "Perhaps better results might have been obtained by cutting down the weight to perhaps 30 K (66 pounds) at the outset," he mused in the conclusion to the discussion of her case. "The question remains whether the pancreatic function is absolutely too low to sustain life, or whether by sufficiently rigid measures downward progress can be halted even at this time." The answer was given in a footnote added in the final revision of the manuscript: "Largely on account of her residence in a city too far away to permit personal supervision and encouragement, this patient finally broke diet, and after a rapid course of glycosuria and acidosis, died in Feb. 1918."[25]

Many of Allen's patients broke diet out of hospital, some sooner than others. Case 1 embraced Christian Science four months after her release, began eating everything at will, and died in a few more months. Case 51, a seven-year-old Polish-American schoolboy, was able to sneak food at home unknown to his parents, and died from it; "the essential cause of trouble lay in the home conditions of an uneducated Polish laboring family." Case 18 was a sixteen-year-old errand boy who adhered to his diet fairly well until summertime when he had a feast of cherries. After that he became uncontrollable and went downhill.[26]

Even inside the hospital the staff had to be constantly on the alert to stop the pilfering of "forbidden food." The most extreme example was case 4, a twelve-year-old boy whose diabetes had already caused blindness when he was admitted. No matter how carefully he was treated, his urine tests on some days would show sugar. It could not be accounted for from his diet. The staff could not understand what was happening:

> It had seemed that a blind boy isolated in a hospital room and so weak that he could scarcely leave his bed would not be able to obtain food surreptitiously when only trustworthy persons were admitted. It turned out that his supposed helplessness was the very thing that gave him opportunities which other persons lacked.... Among unusual things eaten were tooth-paste and bird-seed, the latter being obtained from the cage of a canary which he had asked for.... These facts were obtained by confession after long and plausible denials. The experience illustrates what great care is necessary if records of diabetic patients are to be vouched for as correct.

The gods had their revenge. Thinking the glycosuria was caused by too high a normal diet, the staff cut the boy's normal food supply further and further. It was too late when they realized their mistake. He weighed less than 40 pounds when he died from starvation.[27]

Allen was a stern, cold, tireless scientist, utterly convinced of the validity of his approach. His therapy for diabetes seemed immensely hard-hearted in the extreme cases, and met much resistance from diabetics, their families, other physicians, and other workers at the Rockefeller Institute. Allen defended himself with iron logic. Yes, the method was severe; yes, many patients could not or would not follow it faithfully; yes, in the worst cases it led to death from starvation; yes, all it could do was prolong the lives of diabetics, in some cases for a few years, in severe cases perhaps only a few months. But what was the alternative? All of Allen's experimental and clinical evidence showed that total dietary regulation was the *only* way of prolonging the lives of diabetics. Nobody had a better way. Besides, he claimed, his diet was not impossible to follow: because they were better balanced Allen's diets were often more tolerable than the destructive high-fat alternatives. Most of his disobedient patients had actually been on the more liberal of the series of diets, "and were the sort of persons who would not abide by any restrictions no matter how slight." Some of his most undernourished patients had borne their diet in the most faithful way.[28]

Generally, diabetics on the diet did feel better than those who broke it, the "simple hunger" from careful fasting or dieting being less tormenting than the sick hunger, or polyphagia, of diabetes. Allen's "faithful" patients, even those under an obvious sentence of death, regained a degree of strength and comfort and the ability to enjoy life. "Though always hungry, excessively emaciated, and lacking strength for any real exertion," he wrote of case 60, "some of the noteworthy features are her constant cheerfulness, freedom from infection, and comfort in all other respects. She is able to be up and about, carries on light household duties, and - the point of most importance to her - attends to the bringing up of her child." By her faith and determination, case 60 had won for herself about two extra years of life.[29]

In the final analysis, the only argument against the thorough treatment was the cruelty of prolonging a patient's suffering. "Euthenasia is no more justified in diabetes than in numerous other diseases," Allen argued. "Diabetics who overeat for the deliberate purpose of killing themselves are uncommon."[30] Allen was proud that he was not only keeping diabetics alive longer, but was pioneering in methods of keeping starving people alive; some of his patients were living in stages of inanition not thought possible.

Frederick Allen's determination to apply his methods ruthlessly (to prove his theories absolutely he wanted to be able to control his patients as thoroughly as laboratory animals were controlled) were probably responsible for a decision at the Rockefeller Institute to take away his control of the diabetes clinic. Instead of being the triumph of medical research it appears, the 1919 volume, *Total Dietary Regulation in the Treatment of Diabetes,* actually veils a bitter controversy about the treatment of those hundred cases. The book was later denounced by Allen as an inconclusive, failed study. The diabetologist did not believe there were many shades of grey in medical research, or in life generally. He left the Rockefeller Institute intensely frustrated, served in the army diabetes service during the war, and in 1919 launched a daring bid for personal and professional independence by purchasing the Morristown, New Jersey, mansion formerly owned by Otto Kahn. There he founded the Physiatric Institute, intended to be a prestigious centre for treatment of Americans suffering from diabetes, high blood pressure, and Bright's disease. The fees paid by rich patients for the luxurious facilities in one department of the Institute supported more plebeian facilities in other departments as well as the ongoing research work. One of Allen's rules was that all in-coming patients had to promise their sincere co-operation in the prescribed treatment. As the Institute flourished in 1919-20, Allen worked frantically to pay off his debts and build the resources to support the grand research plan he felt had been frustrated at the Rockefeller.[31]

The total dietary approach to diabetes was the best therapy available at that time. How widely it was actually used is difficult to estimate. In medical schools and among up-to-date practitioners Allen's methods seem to have been universally adopted. But Allen and the other diabetologists often wrote scornfully of the ignorance with which doctors treated diabetes - at worst with opium and over-feeding, at best by handing out printed, out-of-date diets. And even they were better than the patent medicine men who offered nostrums such as Bauer's Antidiabeticum, and the religious people who offered prayer, faith, and Christian Science. At his own professional level, as well, Allen was under attack from several researchers, especially Woodyatt in Chicago, who worked out elaborate theoretical critiques of "starvation" and new justifications for high-fat and fairly high calorie diets.[32] In these clinics, too, the thorough treatment of diabetes was expensive and complicated, involving prolonged hospital stay, careful preparation and weighing of individually tailored diets, elaborate daily tests, and special nursing for children. In prosperous North America diabetes was becoming something of a specialist's disease, with special diabetic wards being set up at hospitals and physicians building whole practices on nothing but the treatment of diabetes.

Other than Allen, the most prominent American specialist in diabetes was Dr. Elliott P. Joslin. A New Englander, a graduate of Yale and Harvard, and student of Naunyn at Strasbourg, Joslin gradually narrowed his medical practice in Boston to diabetes. He was a prolific writer, particularly at the semi-popular level aimed at physicians and the diabetics themselves, and a warm enthusiast. In his writing Joslin tried to put the best face on the diabetic's situation, stressing that it was "the best of the chronic diseases," clean and seldom unsightly, not contagious, often painless, and usually susceptible to treatment.[33]

A friend of Allen and a strong supporter of under-nutrition, Joslin tended to be optimistic about the therapy. He was almost certainly over-optimistic, possibly deliberately so to bolster his patients' morale and his own.[34] It was hard to keep up your spirits to face each day of urging sick people to keep starving. A nurse at the Physiatric Institute remembered how horrifying it was to watch the starving children lying in their beds. "It would have been unendurable," she wrote, "if only there had not been so many others."[35]

Because he tempered his own rock-hard puritanism with warmth and charm and a sense of hope, Joslin may have had more success with his patients than the forbidding Dr. Allen. He was particularly popular with children, some of whom were brought to him because no one else would treat them. When von Noorden came to Boston, Joslin remembered, he shuddered and turned away when shown one of Joslin's skeleton-like diabetic girls. A quarter of a century after the discovery of insulin the doctors were reminded of these pre-insulin diabetics when they saw the pictures of the survivors of Belsen and Buchenwald.[36]

VI

Despite the record of failure, and despite the pessimism of men like Allen, Carlson, and Macleod, attempts to find an effective pancreatic extract continued, "because of the strong theoretical inducements," Allen noted.[37] The most interesting and important of these new attempts involved experiments measuring the effect of pancreatic extracts on blood sugar. High blood sugar, or hyperglycemia, had been recognized for many years as a *sine qua non* of the diabetic condition. Measurements of blood sugar had not usually been involved in diabetes therapy or research, however, because they were very difficult. The chemical tests required to estimate the amount of sugar in the blood called for a lot of blood, usually twenty cc. or more. It was difficult and possibly dangerous to take many of such large blood samples from either humans or animals. As well, methods for testing the sample were time-consuming and so crude that the margins of error in estimating the percentage of blood sugar were very high. It was much more practical, safer, and perhaps more accurate to test the diabetic condition through urine samples alone.

But accurate blood sugar readings would obviously be a useful research tool, supplying a far more reliable guide to diabetes than urine tests. All of the problems and complications and alternative interpretations of glycosuria created by the possibility of kidney disorder could be avoided. If good testing procedures (the lack of which was probably central in E.L. Scott's failure) could be developed, it would be much easier to check short-term fluctuations of blood sugar than to measure, say, the hourly inflow of sugar into the urine. The single most important development in diabetes research, next to Allen's diets, was the rapid improvement between about 1910 and 1920 in techniques for measuring blood sugar. In 1910 a blood sugar test still required 20 cc. or more of blood; by 1920 it could be done with as little as 0.2 cc.[38] The use of blood sugar estimations was soon reflected in the research.

A young American, Israel Kleiner, became interested in pancreatic extracts and blood sugar while working with S.J. Meltzer at the Rockefeller Institute during the time of Allen's researches. Pioneering studies were being done there on the speed with which injections of sugar normally disappeared from circulation (that is, were assimilated by the system). By contrast, in diabetic animals much of the sugar continued to circulate. But when an emulsion of pancreas was mixed and injected along with the sugar solution, the diabetic animal handled it almost normally. Observing this, Kleiner and Meltzer began experiments to see how pancreatic extracts would affect the ability of depancreatized dogs to deal with their system's own excess sugar.

They reported very promising preliminary findings in 1915, but their work was interrupted by the war. In 1919 Kleiner returned to it, running many more experiments. Late in 1919 he published his findings in the *Journal of Biological Chemistry*. Of all publications before the work at Toronto, it was the most convincing.

Kleiner had made solutions of ground fresh pancreas in slightly salted distilled water. These were slowly injected intravenously into depancreatized dogs, with blood sugar readings taken before and after infusion and at later intervals. The 1919 experiments were much easier to do because the new blood testing method (Myers and Bailey's modification of Lewis and Benedict's) required much smaller samples. In both the 1915 and 1919 series of experiments the results were the same and were important: without exception in sixteen experiments the pancreatic extract caused a decline in the blood sugar of diabetic dogs. It was often a very sharp decline, sometimes more than 50 per cent.

Kleiner had not used any chemicals in the preparation of his extract because some of Murlin's recent work suggested that the chemicals themselves, especially alkalis, could artificially reduce blood sugar. He ran checks on the hemoglobin content of his dogs' blood to make sure that the effect he was getting was not just a result of the injected liquid diluting the blood, and checks on the urinary sugar to make sure some strange "washing out" effect was not taking place. Emulsions made from other tissues were injected to see if the effect might be something any ground-up tissue could produce. They caused no significant change in the blood sugar (that they did sometimes cause a reduction in glycosuria indicated the weakness of older methods: "the mere reduction of glycosuria is no proof of a beneficial effect of any agent," Kleiner noted with emphasis).

Kleiner began the "Discussion" section of his paper triumphantly:

> Many investigators have recognized that the best evidence for the internal secretion theory of the origin of diabetes would be an antidiabetic effect of a pancreatic preparation, administered parenterally. The experiments just described show that such a result has been obtained....

His controls had been impressive, his follow-up discussion was a beautiful piece of scientific writing. There was one problem, he reported: the slight toxic symptoms, usually a mild fever, associated with the extract. These symptoms were not particularly marked, and the overall result of the work "indicates a possible therapeutic application to human beings." Before this happened, Kleiner suggested, further knowledge should be obtained. Many other tests could be run. "Finally, the search for the effective agent or agents, their purification, concentration, and identification are suggested as promising fields for further work."

Kleiner did not do any of that further work. In 1919 he left the Rockefeller Institute, and did not return to the problem. The only published comment Kleiner ever made on why he did not continue "and attempt to isolate the antidiabetic factor" was that it was "a long story." As far as can be determined, the university he went to in 1919 did not have the resources to support major animal research.[39]

Another scientist whose work on pancreatic extracts had been interrupted by the war was Nicolas Paulesco, professor of physiology in the Romanian School of Medicine in Bucharest. Paulesco was already a physiologist of substantial achievement and distinction when he returned to an interest in the internal secretion of the pancreas first developed during his student years in Paris in the 1890s. In 1916 he began experimenting with extracts. The Austrian occupation of Bucharest and then the postwar turmoil in Romania delayed his research for four years. Paulesco resumed his experiments in 1919 and published his first results in 1920 and 1921.

Like Kleiner, Paulesco concentrated on measuring the impact of his extract on blood sugar. He, too, reported spectacular decreases in blood sugar after intravenous injections of a solution of pancreas and slightly salted distilled water. He also reported a decrease in urinary sugar and in the presence of ketones in blood and urine. He checked for dilution, controlled with non-pancreatic extracts, and induced fever in his dogs to show that fever itself (which his extract often caused) would not cause a reduction in the sugar content of the blood or urine. He also tried his extract on a normal dog and found that here, too, it caused a reduction in blood sugar.

Paulesco published his earliest findings in his 1920 treatise on physiology, written in French. These and further experiments were described in four short papers published in *Comptes rendus des séances de la Société de biologie* between April and June 1921. A summarizing paper was received by *Archives internationales de physiologie* on June 22 and published on August 31. Paulesco had done fewer experiments than Kleiner, not least because he must have been hampered by the very primitive techniques he was using for measuring blood sugars. These techniques also produced some remarkably low figures, almost certainly based on error. Unlike Kleiner, Paulesco did not set his work and its implications in the context of past and current knowledge. On the other hand his results looked very good, his experiments were more varied than anyone else's had been, and he clearly intended to persist. In his August 1921 paper he mentioned that it would be followed up by "une méthode de traitement du diabète, de l'obesité et de l'acidose, méthode qui est issue de ces reserches expérimentales."[40]

In Germany at the same time, Georg Zuelzer was still trying to find a drug company to take up production of his extract, acomatol. No one in that devastated country was very interested.

VII

In the conclusion to his 1919 study, even while underlining the limits of his diet treatment, Frederick Allen had tried to be optimistic. "The knowledge of diabetes is advancing rapidly enough that even the patient whose outlook seems darkest should take courage to remain alive in the hope of treatment that can be called curative." He must have been discouraged in the next year or two as the most faithful of the cases reported in his Rockefeller study died one after another, with no cure in sight. The idea of advancing beyond diet, perhaps with a pancreatic extract, had been in the back of his mind for some time. In 1921 he began installing facilities for animal experimentation at the Physiatric Institute. He planned to try a new approach to the extract problem when they were ready.[41]

One of Allen's most faithful patients was a young girl named Elizabeth Evans Hughes. She was also his most prominent patient, for her father, Charles Evans Hughes, was one of the most visible men in the public life of the United States. Elizabeth had been born in the New York state governor's mansion in 1907. Her father was later appointed to the Supreme Court, resigned from it to run as the Republican candidate for the presidency against Woodrow Wilson in 1916, and in 1920 became Secretary of State in the administration of Warren Harding. Later he would be reappointed to the Supreme Court and become one of its most distinguished Chief Justices.

One of four children of Charles and Antoinette Hughes, Elizabeth grew up as a lively, intelligent little girl, never very big or strong, but otherwise normal. She had an interesting and exciting girlhood, a beneficiary of all the opportunities open to a family of American aristocrats. It was in 1918, when Elizabeth was eleven or twelve, that something started to go wrong. She would come home from birthday parties, where there had been lots of ice cream and cake, with a ravenous thirst, and would drink glass after glass of water, sometimes two quarts. She was often weak and tired in the winter of 1918-19, and showed increasing tendencies to polydipsia and polyuria. That spring she was taken to Dr. Allen. He diagnosed diabetes and prescribed an immediate fast. Whatever the fasting would do, the diagnosis was like knowing a sentence of death had been passed.

At the onset of her diabetes Elizabeth Hughes was 4′ 11½″ tall and weighed 75 pounds. After the first week's fasting Allen put her on a very low diet, 400 to 600 calories a day for several weeks (with one day's fasting every week), then raised it to 834 calories. He brought her weight down to 55 pounds, then allowed her to rise into the low 60's on a diet going as high as 1,250 calories (350 on fast days). The Hughes family hired a special Joslin-trained nurse to prepare Elizabeth's meals and help her with her tests. Every gram of food she consumed had been weighed beforehand. Sweets and bread disappeared from her diet. She lived on lean meat, eggs, lettuce, milk, a few fruits, tasteless bran rusks, and tasteless vegetables (boiled three times to make them almost totally carbohydrate-free). A birthday cake became a hat box covered in pink and white paper with candles on it. On picnics in the summertime she had her own little frying pan to cook her omelet in while the others had chops, fresh fish, corn on the cob, and watermelon.

Elizabeth disliked Dr. Allen, a square-faced, jowly man who never seemed to smile, never seemed anything but strict. Charles Evans Hughes was one of the sponsors of the Physiatric Institute and had helped Allen with the legal work involved in setting it up; but Elizabeth, who spent several weeks there, found it a horrible place. She disliked her diet, and found the fast days a special nightmare - she tried to plan every minute of these days in advance so she would be distracted from the hunger. She was a vivacious, articulate adolescent, eager for all the experiences life had to offer, and apparently unaware of what was in store for her. Her nurses never told her how serious her problem was. They never told her why friends she had made at Morristown stopped writing or never appeared there again.

She was an obedient little spartan, though, and kept her diet perfectly. She hardly ever showed sugar. Just once, at Thanksgiving, she sneaked into the kitchen and snitched a piece of turkey skin. Her nurse caught her and gave her a severe bawling out. She must *never* take extra food.

Had she been untreated, Elizabeth Hughes would probably have died in the summer of 1919. With stern Dr. Allen's stern diet, her own discipline, and her sheer strength of character, she carried on very well through the winter of 1919-20. She had a difficult time in the spring of 1920, when colds and tonsillitis threw her out of balance, and was often cut back to a diet of less than 500 calories. But she recovered that summer and fall, and at Christmas 1920 weighed in at 62¼ pounds. The winter and spring were bad again, though; by the end of March she was down to 52 pounds. Her diet in April averaged 405 calories. The doctor got her back up to 700 to 900 calories, but her weight was now at a new low plateau, between 52 and 54 pounds. At the age of thirteen, Elizabeth was a semi-invalid. There was great sorrow in the family when one of her older sisters died in 1920 of tuberculosis. While the Hughes family sweltered in Washington in the summer of 1921, Elizabeth enjoyed the fresh air and cool breezes of the Adirondacks. Her condition stayed about the same. In cheerful letters she chatted on about when she would get married and what she would do on her twenty-first birthday. Reading them must have been heart-breaking for Antoinette Hughes. The best medical talent in the world was the Hughes family's to command. But the "curative treatment" for diabetes that Dr. Allen had written about was nowhere in sight.[42]

CHAPTER TWO

Banting's Idea

Frederick Grant Banting, always called Fred, was born on a farm near the small town of Alliston, Ontario, on November 14, 1891. He was of British descent, his grandfather having emigrated to Canada about forty years earlier.* His parents, Margaret Grant and William Banting, were hard-working farm people, devout Methodists, and reasonably prosperous, sober citizens. The youngest of five children, Fred enjoyed a normal farm boyhood in turn-of-the-century rural Ontario, growing up close to nature and with a deep affection for animals. He seems to have been happy at home and to have had a particularly close relationship with his mother. At local schools in Alliston he was a serious-minded but unremarkable student. "We would not have picked him for one on whom fame should settle," his public school principal said some years later.[1]

Banting went on to higher education at the University of Toronto, the province's largest and best university. Located on the shores of Lake Ontario, about forty miles south of Alliston, Toronto was the provincial metropolis and Canada's second-largest city, a thriving community of more than five hundred thousand. Originally uncertain of his vocation – his parents had encouraged him to think of entering the Methodist ministry – Fred dropped out of an arts course towards the end of his first year, and re-enrolled in medicine the next autumn, 1912. The university's faculty of medicine was one of the largest, in terms of student enrolment, in North America, well-equipped, and apparently fairly well-staffed. Its teaching hospital, Toronto General, had recently been rebuilt and was one of the best anywhere. There was a growing emphasis at Toronto on research as a vital accompaniment to the teaching of a medical school.[2]

Banting was an average medical student, more serious and more studious than most, shy, best at athletics. He was tallish, almost six feet, and

*A distant relation, a London cabinet-maker named Banting, caused such a stir in 1864 with his method of reducing corpulence by avoiding fat, starch, and sugar, that the verb "to bant," with such derivatives as "banting," "bantingize," "bantingism," entered the language to refer to weight reduction through dieting.

strong, and when dressed-up could be a handsome young man, with a particularly winning broad smile and an attractive twinkle in his eye. In a less flattering light his face had "horsey" features - it was long and narrow and his mouth seemed to stretch from one side to the other - and in his manners and conversation Fred was very much the unpolished country boy. He could seem intellectually slow; his studiousness was a kind of dogged determination to get through and never won him more than average grades. In his free time he enjoyed most of the male rituals of the university, although he apparently never learned to dance, possessing, as they said, two Methodist feet. Much of his spare time was spent with his girlfriend, Edith Roach, a languages student whose father had been the Methodist minister in Alliston.

Banting's five-year medical course at Toronto was shortened because of the war. The class of "1T7" (Toronto, 1917) took its fifth year in the summer of 1916. "I had five pages of notes on the whole lectures of the fifth year," Banting recalled, writing that he had "a very deficient medical training."[3] Immediately after final results were announced in December every able-bodied member of the class went off to war. Banting, who had been serving part-time in the Canadian Army Medical Corps for two years, was sent to England in 1917. He and Edith became engaged before he left.

After a year working in hospitals in England, Captain Banting was sent to the front as a battalion medical officer. He saw a fair bit of action and received the Military Cross for his courage under fire at Cambrai, where he was wounded in the arm by shrapnel. He had a long convalescence in Britain, returned to Toronto in March of 1919, and was posted to Christie Street Military Hospital before his discharge. During his free time in the army Banting studied to take the examinations leading to the stamp of approval of various medical bodies such as the Royal College of Physicians of London and the Royal College of Surgeons. He seems to have had a deep commitment to his profession, and was gradually developing an interest in research. Before and during his military service he worked with Clarence L. Starr, the brilliant chief surgeon at the Hospital for Sick Children in Toronto, who became something of a hero and medical father-figure to him. As soon as he was free from the army, in September 1919, Banting returned to Sick Children's as a resident in surgery, with a particular interest in orthopedics. Specialization was still fairly primitive in those years, however, and it is misleading to think of Banting as a highly trained orthopedic surgeon. Much of his surgical experience had come from treating wounded soldiers. At Sick Children's he did general surgery.

It is not possible to judge Banting's ability as a surgeon. After he became world famous, and stopped doing surgery, the natural tendency of memory was to say that Fred was a highly skilled surgeon. Evidence of such skill is not contained unambiguously in his insulin notebooks. Banting was certainly experienced, for he had treated more wounds in the summer of

1918 than some peacetime surgeons would see in a lifetime. He was also popular with the sick children at the hospital. But he was not able to win a permanent position at the Hospital for Sick Children. "Surgeons were very plentiful in Toronto. It was my greatest ambition to obtain a place on the staff of the hospital, but this was not forthcoming."[4] Instead, perhaps on the advice of C.L. Starr, and knowing that Edith would be teaching high school in a nearby town, Fred decided to set up a practice in the city of London, Ontario, about 110 miles west of Toronto.

Since returning from the war he had been anxious to marry. He had come home with the veteran's usual minor vices - drinking, swearing, and heavy smoking; but he was still enough of a Victorian boy to believe, with Edith, that a wedding would not be seemly until he was earning money of his own. No self-respecting male Canadian in 1920 would live on his wife's earnings. Fred Banting was twenty-eight years old, a veteran of the world war, a well-trained doctor. It was surely time to settle down, make some money, get married, and have a family.[5]

I

On July 1, 1920, Banting opened an office in a house he had bought on a corner in a residential area of London. He must have known it would take time to build up a practice in a strange city, with whose doctors he had no ties and in a profession which forbade advertising. Even so, he was not prepared for the depressing reality of his situation. Day after day in July, Doctor Banting[6] kept his standard office hours, two to four in the afternoon, seven to nine in the evening, six days a week. He saw no patients at all. Not one. The first customer finally came on July 29. The patient's "illness" was his friends' thirst for liquor in a province where prohibition held sway. Only doctors could dole out alcohol, and then solely for medicinal purposes. "He was an honest soldier," Banting wrote, "who had friends visiting him and he wanted to give them a drink. I gave him the prescription and considered myself rather highly trained for the barkeeping business." Dr. Banting's July income was $4.*[7]

Patients started to dribble in during August, but business was miserably slow. Already in debt from his medical education, Banting had borrowed money from his father to buy the house in which he practised and lived. Every week of medical practice drove him deeper in debt. He tried to save money by cutting out motion pictures and often cooked his meals on the bunsen burner in his dispensary. To while away the time, Banting built a garage and started dabbling with oil paints. He also tinkered with the worthless old fourth- or fifth-hand car he had bought - having paid much

*Banting sometimes told his friends that the patient's problem was syphilis. The two explanations are not mutually exclusive, but do clash with the notation in his account book that his $2 fee was for "baby-feeding."

more than it was worth, he discovered, the kind of realization that reinforced his sense of failure. The car soon failed, too, and had to be scrapped.

Although always a bit of a loner, Banting craved male companionship and female affection. He and a classmate, Bill Tew, who had also begun practice in London, spent "about five nights out of seven" commiserating about the practice of medicine. The camaraderie must have been marred slightly for Fred by the realization that Tew's practice was developing better than his. On weekends he would see Edith, but there was trouble there too. The successful female teacher was making three or four times as much money as the uncertain, insecure male doctor. Apparently they were not as sure of their love as they had been in 1916: the farmer's son who had been to the wars and the minister's scholarly daughter, gold medallist in her class, may have realized they had changed over the years. Edith may have been developing other interests. "I was very unhappy and worried," Banting wrote about that early period in London.[8]

Studying was another way to pass the time. Banting resumed preparation that he had broken off several years earlier to take the difficult exams for a fellowship in the British Royal College of Surgeons. He also got a part-time job in October as a demonstrator in surgery and anatomy at London's Western University. Western's faculty of medicine was small and undistinguished, but had a few good professors and a promising future based on an ambitious building program. As well, Banting could also use the $2 an hour he was paid. He soon began assisting Dr. F.R. Miller, Western's very good professor of physiology, in occasional experiments in cerebral and cerebellar localization.[9]

II

On Sunday, October 31, 1920, Banting spent several hours preparing a talk he had to give to physiology students on carbohydrate metabolism. Neither the topic, nor the associated disease, diabetes, were subjects in which Banting had any particular interest. Stories grew up later about diabetic school chums having had a profound influence on him, and so on, but actually Banting had never treated a diabetic patient and had no interest in the dietary treatment of diabetes. There had been one brief mention of it in his therapeutics lectures at university, an up-to-date suggestion that physicians not be afraid to use Allen's starvation treatment. "I remember seeing one patient only on the wards of the Toronto General Hospital," Banting wrote years later. "I heard of people mostly well on in life dying in coma and believed there was nothing one could do.... There was no such thing as a diabetic in any ward in my surgical experience.... I did not even know that my friend and class-mate, Joe Gilchrist, had diabetes until I had been working on the problem for many months."[10]

His copy of the November issue of the journal *Surgery, Gynecology and Obstetrics* had just arrived.[11] When he had finished work in the evening he took it to bed to read himself to sleep. With carbohydrate metabolism on his mind, he was naturally interested in the leading article in the issue, an analysis of "The Relation of the Islets of Langerhans to Diabetes with Special Reference to Cases of Pancreatic Lithiasis," by Moses Barron.

Barron was an American pathologist who became interested in the pancreas and the islets of Langerhans when, while doing routine autopsies, he came upon a rare case of the formation of a pancreatic stone (pancreatic lithiasis). Rarer still, the stone had completely obstructed the main pancreatic duct. Studying that pancreas, Barron found that while all the acinar cells had disappeared through atrophy, most of the islet cells had apparently survived intact. A review of the literature showed that these observations were similar to those arising when pancreatic ducts were blocked experimentally by ligation. Both experimental evidence, then, and this interesting new piece of pathological evidence, seemed to reinforce the hypothesis, held by many others, that the health of the islets was the key variable in the genesis of diabetes.

Barron wrote up his modest study, presenting it as another bit of evidence in the search for an explanation of diabetes. The work was perhaps of special interest because of the similarity between experimental and clinical cases. He probably submitted it to *Surgery, Gynecology and Obstetrics* because of the interest surgeons had in stones.

Barron's was a useful, not brilliant or trail-breaking study (his next paper, on the relationship between smoking and lung cancer, was a much more important pioneering contribution). His review of the literature was not particularly wide-ranging; his interpretation of his own case was questionable inasmuch as the patient had had islet cell damage and was diabetic (Barron attributed the damage to causes other than the obstruction). The sole importance of Barron's article in the history of medicine is that Fred Banting happened to read it in the evening of a day he had been thinking about carbohydrate metabolism.

Banting's most detailed description of his reaction to the Barron article is in his 1940 memoir, "The Story of Insulin":

> It was one of those nights when I was disturbed and could not sleep. I thought about the lecture and about the article and I thought about my miseries and how I would like to get out of debt and away from worry.
>
> Finally about two in the morning after the lecture and the article had been chasing each other through my mind for some time, the idea occurred to me that by the experimental ligation of the duct and the subsequent degeneration of a portion of the pancreas, that one

might obtain the internal secretion free from the external secretion. I got up and wrote down the idea and spent most of the night thinking about it.[12]

This account should put an end to the story that the idea came to Banting in a dream. Beyond that, it is impossible to re-create the train of Banting's thought as the lecture and the article chased each other through his mind. Speculation about what Banting "must" have thought is hazardous because all his accounts of his inspiration came only after his life had been changed by its consequences. As will be seen repeatedly in this history, Banting was not a precise and reliable guide to the events in which he participated.[13] A clear example of this is the fact that he never afterwards checked his own notebook to find out exactly what he had written at 2:00 a.m. on the morning of October 31. He quoted himself from memory, and always incorrectly. In what is taken as his most authoritative statement of the history of the discovery of insulin, the Cameron Lecture in Edinburgh in 1928, Banting recalled, "I arose and wrote in my note-book the following words –

Ligate pancreatic ducts of dogs. Wait six to eight weeks for degeneration. Remove the residue and extract."[14]

The notebook is in the archives of the Academy of Medicine in Toronto. Banting actually wrote these words:

Diabetus
Ligate pancreatic ducts of dog. Keep dogs alive till acini degenerate leaving Islets.
Try to isolate the internal secretion of these to relieve glycosurea[15]

The obvious comment that Banting didn't know how to spell "diabetes" (and "glycosuria"), let alone treat the disease, is a bit unfair. It will cheer the modern medical student to know that Banting had never been a good speller. He never became one. In any case, the spelling in a person's private notebook ought to be his own business.

The more interesting aspects of the actual notation, as opposed to the remembered one, are twofold. First, the true notation does not contain the word "extract." None of the documents written in the first six months after Banting conceived his idea contain that word. All Banting wrote down was the idea of ligation, waiting for degeneration (he may not have known how long it would take), and then, "try to isolate the internal secretion."

Second, Banting wanted to try to isolate the internal secretion "to relieve glycosurea." He seems to have been identifying diabetes with glycosuria in the traditional way, rather than referring to the newer notion of hyperglycemia as the important condition to be relieved. The possible significance of these points will become clear shortly.

III

The morning after he wrote down his idea Banting mentioned it to Professor Miller at Western. Miller was a neurophysiologist and knew little about research in carbohydrate metabolism. It sounded like a good idea, he told Banting, but surely someone had tried it before. Banting may have asked Miller if it was possible to mount a project at Western to try out the idea. Miller apparently replied that facilities were not available for that kind of work; there were no quarters for large animals, such as dogs, in the old medical building, for example. Besides, Banting ought to consult someone who could be more helpful. Fortunately an expert was close by, in Toronto where J.J.R. Macleod had been professor of physiology since coming over from Western Reserve in 1918. Miller advised Banting to talk to Macleod.

Banting also consulted the professor of pharmacology at Western, J.W. Crane, who knew of no work on the subject. Banting remembered going to the library to look up the literature and finding nothing. This is a surprising statement, even allowing for inadequacies in Western's medical library. It indicates either bad memory by Banting or an inability to search medical literature properly. That night Banting also had a long talk with Bill Tew. Part of his excitement came from his having seized upon the idea as a way out, something worth dropping practice for.[16]

It happened that Banting was going to be in Toronto the next weekend for the wedding of one of C.L. Starr's daughters. At the reception or afterwards, Banting told a number of his acquaintances from surgery about his idea. "I wished to give up practice in London immediately and commence work," he wrote. "They all advised against such a radical move."[17]

Banting saw Professor Macleod, whom he had not known except by reputation, in Macleod's office the next day, Monday, November 8. Shy and inarticulate at the best of times, Banting could not have been at ease. Macleod, on the other hand, was a very senior, very articulate professor, giving up a few minutes of his valuable time as a courtesy to a University of Toronto graduate whom he had never met.

Banting told Macleod that he was interested in doing research work to search for the internal secretion of the pancreas. We do not know how much or how little he knew about previous researchers' attempts to find it (the survey in the Barron article actually said very little). He may have known enough to mention to Macleod, and if not Macleod certainly mentioned to him ("this point immediately came up in our discussion," Macleod wrote later) that many others had tried to prepare an extract of pancreas which contained the internal secretion. They agreed that the problem with such extracts may have been that they also contained the powerful digestive ferments of the external secretion, and that these may have acted to destroy the internal secretion.[18] Banting said that perhaps the use of duct-ligated pancreases would get around that problem, because (as the Barron article showed) the effect of duct-ligation would be to destroy the cells producing the digestive ferments.

"He was tolerant at first," Banting wrote of Macleod at that meeting, "but apparently my subject was not well presented for he commenced to read the letters on his desk." "I found that Dr. Banting had only a superficial text-book knowledge of the work that had been done on the effects of pancreatic extracts in diabetes," Macleod wrote in 1922, referring to their first several meetings, "and that he had very little practical familiarity with the methods by which such a problem could be investigated in the laboratory."[19]

Macleod was confronted with a young surgeon who had walked in virtually off the street, had no significant experience in physiological research, and was talking, haltingly, about a topic he knew about only from standard textbooks and one article. As any conscientious professor in that situation would do, Macleod told Banting that many eminent scientists had spent years, sometimes their whole lives, in well-equipped laboratories working on the problem of the pancreas. They had not even proven conclusively that there was an internal secretion to be found, let alone found it. Research such as Banting proposed could not be undertaken lightly. Indeed it would be "useless" to attempt it without making a full-time commitment for several months.[20]

It is not clear from the documents exactly what Banting was suggesting to Macleod. His later statements that he had asked Macleod for ten dogs, an assistant for eight weeks, and facilities for doing blood and urine tests, are not contained in his or Macleod's 1922 accounts. Banting is maddeningly vague in 1922: "I told him carefully what I had planned...I then repeated my ideas to him."[21]

It may have been that Banting was canvassing the idea of duct ligation as a surgical technique for isolating the internal secretion. Well, what would you do then, after the ligation had caused the pancreas to degenerate? What is the next step? How do you go about proving you have the internal secretion? Macleod might have asked. By making an extract of it and giving it to a diabetic animal, Banting *might* have answered. On the other hand, his notebooks show that some months later, when the work actually began, Banting and Macleod proposed to graft a portion of atrophied pancreas into a diabetic animal as their first experimental approach.[22]

Grafting had been mentioned briefly in Barron's article (extracting had not been mentioned); since Hédon's work in the early 1890s this had seemed the surest way of proving that there was an internal secretion of the pancreas. In 1913 Allen had complained that "deplorably little" had been done in the way of experiments with grafts and pancreatic transplants, implying that it was a very interesting avenue for future research.[23] This may have been the avenue Banting was suggesting to Macleod, or it may have appealed to Macleod as more promising than playing around with extracts. Grafting might have been discussed as a technique for isolating the internal secretion; extracting might have been discussed; both methods might have been discussed.

Whatever technique they were talking about, Macleod started to become interested. Banting remembered repeating his ideas to the professor:

> ...he sat back in his chair with closed eyes for some time. Then he began to talk. He thought that "this might be the means of getting rid of the external secretion." As far as he knew this had never been tried before. "It was worth trying" and "negative results would be of great physiological value." This latter phrase he repeated at least three times.[24]

Speculation is in order here and is permissible because we have some idea of Macleod's knowledge of the literature. Whether he and Banting were discussing grafting or extracting, what must have appealed to Macleod as never having been tried before was the idea of somebody experimenting with degenerated or atrophied pancreas. Now there was nothing new in the idea of producing degeneration or atrophy of the acinar tissues by ligating the pancreatic ducts - all sorts of researchers had done this. Their interest, however, had been almost entirely in measuring the relative amounts of degeneration that took place in the various components of the pancreas, particularly the relative changes in the acinar and islet cells. Nobody, it seemed (except, perhaps, Lydia Dewitt, with her unsatisfactory *in vitro* testing methods), had taken a pancreas in which the acinar tissue had been induced to atrophy and tested to see if it contained the alleged internal secretion. Nobody had either tried to prepare a graft or administer an extract using a fully degenerated pancreas. And yet, theoretically, if there was an internal secretion, and if it did come from the islets of Langerhans, and if it was the acinar cells but not the islets that degenerated after the ducts were ligated, and if two or three other conditions held good, then perhaps some interesting results would follow.

Even if the results were negative, it was the kind of experiment that ought to have been tried long ago, if only for completeness' sake; in that sense "negative results would be of great physiological value," probably valuable enough, for example, to write up for publication. Another possible consideration for Macleod might have been the thought that almost all experiments done in the past, with pancreatic extracts or by any other method, might show different results now that blood sugar could be tested easily and quickly. So Dr. Banting, superficial as his knowledge might have been, halting as his presentation undoubtedly was, had produced a suggestion worth thinking about.

On second thought, there was the great difficulty researchers had commonly had in getting ligation to work so that the pancreas atrophied. E.L.

Scott, it will be remembered, had reasoned the same way as Banting and had wanted to try an extract made from degenerated pancreas, but had found the technique impractical. Scott had been a student; Banting was a surgeon. Macleod may have reasoned that a surgeon would succeed where a student had failed (and, of course, Banting's skills would be particularly important if the discussion was actually centring on transplants or grafts). They may also have discussed the possibility of failure, and/or other ways of attacking the problem of getting rid of the external secretion. Macleod wrote in 1922 that either at this meeting or at a later one, "I suggested freezing the pancreas and then extracting it at the lowest possible temperature with alcohol" - which was essentially Scott's technique. Assuming Macleod's memory on this point is correct (it seems to be supported by Best's 1922 account and was never explicitly denied by Banting), the suggestion passed by as part of the general discussion. As we will see, it became important a year or so later.

Banting had explained that he could not do the work at Western and wondered about the possibility of coming to Toronto to attempt it. Having cautioned Banting about the time it would take, and about the likelihood of negative results - however interesting the idea was - Macleod said that yes, if Dr. Banting wanted to come to Toronto, Macleod would take him into the lab.[25]

The warnings had their effect on Banting. "I was not inclined to give up appointments in Surgery and Physiology in London to get 'negative results of great physiological importance'." He told Macleod he would consider the whole matter carefully. [26]

IV

Before returning to London, Banting explained the whole situation to Dr. Starr, who functioned, Banting told Macleod, as his "father advisor." Starr said he would think it over and write giving advice. Two weeks later Banting wrote Macleod that he was still anxious to do the work, but was still waiting to hear from Starr.[27]

Prodded by Banting, Starr talked with Macleod in December, telling him of Banting's situation in London, and asking him whether there was enough likelihood of "anything real" coming out of the idea to warrant Banting giving up his position in London and coming to Toronto for several months. Macleod probably reminded Starr of the long history of failure in the search for the internal secretion and of Banting's comparatively unimpressive qualifications. Starr apparently told Macleod that Banting was a well-trained surgeon.[28] "He thinks it is very problematical," Starr wrote Banting, "and while he is very much interested in your presentation of the case to him, yet he feels as I do, that probably it would be unwise for you, at this time to give up your work there, and come here to undertake this work. He suggested also that you might possibly come in in the summer and put in a month or two then."[29]

Starr advised Banting to stay in London. He probably believed that Banting was using the research idea as an excuse to get out of his discouraging situation. Surely the thing for Banting to do was to settle down and work hard at building up his practice. As well, Banting already had a useful and promising connection with a medical school. Starr urged him to "stick to your post...I feel you have a great future there if you stick right with it."

Banting took Starr's advice, and settled down to spend the winter in London. He later claimed that he had returned to London from Toronto "more determined than ever to try the experiments. I read widely on the subject of carbohydrate metabolism and even read a little about diabetes. The more I read and thought on the subject and the more subsidiary experiments which I planned, the more impatient I became."[30]

Whatever reading he did, Banting made no entries about carbohydrate metabolism in the little notebook he used to record ideas and research proposals. When Macleod saw him later he was no more impressed with Banting's knowledge than at the first meeting. And in fact Macleod did not hear from Banting again for almost four months. It was not until March 8, 1921, that Banting wrote Macleod saying he would like to spend the second half of May, plus June and July, in Macleod's lab, "if your offer for facilities to do the research still holds good."[31]

Banting was actually considering a number of different things to do with the rest of his life. His practice was picking up, and his income rising, well beyond the break-even point by February. Edith, by all accounts, was encouraging him to settle down with his practice. His work as a demonstrator at Western and assisting Professor Miller had apparently gone well.[32] That, plus his interest in research, may have encouraged him to consider the possibility of full-time university work. He may have gone so far as discussing the possibility with the dean of medicine at Western either before or after his conversations with the people at Toronto; if so, the dean held out no hope of Banting getting the salary he wanted to support himself in a research job.[33] In March Banting was working on several experiments in Miller's lab, none of them relating to carbohydrate metabolism. On the same day that he wrote Macleod about summer plans, he also wrote to C.S. Sherrington, the distinguished professor of physiology at Oxford, asking his advice on an idea he had to study reflex action in the hind-limbs of kittens and dogs.[34]

In his letter to Macleod, Banting suggested that he might start a bit early, coming to Toronto during the Easter holidays "to do" half a dozen dogs so they would be ready for investigation in May. Macleod replied promptly that he would "be glad to have you come up here on May 15, as you suggest, to see what you can do with the problem of Pancreatic Diabetes," but explained that it would not be advisable to do operations over the Easter holidays. Between the end of holidays and mid-May everyone was so busy with exams and winding up the term that no one at the lab would have free time to supervise the animals, "and this supervision, as you know, is of extreme importance in all researches of this character."[35]

Even then, however, Banting had apparently not made up his mind whether he wanted to work in Macleod's or any other lab. His moods undoubtedly varied with the state of his practice and, above all, his romance. Some time during the winter or spring Edith apparently broke off the engagement, returning the ring.[36] At a time like that, Banting wanted nothing more than to get a long way away from his problems. He and Bill Tew talked that winter about joining the medical service of the Indian army and even wrote for details. A bit closer to home, Banting heard of an expedition going to the Mackenzie River valley in Canada's Northwest Territories to drill for oil. They were apparently considering taking a medical officer with them. The head of the expedition lived in St. Thomas, just a few miles from London. About the middle of March, Banting recalled (which would be after he had written Macleod proposing to come to Toronto for the summer), he decided to stake his future on the toss of a coin. "Heads I was to do the research, tails I was to go to the Arctic to search for oil." Three out of five tosses came up tails. The Arctic won.

So Banting took the next train to St. Thomas to see the oil man. "He explained that he was not sure of taking a medical officer, but that if they took one, I could have the job." During March and April, 1921, when Banting is popularly thought to have been waiting with "gnawing impatience and mounting eagerness" to start searching for the internal secretion of the pancreas in Toronto,[37] he was actually waiting for a letter offering him a job as doctor to an oil expedition. A letter finally came saying the group had decided not to take a doctor.[38]

V

"Since nothing presented itself," Banting wrote of his schemes for escaping from London, "I turned the key in my office on the morning of April 26, 1921, parked my suitcase at the station on the way to my last class at the medical college and took the noon train for Toronto."[39] He was committed to the work now, even though Dr. Starr was still advising against it.[40] With no idea how the research would turn out, Banting paid enough attention to Starr that he did not, as some have thought, immediately close out his affairs in London. He kept his house and could have gone back again if the experiments failed.

Banting met again with Macleod to plan the work. Macleod was no more impressed now than earlier with Banting's knowledge of previous research on the problem or of the techniques he might use in the lab. But the work was to go forward. Again, there is no authoritative record of what was said at the meeting, or meetings. "I worked out with Dr. Banting a plan of investigation," Macleod wrote sixteen months later, "the first step of which was to render one or two dogs diabetic by extirpation of the pancreas so that he might make himself familiar with the cause of this condition in animals....At the same time I advised him to tie the ducts in several other animals so that the gland might be suitably degenerated...." Macleod also advised Banting to use Hédon's method of pancreatectomy and gave him references to Hédon's work in the literature.[41]

At one of their meetings, apparently the one in which Macleod gave Banting complete directions for the work, Banting met J.B. Collip, a professor at the University of Alberta who had an interest and some expertise in the study of glandular secretions and the making of tissue extracts. Collip, who had a Ph.D. in biochemistry from Toronto, was passing through on his way for a summer's study in Massachusetts, and was consulting with Macleod about coming to work at Toronto for part of his sabbatical the following year. He planned to work on a very different problem from the one Banting and Macleod were discussing, but would have found the talk interesting. Several years earlier he had published a good summary article on internal secretions and he had recently been giving animals injections of different kinds of tissue extracts, measuring their impact, along with adrenalin, on blood pressure.[42]

Whatever experimental techniques Banting and Macleod planned to use, the results would be measured by tests on urine and blood. Banting had no practical knowledge of how to do this kind of chemical testing - in fact, as his original October 31 note seems to indicate, he may not have been aware of the sophisticated methods of blood testing available to researchers - and would obviously need help. Banting later remembered that he had asked Macleod for an assistant from the beginning, but although one would expect to find it, there is no reference to an assistant in any of the letters they exchanged about the work.[43] Perhaps the matter did not come up until May. Or it might have come up earlier and Macleod, knowing he had student fellows on hand, had assured Banting it would not be a problem.

One day in May, Macleod introduced Banting to the two student assistants he had employed through that winter, Charles Best and Clark Noble. They were fourth-year students in the Honour Physiology and Biochemistry course, picking up extra money as demonstrators and research assistants for Macleod. Both were planning to do a Master of Arts degree with him the next year. Macleod had mentioned Banting in his lectures that winter, Best remembered, saying that he might be coming to Toronto to work on the pancreas. Now Banting explained his hypothesis to Best and Noble.[44] They were going to assist him. It was probably on that same day that Macleod also showed Banting the little room in the physiology department once used for surgical research. Nobody had operated in it for more than a decade, though. As Banting remembered it, the room "contained the truck and dirt of the years."[45]

Macleod held out no false hopes when he talked to Noble and Best about assisting Banting. This kind of research, going after the internal secretion, had been tried many times before and had always failed. "There is always a chance," Noble remembered him adding. Banting later heard, apparently from Best, that Macleod told the two students that the project would probably go up in smoke, but they would at least learn something about surgery from the work; also, as good scientists, "we must leave no sod unturned." Macleod apparently left it to Best and Noble to decide how they would divide the time to be spent with Banting.[46]

The practical problem of arranging the assistance was that in its normal rhythm the University of Toronto went on holidays in July and August. Term appointments and salaries, such as Best and Noble enjoyed, ended on June 30. But Banting intended to work in July, and presumably would need help then. So one of the two students would have to split his summer holidays, taking some time off in May or June and then working in July while the other was on holidays. The contemporary evidence suggests that Best and Noble flipped a coin to see who would go first, and therefore not have to work in July. There is no evidence for the legend in Toronto that the prospect of working with Banting on his wild idea was so unattractive that Best went first because he *lost* the toss. Actually it was the prospect of having a broken summer holiday that was unattractive. Research with a surgeon like Banting would be interesting no matter what the results. So the winner of the toss would work first with Banting. Best won.[47]

VI

Banting was in London for most of the first half of May. On the morning of Saturday, May 14, he presided at an exam for Western's fourth-year medical students. After the exam they gave their demonstrator a box of cigars. Banting "escaped" from London on the next train.[48] Best wrote his last exam on Monday, May 16. Either that day or early the next,[49] Banting and Best cleaned up the physiology department's filthy operating room. They washed the walls and the ceiling. Just as they were about to mop the floor, someone from the floor below complained that water was leaking through. So they cleaned the wood floor of the little room on their hands and knees.

Banting had brought his instruments with him. Towels for the operation had to be borrowed. When everything was ready, on Tuesday May 17, Macleod joined them to begin the first experiment.[50]

CHAPTER THREE

The Summer of 1921

The plan of attack was to start with pancreatectomies on several dogs. Banting could get used to doing the operation, Best could practise his blood and urine tests, and both of them could become familiar with the diabetic condition as it develops after pancreatectomy. These first dogs would soon die from their diabetes. In the meantime, following Banting's idea, the pancreatic ducts would be ligated in several other dogs. They would recover and live more or less normally, but over a period of several weeks their pancreases, unable to secrete juice into the duodenum, would gradually atrophy or degenerate. Then, in the crucial experiment, Banting would re-operate on these dogs, remove the atrophied pancreas, and see if he could somehow use that material, which he believed would contain the internal but not the external secretion, to improve the condition of still more dogs made diabetic from pancreatectomy. Using Banting and Best's original notebooks, it is possible to describe these experiments almost day by day and dog by dog. Most readers should not find it necessary, though, to keep careful track of these dogs or their numbers. Banting and Best sometimes lost track themselves.[1]

I

Banting had probably never done a pancreatectomy, an operation used almost solely in animal research, so Macleod was present to assist and instruct on the first dog. A brown Spaniel female, number 385 in the university's records of dogs used in its labs, was fully anesthetized, strapped to the operating table, and its belly opened by an incision down the middle of the abdomen. Clamps held the abdomen open while the surgeons worked on the pancreas. Removing the pancreas mainly involved cutting it away from the mesentery tissue to which it was attached, while taking care to ligate the many blood vessels which supply the pancreas before cutting through them. Otherwise the bleeding could be uncontrollable. It was important, as well, not to damage nearby blood vessels supplying other organs, or those organs themselves.*

Macleod had decided that Hédon's two-step method of pancreatectomy should be used. The whole pancreas would not at first be removed. Instead, after most of the pancreas had been dissected, a remnant (which Banting called a pedicle), with its blood supply intact, was pulled up through the abdominal wall and sutured or grafted just under the dog's skin. The pedicle gave the animal continued pancreatic function so it would not immediately become diabetic. Instead, the dog could recover from the trauma of the surgery and the incision would heal. About a week later, when the dog's systems had recovered, the researchers could briefly anesthetize the animal, do the second stage by snipping away the pedicle, and then make their observations as the now totally depancreatized, but otherwise healthy animal developed diabetes.[2]

It was not very complicated in theory, and was the kind of operation that could become routine as a careful researcher or surgeon developed experience. The first steps on dog 385 seemed to go well. After the pedicle had been sewn under the skin, the abdomen was closed up, the urethra was enlarged for catheterizing later on, and the dog was allowed to recover. The first operation on May 17 took about eighty minutes.

The next day, Wednesday, Banting and Best began work on their own. The pair had a lot to learn. The first dog they tried to work on died from an overdose of the anesthetic. Another dog survived the anesthetic, but it was a very small animal, and Banting had great difficulty getting at and ligating the blood vessels supplying the lower end of the pancreas. He found himself working in a pool of blood. The dog died after the operation.

They tried again on Thursday, using a larger dog and a longer incision. There was much less bleeding and the dog, number 386, survived. The first dog, 385, was not healing properly, however. It died on Friday. Banting did an autopsy, concluding that he had to be more careful not to interfere with the major blood vessels; he also decided not to sew the pedicle under the skin, but rather to surround it with a "rubber cigarette" so it could be drained from time to time. He cut through the skin of dog 386, which was in bad shape, to allow its graft to drain. No luck. Dog 386 died the next day, Saturday the 21st.

They had worked on four dogs that week. All were dead. There was nothing to do but start a fifth, dog 387, which was operated on that

*It is not clear whether rubber gloves were worn during these pancreatectomies. Physiologists doing experimental work on animals, especially animals like dogs which had fairly tough constitutions, were still rough-and-ready in their methods. Later in the twenties and thirties at Toronto researchers commonly did pancreatectomies without gloves, using their finger-nails to scrape the pancreas away from the splenic artery. Those who remember Banting's own operating techniques from the 1930s contradict each other: some say he followed standard bare hands practice; others say that as a surgeon, who had first worked on humans, he was a stickler for sterilization. Of course the bare hands and finger-nails would have been well scrubbed.

Saturday. Learning from his mistakes, Banting seems to have operated more carefully. He brought the pedicle out to the surface of the skin, placed the finger from a rubber glove around it, and sutured it directly to the surface for easy access and drainage. There was not much bleeding. "Operative prognosis good," Banting noted for the first time. He came in to the lab the next day, Sunday, and must have been pleased to find the dog in fair condition.

With dog 387 apparently healing from the first stage of its pancreatectomy, Banting and Best were ready to start work on the dogs whose ducts were to be ligated. Again they found the going difficult. The first dog they ligated - it was the same kind of major surgery as pancreatectomy except that Banting went into the abdomen to tie the ducts rather than take out the pancreas - died three days later from infection. The second dog recovered and healed, but Banting noted that he was not sure he had actually found and ligated the ducts. They are small and difficult to find; Banting realized he might have just ligated a piece of pancreatic tissue. The third ligated dog survived two days before dying of general infection. There is only a summary note about it: did Banting tear out the missing page in his notebook from sheer frustration? By week's end seven of the ten dogs they had experimented on in the first two weeks of the research were dead.

Banting must have worried about the rate at which he was going through the University of Toronto's dogs. Early the next week a dog appears in his notebooks with the same number as the one that had just died. The double-numbering is occasionally repeated later in the summer. He and Banting did not steal dogs, Best remembered, but bought them on the streets of Toronto for one to three dollars each. He remembered Banting leading one back to the lab by his tie.[3] They would not bother asking how their suppliers procured dogs. If any dog-napping was done to supply the university's animal labs, as Torontonians occasionally feared, and the anti-vivisectionists in the city, whose activities greatly worried medical researchers, regularly charged, at least technically it was not by Banting and Best.

II

There was gradual progress. As dog 387 recovered from its first stage, Best could begin making his tests. They were fairly normal, up-to-date chemical procedures - the Myers-Bailey modification of the Lewis-Benedict method of blood sugar estimation, Benedict's qualitative and quantitative tests for sugar in the urine - but still a long way from the quick, simple, and more accurate tests available today. Best's work involved the use of several reagents, careful preparation and measurement, and various chemical procedures involving centrifuging, filtration, evaporation, and precipitation. The blood testing was the most advanced procedure: as we have seen, it was a relatively new way of measuring the diabetic condition. But Macleod also expected Banting and Best to use more traditional tests of the urine. These involved not just finding the amount of sugar in the urine, but also determining the ratio of glucose to nitrogen. This G:N ratio, also often called D:N (dextrose to nitrogen), was thought to be a particularly accurate reflection of the diabetic state.

Dog 387, the first to be tested, showed normal reactions. Although it had lost most of its pancreas, the presence of the pancreatic remnant was sufficient to keep its blood sugar within a normal range of .085 to .150 per cent. There was no sugar in its urine. Then on May 28 Banting removed the pedicle, making the pancreatectomy complete. The next day the dog's blood sugar had risen to .35 per cent and there was sugar in its urine. The next day the blood sugar was .42 and the D:N ratio was 2.7:1. These figures were both indications of a diabetic state, although Banting and Best may have been a bit disconcerted that the D:N ratio was not higher; they seem to have believed, incorrectly, that it ought to be 3.65:1 in a completely diabetic dog.[4]

Dog 387 died on June 1, partly from its diabetes but probably mainly from the infection in its abdominal wall that Banting found in his autopsy. Still, this was the first experiment to have gone more or less according to plan. Banting already had another dog partially depancreatized, and soon added one more. He ligated the ducts of several other dogs. By the end of the fourth week of experiments, on Sunday, June 12, things must have seemed to be going fairly well. As soon as the two partly depancreatized dogs healed they would be ready for the second stage and more observations of the diabetic condition. Meanwhile there were now six duct-ligated dogs, and a seventh done on June 13, whose pancreases were presumably atrophying according to theory. It would be just a matter of time until they could be opened up and the critical experiments begun.

But what exactly did Banting plan to do? All later accounts, including his own, state that he and Best planned to take out the degenerated pancreases from the ligated dogs, prepare an extract from these, and then administer the extract to other diabetic dogs. As has been mentioned, however, none of the contemporary documents refers specifically to an extract. The clearest, in fact the only, statements of the plan of work are to be found in Banting's original notebook for June 9 and 14. On both these days he seems to have talked over the work with Macleod. The full notebook entries are as follows:

June 9
suggestion.

- Have depancreatised dog c̄ pedicle.
- graft into it remnant of degenerated pancreas

- later remove pedicle
- then remove graft

- Prof. McLeod -
  - 1 put remnant free in peritoneum;
    2 put remnant subcutaneously.
    3 emulsions.
        whole remnant in one shot
    4 repeated smaller shots.
    5 50 gms glucose to totally diabetic dog
        50 gms glucose with whole gland remnant.

June 14. Dr. McLeod's parting instructions

have dogs diabetic c̄ DN ratio constant for 3 days. meat diet.
(1) intra peritoneal graft
(2) subcutaneous graft
(3) whole remnant intravenous injection.
(4) Divided aq. 2 h intravenous
(5) subcutaneous injection.

---

microscopic sections of remnant before and after transplant.

These notes are open to slightly differing interpretations, and it is not clear how the interchange of ideas between Banting and Macleod had developed. In view of the state of research at the time and the widespread interest in grafting as a technique for working with the pancreas (experimental surgery was then in a blush of enthusiasm about the possibilities of grafting) it seems that Banting and Macleod had agreed that the first approach would be to graft pieces of degenerated pancreas into diabetic dogs. The second or perhaps follow-up approach would be to make an emulsion of degenerated pancreas and inject it, in various doses, into the diabetic dogs.[5]

The widely held belief that Macleod set Banting and Best to work and then immediately left town for his holidays is not true. The young men had been at the research for almost a month and seemed to have worked through their early technical problems before Macleod left. He talked over the situation with Banting in June and gave fairly explicit "parting instructions" before leaving. Banting wrote down Macleod's summer address in Scotland. He also noted the summer address in Massachusetts of J.B. Collip, the biochemist who had been present at one of the discussions and knew roughly what Banting intended to do. Banting presumably thought Collip might be someone he could contact for help or advice on short notice.

III

By the third week in June Banting was working completely on his own, for Best, too, was out of the city, having gone off to the Niagara region for ten days of militia training. Charles Herbert Best, "Charley" to his friends, had the world at his feet in the summer of 1921. He had just graduated with his Bachelor of Arts degree from Toronto's Honour Physiology and Biochemistry course, had an interesting summer job with Fred Banting, and knew he would be working again the next year for Professor Macleod while taking an M.A. He was twenty-two years old, a strikingly handsome, tall, blond-haired, blue-eyed, athletic young man. He had grown up on the coast of Maine, where his father, a Canadian with deep family roots in Nova Scotia, was a small-town general practitioner. Charley had come to Toronto in 1916 to finish his high school education and then entered university. He enlisted in the Canadian army after finally finding a homeopathic doctor who certified him as medically fit, not noticing his heart murmur, and had reached England as a sergeant in the artillery when the war ended. He somehow got home quickly enough in 1918-19 to salvage his second year in the "P&B" course at Toronto.

Those last years at the university were a wonderful idyll of fraternity parties and dances, picnics, tennis, riding, golf, semi-pro baseball, and, of course, attention to studies. Best's marks were not outstanding, but they got better the further he went in his course, until in the final year his second-class honours standing was worth a share of the silver medal in his class with Clark Noble. Charley Best was outgoing, sociable, and popular. In choosing him as a demonstrator, and often inviting him to his home, Professor Macleod had obviously picked Best as one of his most promising students. The young man apparently thought he might eventually train as a surgeon and dreamed romantically of going off to practise in South America.[6] He would be accompanied in his adventures by the strikingly beautiful Margaret Mahon, a Presbyterian minister's daughter whose family had recently moved from the Maritimes to Toronto. Charley met Margaret at a sorority party in 1919. They became engaged in 1920, a beautiful couple whose pictures from those years call back all the best nostalgia of the Twenties, like Scott and Zelda Fitzgerald without the crack-up. In fact Charley and Margaret were a more handsome couple than Scott and Zelda.

Best would have enjoyed working with Banting, even though as yet the research had not produced any results. The older man's unsettled life, country ways, and personal insecurity all contrasted sharply with Charley's happy situation; but they were both products of small towns, both veterans, both interested in medicine, sports, and Saturday night dates. Best was learning some surgery from the partnership (Banting some chemistry), and he may have had a special interest in research relating to diabetes. In 1917 one of his aunts, first a nurse and then a patient with Joslin in Boston, had died from the disease.

IV

While Best was enjoying riding with the militia, Banting found work in the lab increasingly frustrating. On Wednesday, June 15, just before Best left, Banting completed his second successful two-stage pancreatectomy, on the second dog to be numbered 386. It rapidly became diabetic. "You ought to get the 3.65:1 D:N on Sat or Sunday," Best wrote to him in their joint notebook as he left for Niagara. Banting did not get the expected ratio. The dog's D:N stayed constant at about 2.5:1 over the weekend and then started dropping in the next week. Later in that week there was more frustration as Banting, apparently operating without assistance, lost another dog from bleeding during the first stage of a pancreatectomy. He apparently went out on the streets and bought a dog to replace it.

The notebooks also show changes and corrections in the figures for urinary sugar, as Banting started using different chemical solutions from those Best had left behind to use in testing. In his 1940 account, nowhere else, Banting wrote that he found Best's solutions to have been anything but standard and the glassware filthy. Banting decided to have his nitrogen estimates done by a friend in the biochemistry building. He must have been even more frustrated when the new, revised figures showed dog 386 with an even lower D:N ratio, well under 1.0. Whatever was happening, Banting apparently decided that Best was at fault.[7]

Best came back to Toronto on Saturday the 25th or Sunday the 26th. He went to see Margaret and then dropped into the lab about eleven at night. "I was waiting for him," Banting recalled in 1940, "and on sight gave him a severe talking to....

> I told him that if he was going to work with me that he would have to show some interest, that his work was totally unsatisfactory, that he lacked accuracy and was too sloppy, and I ended up by telling him that before doing another single thing he must throw down the sink every solution that he had been using, wash every bit of glassware and make up new solutions that were truly "normal".
>
> When I finished thus setting him out he gazed fiercely at me for some moments in silence - he looked very defiant - his fist opened and closed - he was very vexed. I thought that he was going to fight and I measured the height of his jaw. He delayed and I feared that he was not going to fight. Suddenly he turned on his heel, went upstairs

> and I heard rough usage of glassware for some time. He worked all night (for the first time in his life but it was the beginning of many). In the morning he left and I inspected. Everything was spick and span. We understood each other much better after this encounter.[8]

Best apparently never referred to this incident.[9] Banting did in verbal accounts, embellishing it into a kind of fight and saying he had thrown Best's solutions down the bathtub. He tended to embellish, and no one remembered a bathtub in the medical building. Still, the incident probably took place, perhaps as Banting describes it. Crossed-out figures in the notebooks are the circumstantial evidence of Banting's frustration. At all times he was an intensely direct man, quick to anger, ready to fight anyone on any matter of principle. This was not the last time an angry Banting, ready to fight, would confront his research associates in the next eighteen months. By the same token, Fred Banting could shake hands after the air had cleared or the dust settled, and go on to work with a person more closely than ever. His having written "we understood each other much better after this encounter," suggests that it was an important step in the cementing of Banting and Best's partnership.

Best not only cleaned his glassware and re-did his solutions, he had decided to stay on to take Noble's place in July. By all accounts, including his own, Noble fully concurred in the decision. Everyone apparently realized there was no point in a new assistant having to learn all the procedures over again at this stage of the work. Banting thought the students' interests in being near their girl friends (Noble's was out of town) also had something to do with the decision.[10]

V

Best's return and their better understanding of each other did not make the work go better. Creating experimental diabetes in a totally depancreatized dog seemed to them almost impossible. Dogs that healed well enough from the first stage to have their pedicle removed would still not become diabetic; others seemed to become diabetic after the first stage and did not heal. All healing was difficult in the extreme heat and humidity of a Toronto July. Operating in the dingy little room next to the animal quarters, right under the gravel-and-tar roof of the old medical building, was a sweltering ordeal. The animal quarters, by all accounts, were fairly dingy themselves. Banting and Best found they had to help the attendant do the work of caring for their dogs and keeping their cages clean. They worked in heat and dirt and unbelievable stench.

Down two flights of stairs in the physiology department they had been given bench space in room 221, a kind of anteroom between a corridor and one of the main laboratories. Part of it was used as a storeroom for supplies and old apparatus. They saw a lot that summer of Dr. Fidlar, the inhabitant of the main lab, who was using the most complicated apparatus they had ever seen to do intricate experiments on the respiration of one frog.[11]

By early July it had been about five weeks since the pancreatic ducts had been ligated in seven of the dogs. There were two partially depancreatized dogs on hand. Perhaps thinking they were ready for the crucial experiments (or perhaps just to check on the degree of pancreatic degeneration), Banting opened up one of the ligated dogs on July 5. Its pancreas, which should have been markedly degenerated, was apparently quite normal. Something had gone drastically wrong. The duct ligation hadn't worked at all.

That day and the next, in temperatures over 97 degrees fahrenheit, they opened up all seven of the ligated dogs. Five of the seven had normal pancreases. Only two of the dogs showed the expected degeneration. As Banting had feared when he first did some of the dogs, he had failed to isolate and close off all of their pancreatic ducts. The fact that he had used catgut for his ligatures - a stiff material which can easily loosen - had added to his problem. The five dogs were done over again, this time with silk ligatures and sometimes two or three applied at varying tensions.[12] Two of the five dogs died within a day of the operation. So did the two partially depancreatized dogs, suffering from infection in the heat. The week was a disaster.

In fact the whole research program was not far from total failure. Banting and Best had experimented on nineteen dogs. Fourteen had died, no more than two of them according to the research plan. There were five duct-tied dogs left, and only two of them had gone according to plan. Banting and Best took a long weekend off and started over again on Monday.

VI

The pace was a little slower in the second week of abnormally oppressive heat. On Monday, July 11, they did a first-stage pancreatectomy on dog 410, a white short-haired mongrel terrier. While waiting for it to heal they took most of the week off. Charley wrote Margaret on the 14th that he was getting tired of the combination of heat and lab work.[13] The next Monday the terrier's pedicle was removed, completing the operation. The dog did not become very diabetic, its blood sugar ranging from .15 to .22 and its D:N ratio well under 2.0:1. Another dog, 406, a long-haired yellow collie, was started on the 22nd. Tests later in the week seemed to show that dog 410 was becoming more diabetic.

On Saturday, July 30, they decided that the time had come to put the work to the test. The pancreases in the two original duct-tied dogs had been degenerating for seven weeks, surely enough time to destroy all the cells producing the external secretion. That morning Banting removed the pancreas, apparently atrophied, from dog 391 (whose ducts had been ligated on June 7). Whatever plans he had made for transplants or grafts were abandoned in favour of the much easier and quicker method of making an extract of 391's shrivelled pancreas.

"We followed your directions in preparing the extract," Best wrote to Macleod in one of their reports ten days later. They sliced the pancreas up, putting the pieces into a chilled laboratory mortar containing ice-cold "artificial blood" or Ringer's solution, a mixture of salts in water commonly used to preserve tissues. The mortar was put in a freezing brine solution until its contents partly froze. The half-frozen pancreas pieces were then macerated (ground up with sand and a pestle). The solution was filtered, apparently through cheesecloth and blotting paper, to eliminate the solid particles. The filtrate, a pinkish-coloured liquid extract of degenerated pancreas, was warmed to body temperature and was ready for injection.[14]

Reflecting his experience with older techniques requiring larger samples of blood, Macleod had apparently advised Banting that it would be necessary to anesthetize a dog and insert a special cannula or tube each time a sample of blood was to be taken. As Kleiner had before them, Banting and Best discovered that this was not necessary. With a fine needle they were able to get repeated blood samples from veins. After the first few punctures most dogs would lie quietly. The same technique was used for the administration of extract.[15]

At 10:15 in the morning of July 30, 1921, Banting and Best injected four cc. of their extract into a vein of the white terrier, dog 410. Its blood sugar at the time of injection was .20. An hour later, at 11:15, the blood sugar had fallen by 40 per cent to .12. At that time five more cc. were injected. In the next hour the blood sugar barely moved, to .11. Despite another injection, by 2:15 it was starting to rise again, to .14.

To see how the extract influenced the dog's ability to metabolize sugar, Banting and Best decided to give it sugar (20 grams in 200 cc. of water) through a stomach tube. They were not practised in the technique: "Tube first passed into lung dog nearly drowned [*sic*] - completely recovered in 15 min." the notebook records. This was at approximately 2:15 in the afternoon. Three five-cc. injections were made at hourly intervals. As the copy of Banting and Best's chart shows, the terrier's blood sugar rose and the extract did not bring it down. The increase was not as marked as the administration of sugar had caused a few days earlier, though, and very little of the sugar, less than one gram, appeared in the dog's urine or vomit during a five-hour period. After taking a final blood sugar reading at 6:15 on that Saturday night, Banting and Best left the lab.

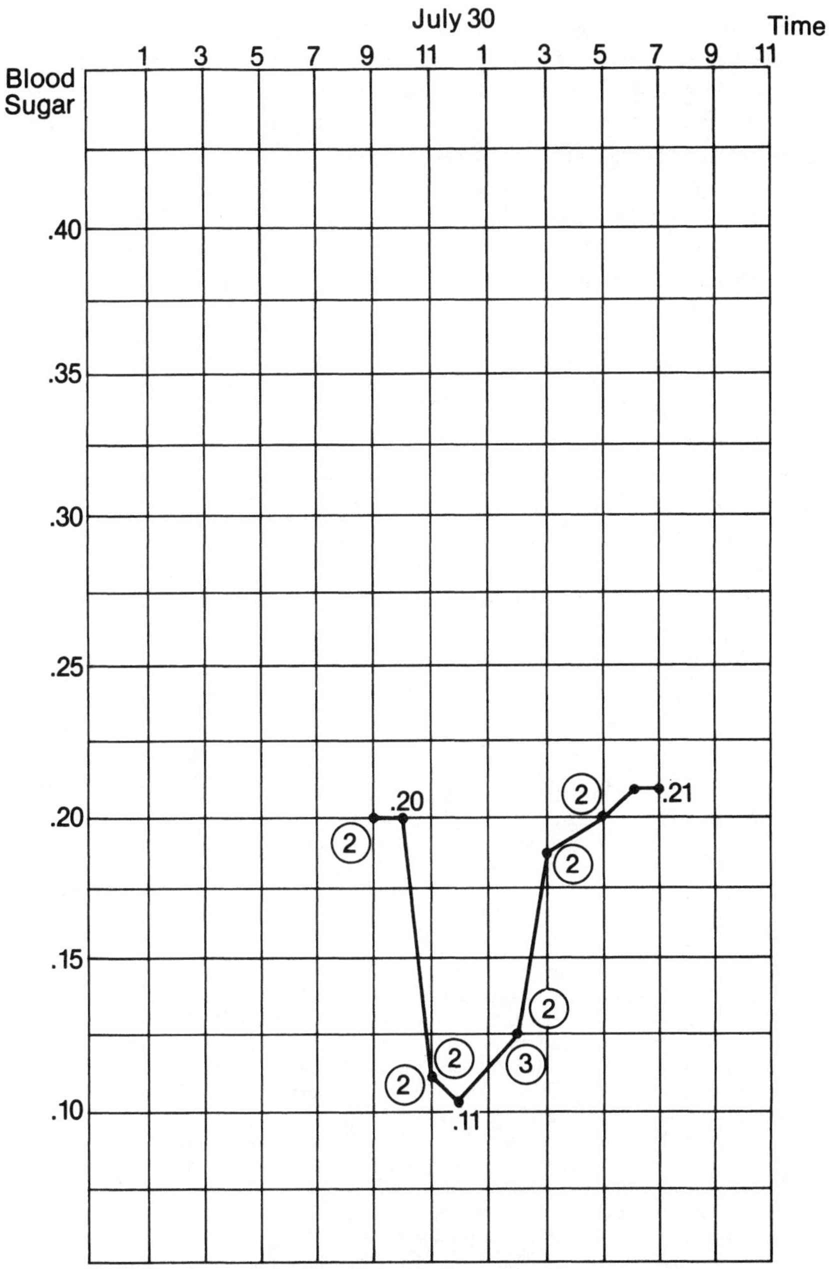

(2) Degenerated gland extract. 5 cc. doses intravenously.
(3) .20 gms. glucose in 200 cc. water by stomach tube.
Dog died July 31st - Cachexia.

(The notebooks indicate that the first dose was actually 4 cc.)

Chart 1: Dog 410, July 30

On Sunday morning they found the terrier in a coma. They took one blood sugar (.15) before it died. No autopsy was done on the dog, the first to receive the extract.* In their first published paper, Banting and Best mentioned that it had appeared to be "entering the cachexial condition characteristic of depancreatized animals which had become infected" when they decided to give it the extract. They also noted that the dog's blood sugar had not risen "to the level usually attained in completely depancreatized animals." Even so, they were encouraged by their result: "The extract seemed to have a marked effect," Best wrote to Macleod a few days afterwards.[16]

On Monday, August 1, they tried again. There was extract at hand from a second duct-tied dog. The only depancreatized dog, the collie 406, was in coma, on the brink of death.[17] Its blood sugar was very high, .50 at 12:10 p.m. Banting and Best injected eight cc. of their extract intravenously. At 1:10 the blood sugar was .42. "Dog able to stand & walk." An additional five cc. of extract brought the blood sugar down to .30 by 2:10, but the dog had lapsed back into coma. It died at 3:30 p.m. The experiment would not look good on paper - two injections on a dying dog, once again no autopsy - and was never written up. To Banting and Best it must have been

*Was it really the first dog to receive Banting and Best's extract? In his 1940 "Story of Insulin," Banting described an earlier experiment, never previously mentioned:

> During Best's absence [late in June] I had tested the effect of the extract of degenerated dog pancreas on a depancreatized dog whose blood sugar was 460 mg percent. Following the injection the percentage fell to almost normal and the condition of the dog improved in an astounding manner. I did not write down the results of this experiment in our notebook. I was so fearful in the first place that this result would not be obtained and when I did get the result I did not think anyone would believe it. This, the first extract tested, was given in the evening and I worked alone all night and finished the readings on the blood sugars in the morning light of the early hours after dawn. No one knew but myself (Banting 1940, p. 28).

The trouble with Banting's description of this incident is that it is impossible to correlate with the known facts about the experimental dogs in June. None of the diabetic dogs during Best's absence had a blood sugar near .46, for example. More important, all of the duct-ligated dogs can be accounted for in the whole series of experiments from May through August. Banting could not have obtained degenerated dog pancreas in June. The only way to make his story fit with known facts about the life and death of his dogs is to suppose that he actually used the fresh pancreas from dog 399, which he took out on June 23. If that happened and gave the result Banting indicates, it underlines the egregiousness of the error mentioned on p. 76 below. More likely, Banting misplaced the whole incident in time and memory. If it took place at all, it was probably early in August, when a supply of extract of degenerated pancreas was available. I suspect that Banting is wrong in saying Best was not present and that the incident was not recorded in the notebooks. My guess is that the incident is the administration of extract to dog 406, their second dog, with briefly sensational results. The germ of truth in Banting's account may be that this experiment was never described in any of the published papers. Another possibility is that he is describing an unnoted administration of extract to dog 409, a control dog which was dying during the night of August 14-15. Banting may have been alone in the lab that night; the dog's last noted blood sugar was .46, and Banting, with extract on hand, might well have been tempted to try once again on this new dog.

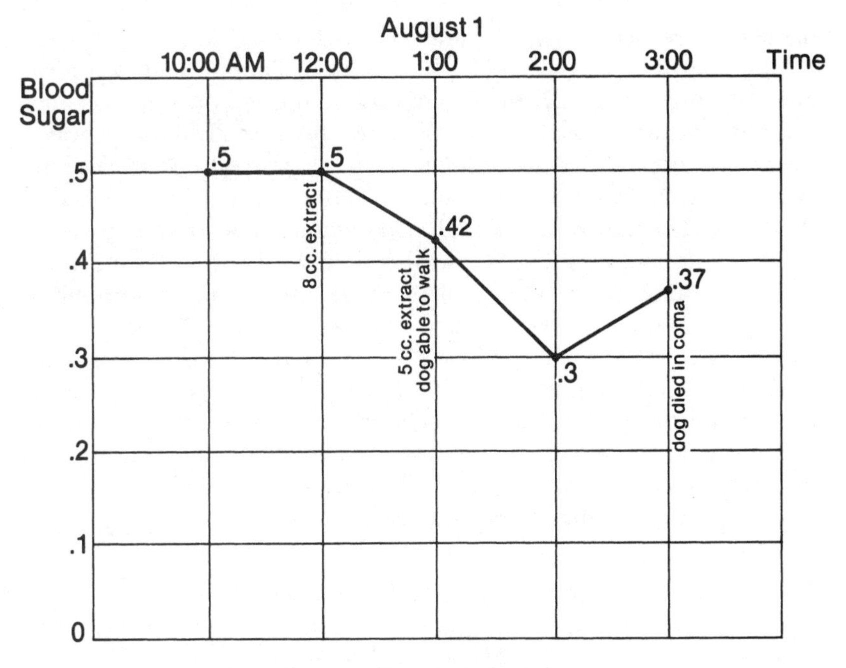

Chart 2: Dog 406, Aug. 1

impressive to see a dog come out of coma, stand, and walk, after an injection of their extract.

They still had more extract, but had run out of depancreatized dogs to try it on. It would take at least a week, perhaps much longer if the same failure rate continued, to get more dogs ready via the two-stage pancreatectomy. Why not speed up the procedure, Best urged, by doing the whole pancreatectomy at once?[18] Banting agreed to give it a try. On Wednesday, August 3, they did a total pancreatectomy, their first, on a young yellow collie, dog 408. "Operation was easiest yet," Banting wrote in his notebook, obviously surprised at how well it had gone. He never again bothered with the time-consuming Hédon procedure.

The next day the collie became the third dog to receive the extract of degenerated pancreas. Five cc. of extract caused its blood sugar to fall from .26 to .16 in 35 minutes early in the afternoon. That evening, another five cc. of extract reduced it from .25 to .18 in half an hour. The dog's general condition, Banting and Best noted, was good. They were still collecting and testing urine, with moderately favourable though slightly puzzling results. The volume of the dog's urine decreased after injections and urination finally ceased altogether.[19] Little attention was being paid to the D:N ratio (which did, however, average around 3.0 for dog 408). More and more as time went by, attention centred on blood sugar readings as the key to the experiments. In their notes on this experiment Banting and Best name their extract for the first time: "Isletin."

Now that they had a live dog to work on, Banting and Best could attempt more varied experiments. On Friday, August 5, they made extracts from the liver and spleen of dogs, prepared them exactly the same way as the extracts of pancreas, and injected them into the collie. Neither caused any significant change in the blood sugar. Injections of "Isletin" later in the day gradually drove down the blood sugar from .30 to .17, and the dog continued to look healthy. "General condition of dog seems much improved...appears more interested in happenings - more susceptible to pain." On Saturday morning they tried boiled extract on the dog. No effect. At midnight, with the blood sugar at .43, they began giving hourly injections of eight cc. of extract.[20] The blood sugar slowly dropped - to .37, .33, .29, and, by 4:00 a.m., .20. At that hour they decided to shoot the works, mixing all the rest of their extracts from the two dogs together, diluting it a bit, and giving the collie a massive injection of twenty-five cc. Result: "Dog suddenly became lifeless and appeared to be dying."

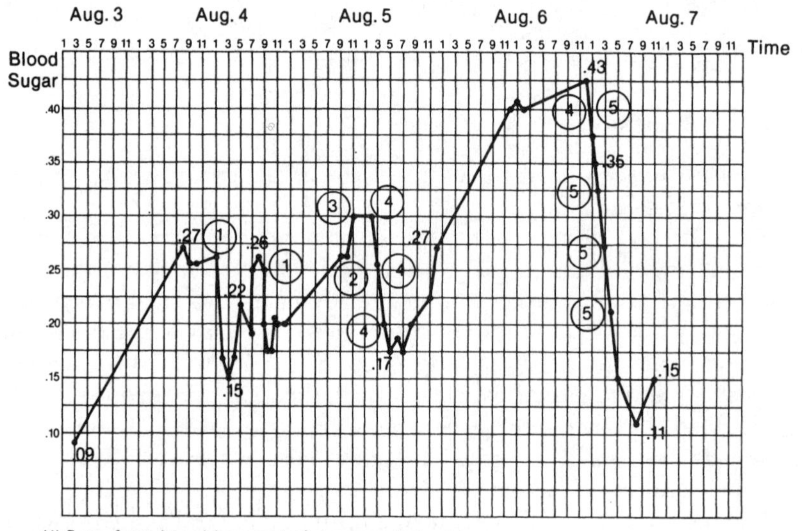

(1) 5 cc. four day old extract of degenerated pancreas.
(2) 5 cc. extract of liver.
(3) 5 cc. extract of spleen.
(4) & (5) extract of degenerated pancreas.
Dog died August 7th - general peritonitis.

Chart 3: Dog 408

Banting and Best thought the injection had caused an anaphylactic shock (similar to an allergic reaction). It might also have been a thrombosis. They tried to revive the dog by injecting large doses of Ringer's solution. It improved slightly that morning but died at noon. The autopsy showed widespread infection, stemming from the operation, which was considered the cause of death. This was the end of the first series of experiments.

VII

Banting and Best were excited by what they had seen. "We got fine results," Best wrote to Margaret on August 8.[21] "I have so much to tell you," Banting wrote Macleod the next day, "that I scarcely know where to begin." The extract "invariably" causes a reduction in blood sugar, he wrote; it improves the clinical condition of the dog; it can be kept active for at least four days, is destroyed by boiling, and extracts of other organs are inactive. But so many new questions came to mind. Banting had jotted them down in his notebook and on index cards, and now listed sixteen of them for Macleod. How could they get the most active form of "Isletin"? What were its chemical properties? Was it actually destroyed by the digestive enzyme, trypsin? Could a diabetic dog be kept alive on the extract? Was it universal in the animal kingdom and in its action? "The whole problem of tissue grafting." The extract's relation to the various forms of diabetes. Its mechanism of action. And other questions. Fifteenth on Banting's list for Macleod - it had been the sixth question to occur to him in his notebook list - was "its clinical application."[22]

"I am very anxious that I be allowed to work in your laboratory," Banting wrote. But he needed more help looking after the animals and needed better operating facilities. He had told Dr. Starr of their problems, and Starr had procured the use of the surgical research operating room for future experiments. They had only two dogs left with ducts tied. Banting wrote Macleod:

> I would like to do about ten so as to have a supply of the extract for you when you return. I have told Dr. Starr all about my results and he advised me to go ahead so I will proceed *slowly* and if I do not hear from you I will take it that I have your permission. Please let me know as soon as possible your wishes. I will not proceed immediately however as I am going to London Ont. to close up my affairs and have them off my mind, and that will give this letter time to get to you.

Whatever friction there had been between Banting and Best early in the summer had disappeared. Banting reported glowingly to Macleod: "Mr. Best has expressed the desire to work with me and I should be more than pleased to have him. His work has been excellent and he is absolutely honest, careful and impartial, and has taken a great interest in the work. He has assisted me in all the operations and taught me the chemistry so that we work together all the time & check up each other's readings."[23]

Even without the early frustrations and failures of the research, it had been a difficult summer for Banting. He had very little money - no more than one or two hundred dollars, he later estimated - no salary, and apparently no prospect of borrowing more from his parents. He earned a few dollars in May and June doing tonsillectomies for one of his friends and later sold some instruments for $25. For part of the summer he lived with his cousin and classmate, Fred Hipwell, minding the Hipwell house while Fred and Lillian and their baby spent a few weeks at a cottage. Then he moved back into the boarding house on Grenville Street where he had lived as a student, paying $2 a week for a little seven-by-nine-foot cubicle. He ate in restaurants, with friends when they invited him, sometimes at the Sunday night suppers of the Philathea Bible Class of St. James Square Presbyterian Church, which he had attended as a student, and sometimes in the lab.[24]

For Charley Best the summer of 1921 was a delightful round of tennis and swimming, golf and baseball, outings with Margaret, and interesting though sometimes frustrating work at the lab. Fred occasionally came on outings with them, dating one of the secretaries in the medical building, but spent much of his time worrying about Edith and his future. Relations with Edith were endlessly complicated - friends read the state of their romance by whether or not the diamond engagement ring was dangling from Fred's watch chain - and his future totally obscure. No one had been very enthusiastic about his research project, he knew, and even Macleod had not expected very much from it. "Worst of all, no one took me seriously," he wrote shortly afterwards. He seems to have taken his own work seriously, hoping from the beginning that he was heading not just for a contribution to physiological knowledge, but for a treatment of diabetes. One day in the summer of 1921 while he was driving Lillian Hipwell and her baby to the beach, he said to her, "Lillian, if what I am working on is a success, I will be a famous man, but then I don't think it will happen."[25]

Now, as his report to Macleod indicated, Banting was encouraged enough by his results to wind up his affairs in London. It would be weeks before Macleod's reaction to their report would arrive, Banting was not being paid for the work, and there is no evidence that anyone had given him any assurance of a paying job of any kind at the university. But he was so sure now that his future lay in Toronto that he was ready to burn his last bridge in London before even knowing how Macleod felt about the work.

VIII

First they decided to go ahead on their own with one more round of experiments, a big push in which they would work around the clock for several days. "It will be quite a crucial test for our Isletin," Charley wrote

Margaret. The plan was to do total pancreatectomies on two dogs, give "Isletin" to one, and compare its health with the other. The dogs, numbers 92 and 409, were totally depancreatized on August 11. About the same time Banting and Best ligated the pancreatic ducts of two cats, and apparently two rabbits; they were getting ready, it seems, for later experiments to see if the extract worked across species.[26]

Dog 92 was a yellow collie. Banting had a difficult time getting its pancreas out. There was a lot of bleeding and he had to ligate major blood vessels normally not involved in the operation. The dog was fortunate to survive, and was still in distress when they began giving it pancreatic extract late that day. Nor did the first injections, at four-hour intervals, seem to have much effect,* except that dog 92's blood sugar, hovering around the .20 level, was staying well below that of the untreated dog, 409. Large doses of remacerated and diluted extract seemed to give excellent results the next day, driving 92's blood sugar down from a diabetic .30 to a low normal .09 in a twelve-hour period.

At 10 a.m. on August 13 the control dog was barely able to walk. Dog 92, on the other hand, was in "excellent condition does not appear tired or sleepy walks about as before operation." A new batch of extract continued to hold down the blood sugar, and while dog 409 continued to get worse, 92 was "feeling fine - rather 'hounded' aspect...but bears herself like a normal dog.... Feeling great...runs around room, frisky." On the morning of August 14 they decided to see if an overdose of the extract would reduce dog 92's blood sugar below normal. It appeared to, thirty cc. in seventy minutes bringing it from .22 to .066. At 1:30 a.m. on August 15 the untreated dog, 409, died, apparently of diabetes. Dog 92, which had become something of a laboratory pet, carried on.

It was an unforgettable time in the lives of the young scientists - working day and night in the lab, snatching a few hours' sleep here and there, frying eggs and heating up steaks over a bunsen burner, Fred's baritone blending with Charley's tenor as they sang war songs while they worked, sitting in the window sills or on the front steps of the meds building in the cool early morning sunshine of a beautiful August day, sitting at an all-night restaurant in the small hours of the morning discussing Isletin's future and their own. After all their troubles the work was coming out so

*It is very difficult to know what effects most of Banting and Best's injections of pancreatic extract *actually* had, as opposed to the *apparent* effects recorded in the notebooks. The impact of intravenous injections of active extract would have been very quick, perhaps only a matter of a few minutes. Many of Banting and Best's blood sugar readings, taken at one, two, or, in this case with dog 92, four-hour intervals, might have missed the period of maximum effect. In some cases, in fact, apparently unfavourable results might have been due to a "rebound" effect of higher blood sugar after the true impact of intravenous injections. A further difficulty in knowing what was actually happening to the dogs' metabolic condition stems from our complete ignorance of the strength of the early extracts. For other problems with Banting and Best's results, see the discussion on pp. 106-8 and 203-11.

beautifully. Everything they tried seemed to work perfectly, so much better than Prof Macleod, who didn't even know they'd gone this far, would ever have expected. This was research. Forget all the bad times gone before. These, for Banting and Best, were the truly memorable days in their search for the internal secretion.[27]

Why not try more experiments on the frisky collie, dog 92? How effective would an extract with a slightly acidic balance be? Or one made more alkaline? The acid extract seemed effective, the alkaline one did not. Extract incubated with trypsin (an active ingredient of the external secretion) was curiously potent. Soon there were no more duct-tied dogs whose degenerated pancreases could be chopped up to make "Isletin." But why quit now, with dog 92 running around the lab friskier than ever? It would come when it was called, lie perfectly still while a blood sample was being taken, then snuggle up to Banting, its head in his lap, while he did the chemistry and talked to it.[28] On August 17 Banting and Best chloroformed a dog and made some extracts of its whole pancreas. Dog 92's blood sugar was .30 at 6 p.m. when it was given ten cc. of whole gland extract. At 6:30 it had fallen to .19 and by 7:00 was down to .17. At 7:30 and 8:30 it had risen, but only to .20. Perhaps Banting and Best were too tired to think clearly.

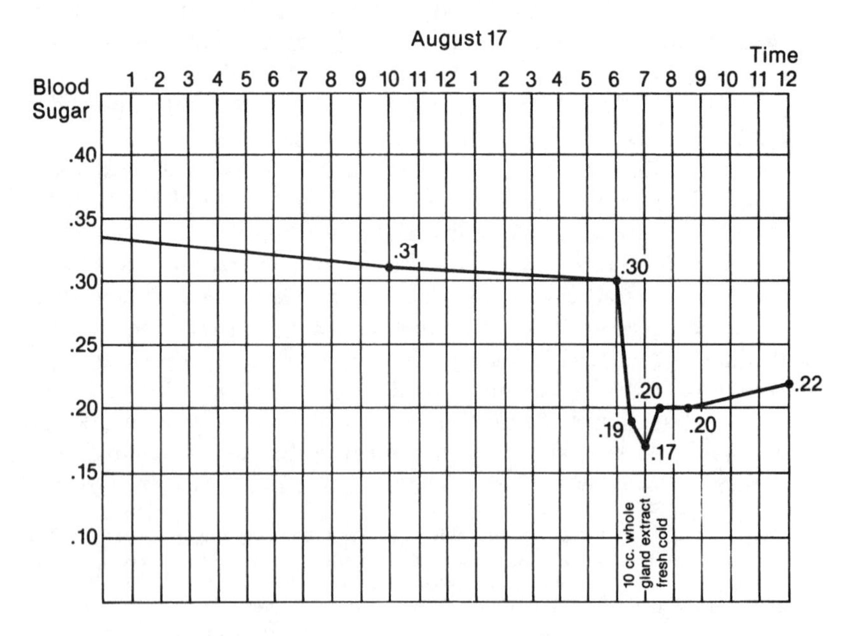

Chart 4: Dog 92, whole gland extract

Perhaps they were too confident of what to expect to notice contrary evidence. Here was fresh whole gland extract markedly reducing blood sugar. It was having as good an effect, perhaps better, than their extract of degenerated pancreas usually had. Banting and Best thought nothing of it. "It is obvious from the chart that the whole gland extract is much weaker than that from the degenerated gland," they wrote in their first published paper.[29] This was not obvious from the chart or the figures; indeed it was not true, and in failing to see this and its implications Banting and Best made a major error. Ignoring clear evidence to the contrary, they continued their pursuit of what we will later see was a faulty hypothesis.[30]

Banting and Best still thought it was necessary to do something to a pancreas to get rid of the external secretion. On the 19th, with no duct-ligated dogs at hand, and the collie now starting to weaken and sicken, Banting hit on another approach to avoiding the external secretion. It was well known that the hormone secretin, formed in the mucous membrane of the duodenum, stimulated the pancreas to produce its external secretion. Why not stimulate a pancreas with secretin until it was exhausted and could produce no more external secretion? Then, containing only the internal secretion, it could be cut out, ground up, and should produce an extract just as potent as those made from duct-ligated pancreases which had taken weeks to degenerate.

It was a complicated procedure, involving a resection of a dog's bowel to obtain the crude secretin, the insertion of a cannula in the larger pancreatic duct to measure the flow of external secretion, the slow injection of secretin for almost four hours until the production of pancreatic juice stopped, and finally the preparation of the extract.[31] But the extract worked marvellously. The collie, 92, had been very sick, unable to get to its feet, suffering from an abscess they had to lance on one of its forelegs. It began getting extract on a Saturday night. By Sunday morning (after another all-night session), the dog was running around the lab again, wagging its tail to welcome the humans, generally in excellent spirits. At two that afternoon, "dog jumped out of cage to floor about 2½ ft lit on fore feet & did not fall," Banting recorded. He never forgot that moment, singling it out for emphasis in accounts of these greatest days of his life.[32]

There was enough extract to try still more experiments. Would simple test-tube experiments show that the extract could "burn" sugar? An earlier experiment on August 6 had given promising results before the extract ran out. It was repeated more elaborately on the 20th, again with apparently good results.* An injection of extract plus trypsin produced no effect on the collie's blood sugar, apparently cancelling out the surprisingly strong result of the similar injection a few days earlier. There was enough extract left for one more injection. It worked well.

*As Banting and Best may have realized later when they chose not to mention their *in vitro* experiments in their publications, the recorded results were inaccurate. The extract could not have had any effect on sugar in a test tube. The temporary thought that it did, of course, would have further convinced Banting and Best that they were on the right track.

On August 22, Banting and Best tried to exhaust the pancreas of a cat by injecting secretin, but the animal died on the operating table. This was one of those times when Fred might have said to Charley, "What the hell?..." They took out the cat's pancreas anyway, made their extract, and tried it on the collie. It threw the dog into profound shock. That was the end of the experiments. Dog 92 lingered for another nine days, its blood sugar gradually rising, its overall condition weakening. It died on the 31st. "I have seen patients die and I have never shed a tear," Banting wrote in one of the more maudlin passages in his 1940 memoir, "but when that dog died I wanted to be alone for the tears would fall despite anything I could do...I hid my face from Best."[33]

IX

About the first week in September, Banting and one of his brothers drove to London to settle his affairs there. In one day, Banting recalled, he sold his house and most of his furniture. The rest he packed into the car he had and into his brother's car.

> That night we slept on the bare floor of the empty house and very early in the morning we took a last look at the place of my hours of misery - and yet it was there that I obtained the idea that was to alter every plan that I had ever made, the idea which was to change my future and possibly the future of others.
>
> One mellows with the years, but I still find it impossible to forget the awfulness, the loneliness and the financial worries that were associated with London. Nor can I forget the feeling of defeat that came over me as I took my final leave on that foggy autumn morning....My car, a little open Ford, would not go fast enough on that morning. I left my brother miles behind. It was a relief to be away and free.[34]

X

They were just beginning to work again on September 6 when Macleod's response to their report arrived. In general he was pleased: the results (to August 8) seemed "certainly very encouraging...definitely positive." He was glad to see that Banting planned to stay on in Toronto, "and you may rest assured that I will do all in my power to help you." In the new anatomy building just being built, Macleod would have new operating rooms. In the meantime Banting could use the facilities Starr had offered, "since I do not wish to spend money at present on the old rooms." He advised Banting to be careful, though, that no one should see him transporting operated animals from one building to another on campus: Macleod was very worried about anti-vivisectionist criticism of animal research, then reaching a peak in Toronto.

There was also much professorial caution in Macleod's reaction. The results were good, but there had to be "no possibility of mistake":

> You know that if you can prove *to the satisfaction of everyone* that such extracts really have the power to reduce blood sugar in pancreatic diabetes, you will have achieved a very great deal. Kleiner & others who have published somewhat similar results have not convinced others because their proofs were not adequate. Its very easy often in science to satisfy ones own self about some point but its very hard to build up a stronghold of proof which others cannot pull down. Now, for example, supposing I wanted to be one of those critics I would say that your results on dog 408 were not absolutely convincing....

Macleod hoped Banting and Best had data on the behaviour of the blood sugar of other depancreatized dogs so they could prove that their results were not merely normal diurnal variations. Did they have similar curves for dogs 406 and 410? Could the large volume of fluid injected into 408 on August 7 explain the drop in blood sugar - in other words, was it a dilution phenomenon? One of J.J.R. Macleod's aphorisms to his students was that one result is no result.[35] He advised Banting to "continue along the same lines without at the present taking up any of the problems which you suggest in your letter." Banting and Best should pay particular attention to the preparation of the extract, setting aside a small piece of tissue from each pancreas for histological examination to confirm that it was really islet tissue that was having the effect. They were to go ahead with the experiment, mentioned in Best's letter, of running two dogs side by side.[36]

They had actually gone ahead with that experiment two days after writing Macleod, had elaborated on it by testing several of the ideas Banting had suggested to Macleod, and had compiled what appeared to be the splendid results on dog 92, of all of which Macleod knew nothing. It was too late to do the pre-injection blood sugar tests on dogs 406, 408, and 410 that Macleod assumed had been done. Sections were, apparently, being taken for histological study, though nothing ever came of it.[37] Macleod's letter, although its encouragement was gratifying, must have seemed somewhat irrelevant.[38]

XI

The experiments Banting and Best began in the week they received the letter had little to do with Macleod's concerns or advice. Except for more careful estimations of D:N ratio (which now seldom rose above 2.0:1), it is not clear what Banting and Best were trying to achieve. One duct ligation was done on the 5th. On the 7th, after receiving Macleod's letter, total pancreatectomies were done on dogs 5 and 9. The operation on dog 5, the notebook records, was done by Best.[39] Both dogs seemed to recover well from the operations.

The next day they tried to make some extract by the quick secretin-stimulation method. In the morning they "failed utterly," as the secretin they injected simply did not stimulate the dog's pancreas. "This afternoon we succeeded in getting about 15 cc. of pancreatic juice in 2 hours when dog died." It was another what-the-hell situation, and although they had no reason to believe that the pancreas was exhausted of external secretion (the earlier stimulation experiment had produced eighty-five cc. of pancreatic juice), they ground it up anyway and considered they had made extract of exhausted pancreas.[40]

That night they first tried the last ten cc. of old extract on dog 9, administering it by rectum in their first change from intravenous injection. It had no effect. Intravenous doses of the new extract seemed to work spectacularly, a total of sixty cc. driving dog 9's blood sugar from .30 at 6:30 p.m. to .07 by midnight. "General condition improved. Injection of extract causes pain." The injections the next day seemed to have little effect. Another rectal injection on the 10th had no effect, but the next intravenous injection caused a sharp drop. Repeating earlier experiments, they again mixed trypsin with the extract, trying to see if the external secretion destroyed the internal. Apparently it did, for the mixture had no effect.

On the 12th they tried to exhaust a cat's pancreas with secretin.* Some juice flowed. They made no attempt to measure it. The cat died after ninety minutes of stimulation. Banting and Best made an extract of its pancreas. Experiments with the cat extract, including injection into dog 9's heart, caused moderate decline in blood sugar, shock, and finally death; the cause of death, they noted, seemed to be poorly ground particles in the extract damaging the dog's veins. The control dog, number 5, which appeared to be moderately diabetic, was killed a day or two later. Why it was killed is not clear. There was considerable pus in its abdomen.

On September 17 they began again on still another depancreatized dog, using extract made after another dubious attempt at exhaustion through secretin.[41] For the first time they tried injecting the extract under the skin - subcutaneously - giving five injections at hourly intervals. The dog's blood sugar did not change (though the fact that it did not rise, hovering around .15 to .18, may have been encouraging). In any case, this method of obtaining extract - from "more or less" exhausted glands, as they put it in

*The two cats, perhaps accompanied by rabbits, whose ducts were ligated on August 11, were never referred to again.

their first paper – was not satisfactory. "No more subcutaneous injection till we get a trypsin free extract," Best noted on September 18. "There is a hole the size of a halfpenny in skin where the injections were given. A superficial vein had been eaten into & there was considerable haemorrhage."

The dog died on the 19th, suffering from infection. No new experiments were begun. The summer's work was over. Banting and Best's companion in the next lab, Dr. Fidlar, also finished his summer's experiments on the respiration of his frog. He liberated it at the exact spot in Grenadier Pond where he had caught it in the spring. On September 21, Professor Macleod arrived back in Toronto from his holidays.[42]

XII

One day in August, worrying about a problem with the bowel of one of his dogs, Banting had dropped in to the office of Velyien Henderson, the professor of pharmacology, who was doing some research on intestinal movements. They naturally began talking about Banting's work, and Henderson took a friendly interest in its progress. As a student Banting had not much like the older, affected professor; Henderson liked to shock the students by using condoms in his pharmacological experiments, would turn an English accent off and on at will, and began lecture series by asking which of his students could wiggle their ears; the students nicknamed him "Vermin" Henderson.[43] Now Banting warmed to the older professor's interests. He saw more of Henderson and at some point in September told him how precarious his financial situation and his future appeared. Sixty years later Henderson's secretary, Jean Orr, remembered vividly Fred Banting coming out of the professor's office, sitting on her desk, and talking about his problems. "He put his hand in his pocket, took out seven cents and put it on the desk, and said, 'There, that's all I have to live on in the world, if I don't get a job.' "[44]

In mid-September the one junior man in Henderson's department was offered a special assignment by the Ontario government. Henderson thought it would be possible to replace him with Banting, and on September 21 wrote the president of the university to this effect. Whether or not he had actually offered the job to Banting by then is not clear. But as Banting prepared to talk over his future with the newly returned Macleod, Henderson had certainly hinted to him that some arrangement might be possible.[45]

Late in September or early in October Banting and Best met with Macleod. His reaction to the news of their experiments is not clear. Macleod remembered that his views on returning to Toronto were about the same as his reaction to the written report. He wanted more work done, and he specifically suggested an experiment to eliminate the possibility that dilution of the blood by the injections was causing the drops in blood sugar.[46]

The more memorable part of the interview came when Banting, probably after relating their problems with working conditions during the summer, demanded four things from Macleod: a salary, a room to work in, a boy to look after the dogs, and repairs to the floor of the operating room. Macleod was reluctant; Banting and Best had already gone through more dogs than planned. What was the point of spending money to fix up an operating room about to be abandoned when the new building opened? The professor told Banting and Best "that if he gave us these things some other research would suffer."[47]

Eleven months later Banting wrote this account of his reaction to Macleod's hesitation:

> I told him that if the University of Toronto did not think that the results obtained were of sufficient importance to warrant the provision of the aforementioned requirements I would have to go some place where they would.
>
> His reply was, "As far as you are concerned, I am the University of Toronto."
>
> He told me that this research was "no more important than any other research in the department."
>
> I told him that I had given up everything I had in the world to do the research, and that I was going to do it, and that if he did not provide what I asked I would go some place where they would.
>
> He said that I "had better go."

Banting expanded on the argument in his 1940 memoir:

> "Alright I am telling you that unless you provide the necessary facilities within twenty-four hours, then I shall leave." Banting walked to the door.
>
> "And where will you go?"
>
> "I don't think it matters a damn to you but I might go to the Mayos."
>
> "Only an advertising institution."
>
> "Or I might go to the Rockefeller Institute."
>
> "They have finished with diabetes research. Allen has been forced to leave."
>
> "The Rockefeller is never finished with research and you know it. And I will be in at this hour tomorrow morning and will continue work if the requirements are met in full. Otherwise I leave."

Macleod apparently relented, thought the matter over, and promised to do what he could.[48] The interview ended.

Best had taken it all in, apparently without saying anything. "I have never heard anyone talk to Macleod as you have," he said to Banting afterwards. The more Banting thought about the interview, particularly Macleod's remark, "As far as you are concerned, I am the University of Toronto," the angrier he became. To Best he poured out his opinion of Macleod.

"Fred Banting began to froth at the mouth," Best remembered. "My recollection of what Banting said to me when he became articulate was, 'I'll show that little son of a bitch that he is *not* the University of Toronto.' "[49]

CHAPTER FOUR

"A Mysterious Something"

"We have obtained from the pancreas of animals a mysterious something which when injected into totally diabetic dogs completely removes all the cardinal symptoms of the disease....If the substance works on the human, it will be a great boon to Medicine."
J.B. Collip, January 8, 1922.

Within a day or two of the confrontation in Macleod's office, all the details had been settled to enable Banting and Best's work to continue. Macleod found a room, big enough for two dog cages and a laboratory desk, that Banting considered quite acceptable. He gave them a part-time lab boy and had the physiology operating-room floor tarred so it could be cleaned properly.[1] Velyien Henderson's opening for a special assistant in Pharmacology solved Banting's employment problem. From October 1, 1921, Banting was on the University of Toronto payroll as a special lecturer in Pharmacology at a salary of $250 a month: in terms of purchasing power - we should multiply by a factor of between ten and fifteen for today's prices - it was a good salary. In view of the "decidedly satisfactory" results they had achieved during the summer, Macleod also arranged for retroactive pay for Banting and Best: $150 for Banting (the new boy), $170 for Best.[2]

Banting's duties in Pharmacology over the winter would be light. Best was staying on as an M.A. student and demonstrator in Physiology. Macleod was cordial and helpful, Banting remembered. "I thought that perhaps I had judged him too harshly." It was a great relief to Banting to have financial support over the winter. His landlady shared his pleasure. He gave her a liquor prescription so she could celebrate.[3]

I

The experiments could go forward again. But in what direction and in whose hands? Banting, we saw, had a long list of directions in which he thought the work should go; the list was so long that Macleod must have thought that Banting, like Leacock's horseman, was trying to ride madly off in all directions. Excited students have this proclivity. Less convinced than Banting that the summer's experiments were definite proof of the isolation of the internal secretion, Macleod advised sticking to the problem at hand so it could be wrapped up to everyone's satisfaction.

It may have been as early as the beginning of October when Banting first suggested to Macleod that either Macleod or others take part in the work. One of the "others" was almost certainly J.B. Collip, now spending part of his sabbatical year working in the Pathological Chemistry department at Toronto. Collip knew about the work, was interested in it, and told Banting several times that autumn how delighted he would be to help. But Macleod advised against expanding the team at this stage. "I pointed out that this being his and Best's research they should independently complete the work as outlined, and that then if the results continued satisfactory I would participate in the further investigations with my assistants." So Banting and Best went back to their dogs.[4]

Actually they first made a detour out to the farm on the outskirts of Toronto that Colonel Albert Gooderham, a local whisky magnate, had given the university to house its Connaught Anti-Toxin Laboratories. These had been founded in 1914 by a professor of hygiene, J.G. Fitzgerald, to produce vaccines and anti-toxins. On October 4 at the Connaught farm Banting and Best tried to ligate the pancreatic ducts of a calf. It died from the anesthetic.[5] Back in the lab, the pancreatic ducts of several dogs were ligated, another dog was totally depancreatized, and extract was made from the pancreas of the one dog, "Towser," whose ducts had been ligated 4½ weeks earlier at the beginning of September.[6] Towser's pancreas was only partly degenerated, so Banting and Best made separate extracts from the degenerated and non-degenerated parts of the pancreas.

The experiments on the depancreatized dog, number 17, a long-haired spotted hunter, were designed to answer several of the questions Macleod had raised about the summer work. To control for diurnal variation the injections were made at the same time every day. Tests after an injection of one hundred cc. of saline solution showed that the extra liquid by itself had no blood sugar reducing effect. Hemoglobin estimations, made before and after injection of the extract, seemed to show no appreciable thinning of the blood. So a dilution phenomenon could be ruled out as an explanation of the results. That was very satisfactory.

The extract labelled "Towser B" seemed very potent in enabling the diabetic dog to utilize the sugar injected along with it. Unfortunately a control injection of sugar alone had the mysterious effect of *not* raising the dog's blood sugar. It was repeated with the same mysterious effect. Not until the third sugar injection did the dog's blood sugar respond properly. Furthermore, Banting and Best's notebooks show that extract B was made from the less degenerated part (the tail) of Towser's pancreas. Injections of extract A, from the most degenerated part (the uncinate process) were much weaker, hardly effective at all. If this experiment showed anything, like the one with whole gland pancreas on August 17 it cast doubt on the hypothesis that a degenerated gland was necessary to produce potent extract. But again, there is no evidence that Banting and Best noticed the problem, for they had managed to get extracts A and B confused.[7] They were probably too puzzled by the strange non-effect of simple sugar injections to check the other aspects of the experiment carefully.

II

Now that Macleod's objections had been met, more or less, what remained to be done? Very little was done through the middle of October – partly because it seemed necessary to wait several weeks before more extract could be produced from the duct-ligated dogs, and partly because Banting seems to have been uncertain about what he wanted to do. He was giving a lot of thought to various possible experiments, writing down his ideas on four-by-six-inch index cards as they occurred to him (unfortunately only now and then jotting down the date on his cards). About the same time, it seems, he and Best were studying the literature more or less systematically, looking for ideas, perhaps also gathering background material as they planned the articles they would write on their work.

Macleod gave them some references. His young secretary, Maynard Grange, doubled as the medical librarian. She was in her mid-eighties and almost blind sixty years later when she told me her vivid memory of Banting coming in one day to look up a book Macleod had recommended. "You know," he muttered to her, "the goddamned little bugger knows everything about this subject." He said it, she remembered, in a tone of grudging admiration.

Banting's index card notes suggest that he had no clear idea where the experiments should go. He was toying with ideas for more test-tube experiments, considering whether injections of pancreatic juice could be used to prove its effectiveness against the internal secretion, noting other methods than pancreatectomy of producing glycosuria, wondering how the extract's effect on the action of the liver in metabolism could be tested. A typical Banting "idea" card, dated October 4, 1921, reads

> Partially depancreatized dogs a la Allan [*sic*] – control diet to see if the tolerance of a partially diabetic dog can be improved by extract – If this occurs a human might be improved by a "course of treatment".[8]

On the back of the card are notes summarizing an Allen article. The importance of the idea, which was not tried, is its underlining of Banting's interest in the possibility of using the extract on humans. He was never a disinterested physiologist looking for the internal secretion of the pancreas, but always a practical researcher looking for a cure for diabetes in humans.

Banting's practical desire to get on with the work, combined with his relative disinterest in medical scholarship, his weak background knowledge, and his inexperience at research, all militated against a careful, thorough study of the literature, including the publications of the others who had gone after the internal secretion. If Banting and Best were aware of the work of Zuelzer and E.L. Scott, for example, they either did not bother to read their articles, which are not listed on their surviving index cards, or they decided there was no need to cite this work in their early publications. They did, however, come across some of the results just published by Nicolas Paulesco.

Sometime between October and December 1921, Best read Paulesco's July 23 publication on the action of his pancreatic extracts. Best summarized its contents on one of the index cards.[9] Paulesco's extract, he noted, lowers the blood and urinary sugar of diabetic animals and definitely reduces the acetone bodies in the urine. Its effect varies in duration and magnitude in proportion to the amount injected. Paulesco also "proves" that his extract lowers the blood sugar of a normal animal. Best also thought it germane to note, however, that Paulesco reported normal blood sugars in his dogs as low as .044 per cent and obtained hyperglycemic readings in his diabetic dogs no higher than .20 per cent. Both figures, Banting and Best knew, were considerably out of line, and this may have cast doubt on the Romanian's methods. As well, Paulesco's animals had been under a volatile anesthetic, chloroform, during his experiments, with the extract injected just after the pancreatectomy; Best may well have realized that the anesthetic's effect on blood sugar would throw all experiments with extracts into question. Moreover, Paulesco did not report either the volume of extract injected or the volume of urine excreted. The index card suggests that Best did not find Paulesco's paper particularly impressive.

One of the least impressive aspects of Paulesco's work, according to Best's summary, was that Paulesco "states that injections into jugular, portal, or mesenteric veins works, but into peripheral veins 'no bon'." Like most English-speaking Canadians then, and now, Best had a very rudimentary reading knowledge of French. In making this note he had misunderstood a key sentence in Paulesco's article. The sentence reads: "Les mêmes effets, c'est-à-dire une diminution ou même une suppression passagère de l'hyperglycémie et de la glycosurie, s'observent aussi lorsqu'on injecte l'extrait pancréatique, non plus dans une veine périphérique, mais dans une branche de la veine porte, par example: dans une veinule mésaraïque ou dans une veinule splénique."[10] Best almost certainly mistook the phrase, "non plus," meaning "not only," for, as he wrote it, "no bon," or no good.

The misreading added to the coolness of the note's summary of Paulesco's work. Banting and Best do not seem to have given much further thought to Paulesco. They did not look up his other publications, and if they reread his work the only result was to compound their misunderstanding by somehow concluding, incorrectly, that Paulesco's experiments showed a less marked effect from second injections.[11] Consequently their brief reference to Paulesco in their first publication in 1922 grossly distorts his work.*

Apart from the mistranslation of French and the other problems Banting and Best may have perceived with Paulesco's work, there are two other possible reasons for the neglect of Paulesco. The most probable is that Paulesco was neglected because, as Banting's cursory reading and notes indicate, the pair neglected almost everyone who had worked on pancreatic extracts before them. Wasn't it obvious that all the precursors had failed to find the internal secretion? If others were working right now on extracts - a probability that Banting and Best, like modern students infatuated with ideas they think no one else has ever had, may not have clearly realized - they surely were not doing as well as the Torontonians. For if they were doing as well, or just a bit better, they would be about to start experimenting on humans. In contrast to the history of much recent scientific discovery, such as the structure of DNA, there is no evidence that the workers in Toronto thought they were in any kind of race or competition with outsiders to be the first to get to the internal secretion. Their intellectual, emotional, and experiential bias was entirely towards the goodness of their own experiments. This impeded their paying attention to the goodness of anyone else's, and encouraged misreadings and mistranslations.

(The result of the sloppiness was harmful not only in creating later misunderstandings and resentments by Romanians and others, but also in the missed opportunities it involved for Banting and Best to plan a rational course of experiments. Had they thought about Paulesco carefully, for example, they might have decided to try their extract on normal animals and to measure its impact on ketonuria in diabetic ones, as he had done. Had they studied earlier workers they might have developed experiments to check for toxic effects of their extract. Had they presented a clear, well-thought-out and productive experimental plan to Macleod in October or November 1921, much later confusion and bitterness about credit might have been avoided.)†

The second probability, partially but not fully contradicting the first, is that Banting and Best saw Paulesco as being unimportant in surmounting the problem at hand, which was to get beyond the stage they had all - Paulesco, Kleiner, Banting and Best - reached of having extracts that suppressed hyperglycemia and glycosuria. Impressive as the blood sugar

*"Paulesco has recently demonstrated the reducing effect of whole gland extract upon the amounts of sugar, urea and acetone bodies in the blood and urine of diabetic animals. He states that injections into peripheral veins produce no effect and his experiments show that second injections do not produce such marked effect as the first."

†This judgment of mine does not go unchallenged. One of the distinguished physiologists who read this passage in manuscript commented, "That's completely pie in the sky. I don't think anyone could have presented such a plan at the time. They were all fumbling in the dark."

evidence was in all three labs, it still did not prove that an internal secretion had been captured. Substances might lower blood sugars and reduce sugar excretion without necessarily permitting the diabetic's system to metabolize its food. Such a possibility must have been on Banting's mind when he made an extensive note on one of his index cards of an article listing a dozen conditions that could lower blood sugar. Among these conditions were several factors, such as shock, moribund states, and the injection of foreign proteins, which could have affected his and Best's dogs.[12] There had to be other experiments, Banting might have reflected, other approaches beyond simple blood sugars and urine tests, to nail down the internal secretion. Paulesco's work, which in any case was less impressive than Kleiner's, appeared to be of no help at this stage of the problem.

Banting kept flirting with the idea of pancreatic grafting. A card dated October 4 has a note on possible kinds of grafts. On Wednesday, October 19, Banting wrote C.L. Starr to say that he had permission from the university's Surgical Research Committee (recently set up by Macleod to head off the anti-vivisectionists) to do "an original investigation of the viability of (1) autogenous, (2) homogenous, (3) heterogenous grafts of pancreatic tissue." He and Best planned to begin work the next day, October 20. "We believe that such an investigation will be of great value," he wrote, "in ascertaining the clinical uses of substances contained in such tissue in the treatment of diabetes."[13]

Banting and Best had dinner together that night.[14] Their notebooks do not show any work done on October 20 or any grafting experiments ever attempted. Instead, on October 24 they began yet another round of injections of their degenerated gland extract into diabetic dogs. The most likely explanation of this sudden change of plan is that Macleod strongly advised the pair to stick to their extract. This may have been the time - alluded to directly in oral sources and ambiguously in the documents[15] - when Macleod told Banting and Best, in effect, that their results were just not good enough and they would have to repeat the experiments to get more and better ones. In view of the likelihood that Banting's proposed grafting experiments would have gone nowhere, this was sound and useful advice.

So more extract was made from degenerated pancreas (not very degenerated, though, for the waiting period after ligation was getting shorter and shorter; in one case it was only eighteen days) and another total pancreatectomy was done to make a dog diabetic. Things did not go well. The dog, number 21, was given twenty cc. of extract at 2:00 p.m. on October 26. At 2:30 it vomited. Excretion of urine almost stopped. The dog's rectal temperature at 3:00 p.m., the only temperature record in all of Banting and Best's notebooks, was a feverish 40C. Neither the first nor a second twenty cc. of extract had any significant impact on its blood sugar. Banting and Best stopped experimenting. The dog died suddenly after a drink of water on the 30th. The autopsy is recorded as showing a ruptured duodenal ulcer. It may have been brought on as the result of a slip of Banting's scalpel during the pancreatectomy.

After most of another week off, Banting and Best began again on November 4, running a control test on one of the ligated dogs to study how it responded to sugar injections. A sugar-plus-extract experiment on yet another depancreatized dog, 26, looked good, except that 26 was a very sick dog. It died on November 10, another victim of a duodenal ulcer and extensive internal bleeding. This is the last dog written up in Banting and Best's first paper; they refer euphemistically to the "early termination of the experiment." The last two experiments had been particularly unsatisfactory, due perhaps to poor surgical technique injuring the animals. Perhaps Banting's heart was not in it.

III

Macleod asked Banting and Best to talk about their work to a gathering of university students and staff at the Physiological Journal Club on November 14. Banting may have been pleased at the thought of presenting his results, which were exciting regardless of the most recent experiments, on his thirtieth birthday. A notice of the meeting lists Banting and Best as speaking on the subject of "Pancreatic Diabetes."[16]

The importance of this talk has been grossly exaggerated, to the extreme of the Historic Sites Board of Canada stating on its major commemorative plaque on the University of Toronto campus that the meeting marked Banting and Best's "public announcement of a therapy for use in the treatment of diabetes mellitus," hence the discovery of insulin. This is nonsense. The session was an informal presentation to a semi-private university group. Banting and Best had not yet finished writing their first paper describing their experiments. When it was finished, a week or more later, the authors concluded that it was "very obvious" that the results of the experiments through November 10 "do not at present justify the therapeutic administration of degenerated gland extracts to cases of diabetes mellitus in the clinic."[17] At the Journal Club meeting on November 14 Banting and Best gave a preliminary report to some interested colleagues and students, possibly a few outside visitors, on their work-in-progress.

Banting liked to think and write late at night. At 12:15 a.m. on November 14 he noted in a diary the coming of his thirtieth birthday. His ambition was to write one article per year for five years - "I have my first paper well under way." The questions of the moment, he thought, were whether to study for his fellowship in the Royal College of Surgeons, whether to leave surgery for experimental physiology, and whether to marry. "Time alone I suppose will only solve these problems. - At the present it behooves me to study & work at the internal secretion of the pancreas. & if possible isolate it in a form that will be of use in treating Diabetes."[18]

There are no records of the presentation or the discussion at the Journal Club meeting later that day, but Banting's and Macleod's 1922 statements disclose two important consequences of the meeting. The first arose from a misunderstanding. Banting apparently had asked Macleod to introduce them. Best would show charts of their dogs while Banting talked about their work and its relation to that of other investigators. To Banting's dismay, Macleod in his introduction said all the things he, Banting, had planned to say about earlier research. Banting was inexperienced as a speaker, nervous and inarticulate, and could not have adjusted easily to the surprising introduction. His natural reaction to the misunderstanding would have been to become angry - and to notice, as he stressed in writing about the meeting a year later, how often Macleod was using the pronoun "we" in describing the work. His mood could not have been helped after the meeting when he learned that students were talking of the remarkable work of Professor Macleod.[19]

Banting chose to say nothing to Macleod about his feelings. It was two months and many events later before Macleod heard of Banting's sensitivity about the Journal Club meeting. "Had I been told of this attitude of Banting at the time," he lamented in September 1922, "it would have served to warn me of his peculiar temperament and of his entirely unwarranted suspicions...."[20]

The more constructive consequence of the meeting came as the result of a suggestion by Dr. N.B. Taylor in the discussion after the presentation. He thought that a convincing demonstration of the extract's effect would be to show that regular administration of it could prolong the life of diabetic dogs. When Banting, Best, and Macleod discussed the future course of the work the next day, Macleod suggested they try this longevity experiment. They agreed.[21]

IV

But that agreement brought what had always been a bedevilling question to the fore. Where would the extract come from for such an experiment? The duct-ligation method of obtaining extract was slow and cumbersome and expensive at the best of times. It involved delicate operations, many dogs, and a four-to-seven-week waiting period. At this time, November 15, 1921, when there was at most only one duct-ligated dog on hand, Banting and Best faced the depressing prospect of being able to do next to nothing for a month or two while waiting to obtain extract. The one short-cut they had tried, injecting secretin to exhaust pancreas of its external secretion, had not worked at all satisfactorily. This problem with the supply of extract, then, was a crippling limitation on the work. Indeed, in a larger sense the supply problem threatened to be an over-riding barrier: there would probably never be a practical clinical use of the internal secretion of the pancreas if duct ligation and degeneration was the only way of capturing it. There had to be a better way of obtaining pancreatic extract.

Banting thought about these problems late into the night on the 15th. His reading had given him some clues. Laguesse had found that in the pancreas of new-born and foetal animals the islet cells are more plentiful in relation to the acini than in mature animals. This should mean, Banting reasoned, that their pancreases produced abundant quantities of internal secretion. Perhaps he and Best should try to obtain an extract from new-born animals. But then, Banting realized, there might be a possibility that in the *foetal* pancreas, as opposed to the new-born, the internal secretion might exist and the external secretion not be found. Other internal secretions, such as adrenalin, were present in early stages of foetal development. Since digestion does not begin until after birth, it was likely that the external secretion was not potent in the foetus. Further interesting evidence was an article about Carlson's work in Chicago in 1911 in which he and Drennan found that a depancreatized pregnant dog did not become diabetic until after delivery. They postulated that the foetal pancreas must supply the necessary deficiency. If it did have that kind of potency, and did not contain destructive pancreatic juice, perhaps the foetal pancreas could be used to make an effective extract. This idea "presented itself" to Banting at about 2 a.m. in the morning of November 16.[22]

Banting first thought of obtaining foetal pancreases by producing abortion in dogs. Then, surely realizing how cumbersome that procedure would be, the farmer's son remembered that growers often bred their cattle just before slaughter to make them better feeders and fatter. There would be plenty of calf foetuses available at the slaughter-houses. The next morning he and Best went to the William Davies Company's abattoir in northwest Toronto, cut out the pancreases from nine calf foetuses, and brought them back to their lab.

They prepared an extract by their usual method of macerating the tissue in ice-cold Ringer's solution and filtering.[23] Dog 27 had been depancreatized on the 14th. Early on November 17, showing a blood sugar of .30, the dog was given an intravenous injection of five cc. of the foetal calf extract. Forty-five minutes later its blood sugar had fallen to .20. It got two more injections that day, and the next morning a ten cc. injection reduced its blood sugar from .175 to .08 in one hour. Twenty-four more hours and its urine was, as they underlined in their notebook, *sugar free*. Extracts of foetal pancreas worked. There would be no more duct ligation, no more shortages of extract. The abattoirs could supply all the foetal pancreas the labs needed. As Banting and Best wrote a few weeks later, this was the beginning of a "new era" in the work.[24]

V

For you, the reader, perhaps starting to tire of dogs and extracts, the first month of the "new era," described in the rest of this chapter, is the most technical part of the history of the discovery of insulin. I cannot present an accurate record without the technical detail, and you would be unwise to skip this important material. Try not to worry about the individual dogs and their blood sugars, but instead notice the pattern of development of the research problem and the achievements and failures of Banting, Best, and the others.

The longevity experiment was begun on dog 27, which was given one or two injections of foetal calf extract daily. A second depancreatized dog, 33, was used to test various doses of the extract, which Banting and Best were now determined to make in its most potent and effective form. No name was being used for the extract, "Isletin" having not been mentioned since early August. The job at hand was to capture the "active principle" of pancreatic extracts in some form that could eventually be tested clinically.

The first improvement in extract preparation involved a final filtration through an unglazed porcelain, or Berkefeld, filter, which trapped clumps of bacteria, assuring sterility although apparently reducing potency. Heating or boiling the extract, Banting and Best found, seemed to destroy the active principle. With the Berkefelded extract it seemed useful to return to trying subcutaneous rather than intravenous injections; they would spread the extract's action over a longer period of time and be less likely to cause shock. On November 23 Banting experimented with a new blood sugar test, the Shaffer-Hartman method, probably introduced to them by Collip, who had learned it in the United States that summer. More significant is a brief note in Banting's handwriting the same day: "One of us (FGB) had 1½ cc Berk. ext. subcut. No reaction." This is the only record of the pair's first tentative...what the hell...why wait?...only a little bit...use of their extract on a human being. "No reaction" meant no harmful effects; they did not try taking their own blood sugars.

Banting and Best were starting to pay attention to the quantities of pancreas and Ringer's solution being used,[25] and were experimenting with more concentrated doses of the extract. The addition of tricresol, used to preserve diphtheria anti-toxin, did not seem to interfere with the active principle.[26] They were beginning to have trouble getting blood from dog 33's veins,[27] but tests continued through to the end of November. The notebooks record fairly low blood sugar readings on the 27th (.048 at 2:30 p.m., followed by an injection of extract half an hour later; .05 at 7:30 p.m.) without any observations on the dog's condition. Administration of extract through a stomach tube – to see if it might be absorbed through the gastric mucosa – produced no effect. As earlier experiments had shown, only injections seemed to work.

In the meantime the longevity experiment continued on dog 27. Although few observations of its condition were recorded, the experiment seemed to be going smoothly.[28]

VI

Banting and Best finished their first paper in late November, reporting the results of their work through November 10. Macleod apparently advised the pair on its format and helped polish the final draft. When the manuscript was finished, Macleod recalled ten months later, "Banting asked me if I wished my name to appear along with his and Best's. My reply was that I thanked them but could not do so since it was their work and 'I did not wish to fly under borrowed colours'."[29] Under the bold title, "The Internal Secretion of the Pancreas" (by F.G. Banting, M.B., and C.H. Best, B.A.), the paper was sent to the prestigious *Journal of Laboratory and Clinical Medicine,* published in St. Louis. It was accepted for the February 1922 issue.

The paper was a reasonably straightforward description of the work, set in the context of the background reading the pair had done. Much of the travail and disappointment of the early summer's experiments was omitted, as were one or two particularly badly done experiments. The article does not omit the misreading of Paulesco's work and the misinterpretation of the first experiment with whole pancreas. Like all of Banting and Best's joint papers, the article contains minor factual errors. Figures given in the text and the charts sometimes disagree with each other and/or with figures in the notebooks. The description of the last experiment is particularly bad: the charts show one set of figures for volumes of extract injected and duration of duct ligation; the text contains a second set; the original notebooks contain a third set.

The key sentences of the paper are a sweeping summary and claim:

> In the course of our experiments we have administered over seventy-five doses of extract from degenerated pancreatic tissue to ten different diabetic animals. Since the extract has always produced a reduction of the percentage sugar of the blood and of the sugar excreted in the urine, we feel justified in stating that this extract contains the internal secretion of the pancreas.

That summary is inaccurate, representing enthusiasts' tendency to put a totally favourable gloss on their results. It was simply not true that Banting and Best's extracts had *always* produced a reduction of the percentage sugar of the blood and of the sugar excreted in the urine. Sometimes the extracts had not worked at all; other times their effects had been inconclusive; a few times the necessary tests had not been done. It was not a long series of specific results, all of which were clearly successful, that was impressive. Such a series did not exist. Rather, it was the overall pattern formed by the experiments. By my rough (because partly subjective) estimate, compiled from the notebooks and charts, Banting and Best's first 75 injections of extract of supposedly degenerated or "exhausted" pancreas, using nine dogs, produced 42 favourable results, 22 unfavourable ones, and 11 inconclusive observations.[30] This is an impressive statistical picture in its own right, impressive enough to justify the work on the one hand and explain the researchers' overly enthusiastic claims on the other. In the face of so many good results, the tendency was to forget or ignore the bad ones.

Whether the pattern of results justified the claim to have captured the internal secretion of the pancreas is another question. There is no doubt that Banting and Best thought it did, but as with all claims of discovery the difficult job was to convince other people. The effectiveness of Banting and Best's results in inducing conviction would be clear only after publication or other public presentation. And, as we will see, it would be well after publication before anyone asked whether the results verified the researchers' subsidiary hypothesis about the external secretion being destructive of the internal. Finally, despite their claim that their extract contained the internal secretion, Banting and Best specifically said they did not yet have a therapeutic agent. The clinical condition of their dogs had "always" distinctly improved after administration of the extract, they reported, "but it is very obvious that the results of our experimental work, as reported in this paper do not at present justify the therapeutic administration of degenerated gland extracts to cases of diabetes mellitus in the clinic."

Before actual publication, there would be an occasion for public presentation of the work at learned society meetings held during the Christmas holidays. Macleod was a member of the American Physiological Society. When he received the call for papers for its annual meeting, to be held in New Haven, Connecticut, he suggested to Banting that a report on the work be presented. According to Macleod, Banting asked that Macleod's name should appear on the report to draw attention to it. "I agreed to this arrangement but stipulated that Banting and Best should both attend the meeting and should themselves present the report."[31]

Macleod's attitude to Banting and Best's work at this time (towards the end of November) is a bit obscure. While giving them advice and help he was not actively directing them, and had not yet added to their resources. There was no secrecy about the work, however, and at least one inquiry came from outside. At a meeting of the Southern Medical Association in Arkansas, Elliott Joslin talked with Dr. Lewellys Barker of Johns Hopkins Hospital, a Canadian and a University of Toronto graduate who had recently been back for a visit. Perhaps Barker had attended the Journal Club meeting; perhaps he had been talking with Macleod. He had learned what was going on at Toronto and mentioned it to Joslin. Joslin wrote Macleod asking whether he had published or was about to publish anything on the work. "Naturally if there is a grain of hopefulness in these experiments," Joslin added, "which I can give to patients or even can say to them that you are working upon the subject, it would afford much comfort, not only to them, but to me as well, because I see so many pathetic cases."

"It is true that we have been doing work on the influence of Pancreatic extracts, which has yielded most encouraging results," Macleod answered on November 21.

> But I would rather hesitate to attempt the application of these results in the treatment of human diabetes until we are absolutely certain of them. Dr. Banting and Mr. Best who have been doing this work, are to report their findings at the meeting of the Physiological Society at New Haven, by which time we expect to be in a position to come to a definite conclusion. I may say privately that I believe we have something that may be of real value in the treatment of Diabetes and that we are hurrying along the experiments as quickly as possible.[32]

Another person Barker told about the Toronto work was Dr. George H.A. Clowes, the research director of Eli Lilly and Company, a pharmaceutical manufacturing company located in Indianapolis, Indiana. Clowes took no action at the time, but resolved to be at the New Haven meeting to hear the presentation.[33]

VII

Something suddenly went very wrong with the longevity experiment on the afternoon of December 2. About four hours after an injection, dog 27 began showing symptoms of "a peculiar nature" ("Convulsive twitchings - retraction of head - unconscious for several hours. Salivation & frothing - seemed to improve for a time, then repetition of symptoms").[34] It seemed improved the next day, but ninety minutes after injection the symptoms began again and were more severe. This time the dog did not recover. It died that night, bringing the longevity experiment to an abrupt end. Banting and Best recorded the death as due to an anaphylactic-like reaction.*

Instead of prolonging dog 27's life, the extract had killed it. The idea of

*Was the death perhaps caused by an as yet unrecognized hypoglycemic reaction? Banting and Best thought it was shock, for a double dose of the same extract produced no reaction when injected into a normal dog. Later, however, they thought it might have been hypoglycemia. It does not seem possible, on the basis of the evidence extant, to know what had happened to the dog. No autopsy was done, apparently because Banting and Best just wrote it off as a failed longevity experiment. Banting 1923, A, B.

running a longevity experiment still seemed like a good one, though, and on December 6 the pair decided to convert the test dog, 33, later known as Marjorie,[35] to longevity. She had been depancreatized on November 18. Another experimental dog, 23, was started, pancreatectomy by Best.

VIII

The first step towards another important breakthrough also came on December 6 when Banting and Best decided to try using alcohol in the preparation of their foetal calf extract. Macleod had suggested the idea months earlier, and it had been used by both Scott and Zuelzer. In 1922 Best wrote that trying alcohol was a "fairly obvious" idea, which occurred to Macleod, Banting, and himself independently.[36] The trouble with an aqueous saline solution of extract, they seem to have thought, was that any attempt to concentrate it to get at the pure active principle by boiling away the water also seemed to destroy the active principle. Alcohol evaporates at much lower temperatures than water. It may also have been used in the hope that it would dissolve and remove some of the contaminating impurities in the solutions. Banting and Best ground the foetal calf pancreas up in alcohol, filtered the mixture to get out the solids, and then evaporated the alcohol by a technique Macleod had shown them of using a current of warm air flowing over porcelain dishes containing the solution. The dry residue was next redissolved in a saline solution. It was given to dog 23 on Wednesday, December 7, and worked well. Injections given to dog 33 the next day were less potent, but moderately successful.[37]

The realization that the active principle in a foetal pancreas was soluble in aqueous alcohol led Banting and Best to wonder whether they could get a similar result from a fresh adult pancreas. On December 8 they did a pancreatectomy on dog 35. Instead of throwing out its pancreas, they cut it up into slightly acid alcohol, macerated it, and allowed it to stand for forty-eight hours. The solution was filtered, the alcohol was evaporated off in the warm air current, and the dry residue was redissolved in saline. On Sunday morning, December 11, dog 35 was given six cc. of extract of its own whole pancreas. Its blood sugar dropped from .38 to .18 in four hours. Whole pancreas extracted with alcohol worked. Here was another major advance. No more degenerated pancreas. No more foetal pancreas. Now the research could go forward using cheap, easily obtainable supplies of fresh whole pancreas.

IX

Banting had continued to press Macleod for help, several times asking if J.B. Collip could join the team to work on the biochemistry of pancreatic extracts. Collip wanted to help. Working several blocks away in the pathology building on the grounds of Toronto General Hospital, he saw Banting and Best every few days, took a great interest in their experiments, and often left with the comment, "Well, if I can be of any assistance let me know."[38] Many years later, Dr. E.E. Shouldice, who had been working in the same lab with Collip that fall, remembered the biochemist's anxiety to get to work on Banting and Best's extract. Collip said, Shouldice remembered, that it would take him about two weeks to purify their crude extracts.[39]

James Bertram Collip was well qualified to take part in the work. At the age of twenty-nine in 1921 he was a year younger than Banting, but had far, far more experience at medical research. Born in Belleville, Ontario, a florist's son of British descent, Collip had taken his B.A. in Honour Physiology and Biochemistry at Toronto in 1912. He had gone straight on in biochemistry at Toronto, earning his M.A. in 1913, his Ph.D. in 1916. In September 1915 he had begun work as a lecturer in physiology at the fledgling University of Alberta in Edmonton. By 1920 he had been promoted to full professor in charge of a new Department of Biochemistry.

Collip was an active researcher, working on a variety of problems most of which involved blood chemistry. By the end of 1921 he had a respectable list of twenty-three academic publications, including a very good 1916 summary article on internal secretions, a subject in which he had a longstanding interest. In 1920 he spent some time making and injecting tissue extract, accompanied by adrenalin, and studying the resulting effects on blood pressure. Although he had had to shoulder a heavy burden of teaching at Alberta, Collip's deepest commitment was as a research scientist. Happily married and with a young family, he was a shy, sensitive, and boyish-looking young scientist, who loved nothing more than long stints in the lab, preferably late into the night, trying out this or that mixture of ingredients to produce a desired, or perhaps an undesired, physiological reaction. He was being supported on his sabbatical leave by a Rockefeller Foundation Travelling Fellowship and a temporary appointment as assistant professor at Toronto.[40]

Macleod finally agreed to Banting's requests and invited Collip to help Banting and Best in the development of their extract. They needed help not, as legend has it, because they were floundering around going nowhere, but because the pace was speeding up thanks to the development of means to produce large amounts of extract. There was so much to be done. Banting wanted it done quickly. If there was any objection to Collip joining the work, it came from Best, who apparently felt slighted - a student of biochemistry seeing a full professor and Ph.D. come onto the team. "I was opposed to Collip's participation in our work for obvious and selfish reasons," Best said in an unpublished address in 1957, "but Fred Banting persuaded me not to protest too vigorously. This was also for obvious reasons – i.e. the urgent need of our antidiabetic material for clinical use."[41]

The exact date in December on which Collip started work is not known. It would be helpful to know it because of the terrific pace of discovery in December, as well as the later dispute about who did what when. Collip appears to have been at work on pancreatic extracts by Monday, December 12. It is important to note that Banting and Best appear to have begun making alcoholic extracts of fresh whole pancreas – at least the one canine pancreas tried on December 11 – before Collip joined the group.

His first contribution to the research may have been to add a rare and totally unintentional comic note. The very first extracts he made, using alcohol and fresh whole glands obtained from the abattoir, apparently did not work. Collip went to see Banting and Best:

"There is something wrong with this whole piece of work," Banting in 1940 remembered Collip as saying.

"What makes you think that?" asked Best.

"Well, I made some extract and did not get the results which you got."

After much discussion of Collip's methods, it turned out that he had told his lab boy to go to the abattoir and ask for sweetbreads. Instead of getting pancreas, which at that abattoir was being ground up for fertilizer, the boy was bringing back thymus or thyroid glands.

It was a good anecdote, one Banting liked telling to denigrate Collip's work. Perhaps it is a true story, for an entry in Banting and Best's notebooks later in December shows them using, as a control, an extract of "Protein free Thymus à la Collip ± Pan."

X

Collip's notebooks have not been found, so his work has to be pieced together from several later accounts, all of which are less detailed than Banting and Best's records.[42] There is no doubt, however, that, coming into the work just at the stage when Banting and Best had found that alcoholic extracts of whole pancreas lowered the blood sugar of diabetic dogs, Collip quickly began making important findings.

He immediately began making extract from whole beef pancreas. Then he followed up a suggestion Macleod made in the presence of the other three researchers. Why not try the extract on the blood sugar of rabbits, Macleod wondered, particularly those made "diabetic" experimentally. There were several ways of inducing hyperglycemia in rabbits, all of which would be cheaper and easier than depancreatizing dogs. Possibly because he was working a long walk away from the dog quarters in the medical building, Collip used rabbits from the beginning. As soon as he got his sweetbread confusion straightened out, Collip found that pancreatic extracts were effective on rabbits. And not necessarily diabetic rabbits, but just plain rabbits, perfectly normal ones. Extract lowered their blood sugar from normal to below normal just as Banting and Best's extracts lowered the blood sugars of diabetic dogs from above normal to normal and below. This observation had immense practical importance, which Collip realized immediately, in giving the group a quick, easy way of testing the potency of a batch of extract. Its strength could be measured by its effect on the blood sugar of a normal rabbit, procured quickly and easily from a vein in its ear, tested by the new Shaffer-Hartman method.[43]

Collip also began work on depancreatized dogs. On Friday, December 9, Banting and Best had run a strange experiment on dog 23, recording the effect of an injection of extract on its blood pressure and blood sugar while it was under anesthetic. It is one of the least satisfactory of Banting and Best's experiments: the dog was apparently dying of infection and figures given in the published account bear little relation to those in the notebooks.[44] There was nothing significant in the blood pressure performance, but there was a surprising failure of the dog's blood sugar to respond significantly to the injection of apparently potent extract. Why did the extract not seem to work on a dog under anesthesia?

It was apparently the discussion of this experiment among Banting, Best, Macleod, and Collip which brought the liver/glycogen issue to a head. The discussion was along the following lines: An important process in carbohydrate metabolism is the conversion of glucose into glycogen in the liver. In the diabetic condition very little of this conversion takes place, so the glycogen content of the liver is low. Banting had wondered from time to time whether the pancreatic extract would enable the diabetic dog's liver to start storing glucose. This would be an important demonstration of its potency as an anti-diabetic agent.

Why wasn't the extract effective on the anesthetized dog? It was known that volatile anesthetics inhibited the glycogen-forming action of the liver in normal animals. Perhaps the extract did not work on the anesthetized diabetic dog because its liver was similarly inhibited by the anesthetic. If that were so, the reasoning went (apparently during discussions the four were having over lunch daily), it might be that the key to the extract's overall effect was in the liver. Whether or not that was true, it was certainly time to turn to the liver and find out what the extract did to it.

Collip undertook to measure the extract's effect on the glycogen-forming function of the liver, as well as to make observations on the extract's effect on the amount of ketone bodies excreted in diabetes. He first confirmed, or thought he did, that the extract had no effect on ether-anesthetized normal dogs as well as ether-anesthetized diabetic ones, reinforcing the hypothesis that the liver was critical to its action.[45] On Tuesday, December 13, Banting and Best depancreatized a large Airedale in the surgical operating room and Collip took it to his lab in the pathology building to await the development of diabetes.[46] Meanwhile he continued working on his rabbits and tinkering with batches of extract.

About this time the team had agreed that another highly desirable experiment to test the potency of the extract would be to measure its effect on the respiratory quotient of diabetic subjects. This measurement of the ratio of carbon dioxide breathed out to oxygen absorbed was thought to be a reliable guide to whether or not carbohydrates were being burned in the body. It involved complex apparatus and gas analysis, however, and it was agreed to delay this work until January when another researcher would be available to work on the problem with Best.[47]

XI

While Collip was starting work in his lab, Banting and Best spent the week of December 12-16 working on their newly discovered extract of whole pancreas. That first injection of whole dog pancreas had worked on December 11. On the 12th an alcoholic extract of whole pancreas seemed to work when administered through a stomach tube (apparently causing a blood sugar reduction from .42 to .28 in four hours). Whole cow pancreas injected intravenously also seemed potent. On December 13 and 14 Banting and Best administered extracts of liver, spleen, thyroid, and thymus, all made the same way as pancreatic extracts, to the test dog. None of them was effective. To try to make their pancreatic extract purer, they experimented with dialysis (the use of a semi-permeable membrane to filter out small molecules from colloids in a solution) and also found they could wash the residue with toluene after alcoholic evaporation to rid it of more of the fat-like lipid impurities. On the evening of December 15 they gave their second talk on pancreatic diabetes, addressing a group at Toronto General Hospital.[48] On either that day or the next[49], dog 35 was given by injection a piece of dried extract the size of a match which had been washed twice in toluene and redissolved in ten cc. of saline. In four hours its blood sugar went from .37 to .06, a spectacular result.[50]

Then things went downhill. At week's end Banting and Best made up the largest batch of extract yet, some seven hundred cc. of tissue and pancreatic juice with five hundred cc. of acid alcohol. They tested it on December 18 and found it had no potency whatever. They concluded that they had probably used too much acid. On the 19th or 20th another batch was made up, using just alcohol, no acid (the notebooks tend to be vague on the quantities used in these mixtures). A subcutaneous injection from this new batch had no effect. A little bit of acid was added to the final saline solution of extract and an intravenous injection was tried. The only effect was to cause the dog to vomit.

They still had some extract on hand that they knew to be potent. Perhaps in reaction to the week's disappointments, they decided, apparently without telling anyone, to try it on a human diabetic. One of Banting's classmates, Joe Gilchrist, had become diabetic in early 1917 a few months after their graduation. He took the Allen treatment and was able to carry on with gradual downhill "progress" until October 1921 when a bout of influenza shattered his carbohydrate tolerance. All the symptoms returned and he began to deteriorate rapidly.[51]

An index card in Banting's papers is the only record of Banting and Best's first "clinical" test of their extract.

Clinical Use

Dec. 20. Phoned Joe Gilchrist -
gave him extract that we knew to
be potent. - by mouth - empty stomach

Dec. 21 - no beneficial result.[52]

It was not yet known that no pancreatic extracts ever work by oral administration (they are, in fact, "digested" by the proteolytic digestive enzymes). Only a few days earlier, on December 12, Banting and Best had had that apparent success giving extract via a stomach tube to one of their dogs. It might work on Gilchrist, they must have thought. But it was still too risky to inject the extract into the blood stream or under the skin of a human.

XII

While Banting and Best were throwing out batches of impotent extract and seeing their potent extract have no potency on a human diabetic, Collip's experiment was proceeding smoothly. He was making his own extract, using a technique similar to Banting and Best's but with various improvements. Instead of evaporating the alcohol in a warm air current, he used a laboratory vacuum still. He did not evaporate all the liquid in the pancreas/alcohol solution, as Banting and Best did, but instead reduced it to about one-fifth of the original volume, giving him a suspension of fine particles in a clear straw-coloured liquid. This was filtered, leaving a liquid filtrate and a residue of solid particles.[53] On December 20 Collip was ready to administer pancreatic extract to his diabetic Airedale. First he injected fifteen cc. of the liquid filtrate he had prepared. The Airedale's blood sugar dropped from .309 to .217 in two and three-quarter hours. Then Collip injected a solution made with the solid particles left after that last filtration. It proved more potent, ten cc. dropping the blood sugar to .085 in 65 minutes and .051 in a further two hours and ten minutes.

The lesson Collip learned in terms of extract preparation from testing both filtrate and residue - we will see its importance later - was only a bonus in terms of the main aim of the experiment, which was to test the extract's effect on glycogen formation and ketonuria. As the Airedale became diabetic, Collip had been carefully measuring the amount of ketone bodies in its urine. On the 21st, after the injections of extract, the dog's urine became completely ketone (and sugar) free. In their notebooks for August 5 and 7, Banting and Best had jotted down two casual, perhaps retrospective observations about their extract causing "acetone bodies" to disappear; they did not mention the subject in their first paper. Collip's experiment was the first measured demonstration in Toronto that the extract could abolish ketosis.[54]

There was much more to be demonstrated in the experiment. Through the 21st and into the 22nd Collip continued to give periodic injections of extract, while allowing the dog to consume glucose and milk freely. He was hoping that the extract would enable the diabetic dog's liver to start making glycogen again from the carbohydrates it was consuming.

Banting and Best spent the 21st and part of the 22nd experimenting on normal animals for the first time. On the 21st their extract had no effect on a normal dog. On the morning of Thursday the 22nd another batch had no effect. They tried a normal rabbit. No effect. On that discouraging note, seven failures in a row that week, they quit for the day and for the Christmas holidays. Their notebooks end. The one experiment they kept going over Christmas was the longevity test on dog 33.

Collip stayed in his lab that afternoon to complete the glycogen experiment. At 6 p.m. he chloroformed the Airedale, cut out its liver, and measured it for glycogen. An untreated diabetic dog's liver would not have very much, no more than about 1.5 per cent at the very most. This dog's liver was full of glycogen, so full it could hardly be measured. The liver was an incredible 25.6 per cent glycogen, Collip recorded.[55] This was a result beyond anyone's expectation – a crystal-clear demonstration that the extract enabled a diabetic animal's liver to form glycogen. "There was thus afforded definite proof," Collip wrote later, "of the restoration...of a function which was definitely known to be lacking in the diabetic state."[56]

Collip did not see Banting and Best until after Christmas when they all travelled by train to New Haven together for the meeting of the American Physiological Society. He had good news to tell them about the glycogen experiment. Great news, the most solid evidence yet that the group was on the right track, heading towards triumphant success. The "mysterious something," as Collip described the active principle of their extracts a few days later, worked against diabetes.[57]

Banting and Best must have been pleased by Collip's news. But they must also have been a bit chagrined that it was Collip who had achieved so much with their extract just when their own attempts to make it work at all had resulted in a week of total failure.

CHAPTER FIVE

Triumph

Most of the important people in North American diabetes research came to the Friday afternoon, December 30, session at the American Physiological Society conference at Yale University in New Haven. The program announced a paper by J.J.R. Macleod, F.G. Banting (by invitation), and C.H. Best (by invitation) on "The Beneficial Influences of Certain Pancreatic Extracts on Pancreatic Diabetes." Among those present were Allen, Joslin, Kleiner, Scott, Carlson, and the Eli Lilly research director, George H.A. Clowes. It had been arranged that Macleod would chair the meeting and Banting would give the paper.

Everyone who ever described that session (there are no formal records of it, only a half-page abstract of the paper, very similar to Banting and Best's first long article, published a few months later in the A.P.S. Proceedings)[1] remembered that Banting was nervous and spoke haltingly. The best account is Banting's own (1940): "When I was called upon to present our work I became almost paralyzed. I could not remember nor could I think. I had never spoken to an audience of this kind before - I was overawed. I did not present it well."[2]

I

The audience was an experienced, tough, and critical group of experts. There would have been many searching questions even if Banting's presentation had been good. As it was, after Banting spoke, Allen, Kleiner, Carlson, and others, all raised points about the work. As Macleod recalled the meeting nine months later, "it was evident that he [Banting] had not succeeded in convincing all of his audience that the results obtained proved the presence of an internal secretion of the pancreas - the primary object of the work - any more definitely than had those of previous investigators."[3]

Macleod found himself in the unhappy position of seeing a presentation of highly promising research from his own lab, to which he had allowed his name to be attached, fall flat. It was a situation all scholars dread when they are students and fear for when their own students give their first papers. There was only one decent thing Macleod could do in the circumstances. Instead of asking the hapless Banting to respond to the criticisms, which would be like throwing the lamb to the wolves, Macleod came to his defence by joining the discussion himself. He tried to answer the critics, "laying stress," he wrote, "on the frequency of direct relationship between the injections and the lowering of blood sugar and on the prolongation of life of two treated animals." Elliott Joslin wrote thirty-five years later about the meeting that "Banting spoke haltingly, Macleod beautifully."[4]

Knowing of the work Allen, Scott, Carlson, and Kleiner had done, it is not difficult to imagine the questions they posed to the Toronto people. The most obvious, dealing with the extract's "beneficial influences," would be whether it also had toxic effects. Did it cause fever, for example? A hard question to answer in view of there being no temperature records for Banting and Best's dogs (except one reading, showing fever). Were there other reactions? Well, the dogs did sometimes react to the extract.[5] And if the longevity experiment, not yet complete – for dog 33 had been going just under five weeks – was being discussed, it had to be admitted that the first attempt at longevity, with dog 27, had ended abruptly when the dog had died of severe reaction to the extract.

And what was the precise condition of these dogs? Best's one memory of the meeting was of Anton Carlson mentioning that his depancreatized dogs sometimes lived for several weeks. Best interjected the comment that if this was so he had probably not taken out all the pancreas. "Young man, you might be right!" Best remembered Carlson responding.[6] It was probably a two-edged comment, implicitly raising the question of the completeness of Banting and Best's operations. Was it their extract or was it pancreatic remnants that kept their dogs alive? Were routine autopsies adequate to prove total pancreatectomy?[7] What about the D:N ratios on the experimental dogs? Another embarrassing question, difficult to answer.

A host of other questions could have been raised. How soundly based in the literature, for example, was the theory of selective atrophy after ligation, which had been so important at the start of the work? How sure were Banting and Best that the early pancreases they used had actually been fully atrophied? How could a fully atrophied pancreas supply as much extract as the figures indicated? Was it really clear that the proteolytic digestive enzymes were the problem in preparing pancreatic extracts?[8] Many researchers, including at least three members of the audience, had made extracts that reduced hyperglycemia and glycosuria. Where exactly had Toronto gone beyond them? And so on. Few of these questions, if they were asked, could be answered satisfactorily by the Toronto team. What they could do, as Macleod did, was to keep drawing attention back to the blood sugars and the dogs' survival. The extract reduced blood sugar; it apparently had kept two diabetic dogs alive (dog 92 back in August, and dog 33 now in December). Macleod might also have referred to the experiments in progress on the new whole gland extract; of course they had to be repeated, but just before Christmas there had been some exciting findings about ketonuria and glycogen formation. The work was going ahead vigorously on several fronts and there would be further reports in the near future.

Joslin remembered the overall reaction to the session as being "little praise or congratulation, and a moderate amount of friendly but serious criticism of the work."[9] Judging from surviving correspondence, the experts took a cautious interest in the Toronto work. They might be onto something up there in Canada; we'll look forward to hearing more. After the meeting E.L. Scott walked back to the hotel with Macleod, discussing his 1911 work. A few weeks later he sent Macleod details of his extraction methods, commenting that his extract would not have been likely to cause such sharp reactions as Macleod had described, but on the other hand it was never put to the "severe trials" they were using in Toronto. Writing Macleod about the same time on another matter, Frederick Allen added a last paragraph about their mutual interests:

> I hope your work with the pancreas extract is progressing satisfactorily. With the beginning of our animal experimentation here, I shall probably go ahead with plans I have had for a long time, in the direction of an extract. The methods in view are totally different from yours. You not only have priority, but, if you have solved the initial difficulties, your method is better than mine could ever be. I merely thought out my method as a means of escaping those difficulties, and it may have some value for other purposes at least, so I shall probably give it a trial. It is high time we had some treatment beyond mere diet, though I recognize the difficulties in the way of a practical application of any extract.[10]

There was one exception to the experts' wariness. When Macleod returned to his New Haven hotel room after the session he got a phone call from George Clowes, the Lilly research man, who said that he thought the evidence was convincing and asked whether Eli Lilly and Company could collaborate with the Torontonians in preparing the extract commercially. ("It is true that Banting presented his material somewhat haltingly and certainly very modestly," Clowes wrote in 1948. "However, anyone who was at all cognizant with the subject must have realized that a great discovery had been made and that provided the work could be brought to fruition there was every prospect that an important means of treating diabetes would be developed.") Clowes talked the matter over with Banting and Macleod. He was told by Macleod that the work was not sufficiently advanced for commercial preparation. Clowes' suggestion would be borne in mind.[11]

The person most disappointed with the New Haven session was Fred Banting. However good or bad the reaction to the work had been, it was obvious to him and to everyone else how badly he had failed in presenting it. Instead of his idea and his experiments culminating in a great personal triumph, he had endured an embarrassing, humiliating afternoon. And this after all the frustrations of the week before Christmas, when his and Best's work had gone so badly and Collip's so well.

It particularly rankled that Macleod had stepped in and expressed himself so smoothly. More than smoothly - almost proprietarily. Instead of feeling grateful that Macleod had bailed him out, Banting decided that the professor had gone too far. The bugger had kept using that word "we" even though he had never done a single experiment, "nor had he contributed one idea of value except estimation of haemoglobin...I was the only one who gave a paper to the Physiological Section who was not asked to respond to his paper." Who was the chairman who had not asked Banting to respond? Macleod. Come to think of it, whose name was first on the program? Macleod's.[12]

Then there were all the earlier events to consider: Macleod's discouraging comments at the beginning of the work; that quarrel in his office early in the fall; his having said so much, using "we" all the time, at the Journal Club meeting; the interest he was taking in the work now; the good results Collip was getting (and apparently reporting to Macleod)[13]; and then everything he had said this afternoon. It seems to have been during his emotional turmoil after the New Haven meeting that Banting first decided Macleod was trying to take over the work, trying, in fact, to steal his results. His memory is not to be trusted for particulars, but Banting's 1940 account of the train trip back to Toronto leaves no doubt that the meeting triggered a personal crisis:

> I did not sleep a wink on the train that night - I did not even go to my berth but sat up in the smoker condemning Macleod as an imposter and myself as a nincompoop. I decided that I must first learn to write clearly, precisely, legally, explicitly and then be able to talk convincingly, freely and unhesitatingly. I knew Macleod for what he was, a talker and a writer. Apart from his pen and his tongue he would not even be a lab. man for he had no original ideas, he had no skill with his hands in an experiment. He only knew what he read or was told and then he could rewrite or retell it as though he were a scientist and a discoverer. It was foolish to spend weeks and months working night and day at experiments and then have them told beautifully by someone else who had the art as though they were his ideas and works.[14]

Macleod knew nothing of Banting's feelings. In early January he wrote to a colleague that the New Haven meetings had been "in every way a great success, the discussions being particularly interesting." About this time Banting began telling his friends that Professor Macleod was stealing his results.[15]

II

Back in Toronto in the New Year the work advanced about as quickly as relations among the workers deteriorated. Our documentary sources also deteriorate somewhat, for no Banting and Best notebook has been located for the period between December 22, 1921, and February 13, 1922. This may be because they did very little work together. The longevity experiment continued on dog 33, Marjorie, but relatively little is reported in any publication about her condition in January. The dog had apparently been getting a daily injection of six cc. of whole gland extract over Christmas.[16] They discontinued this on January 4, measured the sugar in the dog's urine (but not its blood sugar), and observed its general condition. When it seemed to get worse, they resumed injections on January 8.[17] There are no records of Banting and Best doing any other experiments on dogs in January.

It is not clear what role they were to have in the ongoing work now that Macleod had turned his whole lab over to the problem. A rough division of labour appears to have been worked out informally around Christmas-time.[18] Clark Noble was added to the group to help work with the rabbit testing and the glycogen experiments. Best and Dr. John Hepburn were to do the respiratory quotient tests (this would be Best's M.A. thesis project) when the apparatus could be set up. Collip was to try to purify the extract to see if he could get it pure enough for clinical trials. While waiting for his respiratory quotient experiments to begin, Best seems to have done the preliminary work collecting pancreas and taking it through the first stages of extraction before passing it to Collip.[19] Banting appears to have done whatever surgery the group required. He may have expected to play an important role in the impending clinical tests because he was the only practising - more or less - physician in the group.[20] It was well understood that all results were being pooled in what had now become a team effort.*

Collip was working hard and enthusiastically at his several problems. It

*This was probably taken for granted, as it would be by any scientists in a similar situation. But Banting in his 1922 account suggests that he, Best and Collip had explicitly agreed to tell their results to each other. In his 1940 account he elaborates as follows:

> Collip, Best and I stayed together [at New Haven] in the same hotel and were together a good deal. Naturally we talked incessantly about the work. There were so many problems that were opened up and demanded immediate investigation that it was agreed that Best and I take Collip into partnership - and that we should pool all results and share alike in all publications; that all results should be confided to each and all of us. It was a gentleman's agreement and we shook hands

was probably in the first week of the New Year that he made a series of vitally important observations. When he first began injecting extract into normal rabbits he had noted how very hungry they became as their blood sugars fell, some of them avidly eating paper or wood shavings. As he started using more potent batches of extract, the rabbits would occasionally go into convulsive seizures. Their heads snapped back, eyeballs protruding, limbs rigid, they would violently toss themselves from side to side, then collapse into a kind of coma, lying still on their sides and breathing rapidly. The slightest stimulation, such as a shaking of the floor, would set them off again. Sometimes lying on its side the animal's limbs would move rapidly, as in running. The convulsions would recur every fifteen minutes or so until in most cases the rabbit died, rigor mortis setting in immediately.[21]

A pre-doctoral fellow in the pathological chemistry department, O.H. Gaebler, remembered witnessing Collip's reaction to the first appearance of the convulsions. Collip's first thought, according to Gaebler, was that the extract must have toxic properties to cause the reaction. "On second thought, he took a blood sample and set it aside, emptied solid glucose into water, shook it about, and injected it. The rabbit recovered shortly. Subsequent analysis of the blood indicated virtual absence of glucose. It all looks simple now, but it was the most thinking per square meter per minute that I have seen."[22]

Collip had been dealing with the hypoglycemic reaction, now called "insulin shock," which develops when blood sugar falls below certain levels. He had learned the remarkable way in which sugar clears up the condition, the symptoms quickly disappearing as the blood sugar rises again. Clark Noble learned of the same phenomenon about the same time when Joyce, the animal keeper, told him of coming in in the morning and finding rabbits dead or convulsive (in one version of this story a rabbit was stuck in the ventilating system). Noble was doing blood sugars on the animals, finding them very low, when Macleod came in, took the pipe out of his mouth, and said, "Ah, Noble, very interesting. Did you give them glucose?" He had not, did now, and half a century later thought he and Macleod had been the first to see its effects.[23] Macleod had probably earlier learned what to do from Collip. Or he and Collip had both been alerted to the phenomenon and its antidote by a paper published earlier that year by F.C. Mann and T.B. Magath reporting their observations of hypoglycemic shock after hepatectomy (removal of the liver).[24]

Collip was meeting Macleod almost every day for lunch now (neither of the two professors would have gone out of their way to socialize with the

---

on it. I had no suspicion at that time but I believe Best had. Collip was to work on refining the extract - Best and I were to test the extracts and continue the physiological investigations."

inarticulate and probably increasingly sullen Banting) and telling him of his very satisfying results.[25] Just how satisfying they were to Collip is clear from one of the group's few surviving letters of that winter, a January 8 report Collip made to the president of his University of Alberta, H.M. Tory, on the use of his sabbatical time:

> I will never regret having decided to spend a year near Professor Macleod. At the recent Conference at Yale he stood out most obviously as the leading man present. Last spring the old problem of diabetes was again taken up for re-investigation in his laboratory. During the summer such encouraging results were obtained by Dr. Banting and Mr. Best that in the fall the scope of the work was much enlarged. I was given the chemical side and a good part of the Physiological to push along with.
>
> I planned a series of experiments the results of which when obtained gave me a direct lead to the solution of the basic functional derangement in diabetes. The crucial experiment was tried out just before the Christmas break and the results were so striking that even the most skeptical I think would be convinced. I have never had such an absolutely satisfactory experience before, namely going in a logical way from point to point into an unexplored field building absolutely solid structure all the way. However to make a long story short we have obtained from the pancreas of animals a mysterious something which when injected into totally diabetic dogs completely removes all the cardinal symptoms of the disease. Just at the moment it is my problem to isolate in a form suitable for human administration the principle which has such wonderous powers, the existence of which many have suspected but no one has hitherto proved.
> If the substance works on the human it will be a great boon to Medicine, but even if it does not work out a milestone has at least been added to the field of carbohydrate metabolism.
>
> Professor Graham was in my laboratory today discussing the whole matter and in the course of a few days time we hope to have had a clinical test made. If it works we will turn over in all probability the formula to the Connaught Anti-Toxin laboratories for manufacturing purposes.
>
> To be associated in an intimate way with the solution of a problem which for years has resisted all efforts was something I had never anticipated. I only wish that the various papers which will be published on this work were coming from Alberta rather than Toronto.[26]

III

Fred Banting's dissatisfaction with the state of affairs in the lab had not eased. Macleod had become the quarterback of the team. Collip seemed to be doing all the running with the ball. Collip expected soon to have an extract ready to try on humans. Nothing is more evident from Banting's notes and ideas, going right back to October 31, 1920, than his belief that the real test of his work would be the one done on a human diabetic. He was determined to participate in the first clinical trial.

He might not. It was Collip, not Banting or Best, who had the job of preparing the extract which would be used in the clinic. Surely Dr. Banting would administer it, though. Not necessarily, for he had no standing at Toronto General Hospital, the university's teaching hospital, where the trial would take place. In any case, he wanted the first test to be of extract he and Best had made, not Collip, and he began urging Macleod to let him try on a human the extract he and Best were using on dog 33. To clear himself to do this, or perhaps to be in on the testing of Collip's extract, Banting apparently applied to Professor Graham at the Department of Medicine for a temporary appointment in that department to make possible his testing pancreatic extract on humans at the hospital.[27]

Duncan Graham, an Ontario-born Scotsman, trained in Toronto, the United States, Britain, and Germany, had recently become the Eaton Professor of Medicine at Toronto, one of the first appointments made under the controversial "full-time" system. Graham was a tough cookie at all times, but particularly so when it came to protecting patients in the hospital wards under his control from premature experiments or investigations.[28] He decided that Banting, a surgeon who was not currently in practice, had no qualifications to experiment on his patients. Banting remembered Graham saying to him, on either this or a later similar occasion, "What right have you to treat diabetics? How many of them have you ever treated?"[29] Not easy questions for Banting to answer.

Banting was nothing if not persistent, and by now must have been desperate to stop what could have only seemed more and more like some nightmarish conspiracy - Graham and Macleod were the best of friends - to push him out of the picture. We know nothing of the arguments Banting used in persuading Macleod to let him and Best try their extract on a human. Perhaps he claimed it would only be fair to give them the first chance. Perhaps he convinced the professor that the extract being used on dog 33 was not having toxic effects. Perhaps Macleod thought it wise to give Banting the reassurance he seemed to want (Macleod did not yet know of Banting's belief that he, Macleod, was trying to push him aside, but he had noticed, after a meeting of the Journal Club at which ketonuria was discussed, that there was a "strain" between Banting and Best on the one hand and Collip on the other).[30] Perhaps Banting just wore Macleod down. Whatever the reasons, Macleod relented and agreed to intercede with Graham to make possible a clinical trial of extract prepared by Banting and Best.[31]

The patient chosen to receive the extract was a fourteen-year-old boy,

Leonard Thompson. Thompson was a public ward patient (i.e., a charity case) in the diabetic clinic Dr. Walter "Dynamite" Campbell had founded a few years earlier at Toronto General Hospital under Duncan Graham's supervision. Leonard's diabetes had been diagnosed in 1919. Allen therapy was tried. By December 1921 the boy was reduced to skin and bones. As a favour to his doctor, Campbell agreed to pull strings to have him admitted to the General Hospital rather than the Hospital for Sick Children, since the latter had no diabetic clinic. To arrange this he had Thompson's father take the boy to Duncan Graham's office. When the father walked in, carrying the boy, Graham's secretary, Stella Clutton, was horrified. "I've never seen a living creature as thin as he was," she told me sixty years later, "except pictures of victims of famine or concentration camps."[32]

Leonard Thompson weighed 65 pounds on admission to hospital on December 2. He was pale, his hair falling out, abdomen distended, breath smelling of acetone. He was dull and listless, content to lie in his bed day after day. "All of us knew that he was doomed," a senior medical student in the hospital recalled.[33] Campbell tried various adjustments to his hospital diet, finally settling on a regimen totalling 450 calories daily. When the boy continued to worsen, Campbell told his father that unless Banting and Best's new extract had some effect the result was inevitable. The father agreed to let them try the extract on Leonard.[34]

Best made some extract by the process worked out in December. Whole beef pancreas was ground up in an equal volume of slightly acid alcohol. The solution was filtered, most of the alcohol was evaporated off in a vacuum still, the solution was washed twice with toluene, and the remaining watery solution was sterilized with a Berkefeld filter. Banting and Best tested the extract's potency on a dog. They may have given each other injections to make sure it was safe for humans; if so, there was only a little redness in their arms. The next day they took the extracts across the street to Campbell's clinic on Ward H of Toronto General Hospital.[35]

Campbell remembered the extract as being "a thick brown muck" in appearance.[36] The actual injection was made by a young house physician, Ed Jeffrey. In the afternoon of January 11 he injected fifteen cc. of the (presumably diluted) extract into Leonard Thompson, seven and a half cc. into each buttock. The quantity chosen was one-half the dose it was thought would have a definite result on a dog of equal weight.[37] The only detailed description of the scene that day is in Banting's 1940 memoir:

> We went to the hospital and remained in the corridor while a houseman injected it into the patient. We had advised Campbell concerning the time for taking samples of blood for blood sugar estimations and also concerning specimens of urine. We waited around for the first specimens and could hardly contain our suppressed excitement. This was in reality the first human diabetic to be treated.

When the specimen of urine arrived we were told that it would be tested in due course. We asked for a small sample, a few drops, but we found that the whole sample was the property of the hospital, that all specimens would be done together along with samples of blood and that we would be given the results on the following day. There was a cool atmosphere about the place but there did not seem to be anything to do so we went back to the laboratory.[38]

The result of the injection, as reported in a publication signed by Banting, Best, Collip, and Campbell, was as follows: Leonard Thompson's blood sugar dropped from .440 to .320. The twenty-four-hour excretion of glucose fell from 91.5 grams in 3,625 cc. of urine to 84 grams in 4,060 cc. The Rothera test for ketones continued to be strongly positive. "No clinical benefit was evidenced."[39] A sterile abscess, caused by the impurities in the extract, developed at the site of one of the injections.[40]

There is substantial evidence that one or two other patients also received injections of Banting and Best's extract,[41] but there are no records or detailed references in the publications. Some of Banting's accounts in the 1920s suggest that Thompson was the one of three patients on whom the extract had some noticeable effect. Many years later Walter Campbell told Robert Noble (Clark Noble's brother) that Thompson was the only one on whom they had even bothered to do blood sugars.[42]

Banting and Best's extract had failed. Of course a good face could be put on the results: the 25 per cent decrease in the blood sugar, the reduction of glycosuria (and Banting put a better face on the results in his Nobel Prize lecture by talking of a "marked reduction" in blood sugar and saying that the urine had been rendered sugar-free).[43] But there was the overwhelming fact that the extract's actually very modest impact did not outweigh the reaction it caused. Even though Leonard Thompson was a very sick diabetic boy, the doctors decided not to give him further injections of Banting and Best's extract. It was "absolutely useless for continued administration to the human subject," Collip wrote in 1923, in a mood, which we will come to understand, of considerable bitterness. Banting himself accurately summed up the situation after January 11 when he wrote in his published "History of Insulin" that "These results were not as encouraging as those obtained by Zuelzer in 1908."[44]

Banting may not have known at the time that the records of Toronto General Hospital listed Thompson as having received "Macleod's serum." When he found out about it, he did not appreciate the irony.[45]

IV

Macleod had probably made a serious mistake in bowing to Banting's pressure for a clinical test. It was a kind of crossing of the Rubicon (which in the geography of insulin was Toronto's College Street, running between the hospital and the university), from the clinical side of which there could be no real withdrawal. That boy was in the hospital dying. The impure extract had been a little bit effective. The pressure must have increased on Collip to come up with better extract, fast. He was working very long hours.

As he worked, Collip could not have been at all happy about the behaviour of Banting, who seemed to have undercut the group's arrangements by turning the purification problem into some kind of competition between Banting and Best on the one hand, and Collip on the other. What was the point of this? Especially because there was no true competition, for the trio had already pooled its methods. In making their extract for Thompson, for example, Banting and Best had apparently adopted the improvements Collip had worked out in December, notably the use of a vacuum still and the technique of not evaporating off all the alcohol.[46] What were Banting and Best up to in testing that extract on Thompson? Were they hoping to take credit away from the other members of the team, hoping to say they made the extract first used on humans? If so, would they acknowledge that even this process relied on contributions by others?

Collip might have been very angry at the breach of the experimental plan, perhaps regarding it as a breach of faith or trust. Both Macleod and Collip might well have regretted the scientific blunder and embarrassment of premature testing. Relations between laboratory experimenters and clinicians are seldom without stress. At that time in Toronto, with Duncan Graham's particularly strong views about clinical experimentation, and in a climate of deep public suspicion (caused by a struggle over appointments) about the university's relationship with Toronto General Hospital,[47] Macleod must have found the Thompson test something of a humiliation.*

He found it had another disastrous consequence when a day or two later Best came into his office with a reporter from the Toronto *Star,* the city's dynamic evening newspaper which was just entering a period of all-out enterprising reporting. In fact it was a sign of the *Star*'s enterprise that the reporter, Roy Greenaway, had somehow found out about the test on Thompson and was about to scoop the world on this new treatment for diabetes. He had found his way to Best, who thought the best way to handle the situation was to give him to Macleod. Macleod was appalled at the prospect of the impact of premature publicity, especially on diabetics

*Macleod and Collip may have been deeply disturbed for a further reason if my ordering of events is incorrect in placing the discovery of the hypoglycemic reaction before January 11. If Collip only discovered it after January 11, as is possible, then he and Macleod would have been appalled at the thought of a clinical test having been held before animal tests revealed the hormone's potentially fatal side-effect. Collip's accounts, 1923K,L, go out of their way to stress how thoroughly the working of the extract had been investigated before it was used successfully on humans.

desperate for treatment. He probably urged Greenaway not to publish anything; Greenaway agreed that he would emphasize that the work was preliminary. He did, more or less. The article, appearing on January 14, emphasized Macleod's cautions. "We've really no hope to offer any one at all as yet," Macleod was quoted as saying. "We don't know anything yet that would warrant a hope for cure. But we are working intensively at the thing with a hope that some day we may be able to help on a little bit." Last summer's experiments had not been new by any means. Hundreds of people all over the world had been working on the problem of sugar and the blood. "At New Haven we were able to report results that were more definite; that was all. We are working very conservatively striving to awaken no false hopes."[48]

To Fred Banting everything about the article, which barely mentioned Best and himself, was a distortion. To understand why, reread the last paragraph from Banting's point of view. Think about his situation on January 14; think about Macleod's use of "we." Banting's near paranoia about Macleod is surely understandable.

Macleod probably did not give Banting's sensitivity any thought, concerned as he was about the trouble the publicity was bound to cause. The *Star*'s report was picked up by other papers. Within a few days letters started to arrive from diabetics and their relations asking about the extract.[49]

Macleod's innocence of Banting's suspicions ended a day or two later when Duncan Graham came to see him to report a conversation he had just had with Banting. Banting was accusing him, Macleod, of stealing his work, Graham said. He had been scattering these accusations freely for some time and now wanted to see Macleod.

Macleod's first reaction was not to take it seriously. But Graham "impressed me with the serious character of these charges of Banting, and as Banting had been discussing this matter with other people advised me to take steps to put things right." Macleod immediately went to C.L. Starr, whom he knew Banting trusted. Starr agreed to see Banting. Starr and Banting had a talk; possibly Starr and Macleod had a second talk (Banting's and Macleod's recollections of the comings and goings do not agree). Banting was instancing both the New Haven session and the *Star* interview as evidence of Macleod's bad faith, and apparently was able to call on others who had been at New Haven to at least corroborate Macleod's dominance of the meeting.[50]

Banting and Macleod finally met. Each claimed later that the other apologized. Macleod told Banting he regretted having taken over the discussion at New Haven, but did it to emphasize the real value of the work. Banting agreed he had misunderstood Macleod's action, "and assured me that he would do his utmost to undo the harm he had done among his friends." The two apparently agreed on a *modus vivendi*, described by Macleod:

> Dr. Banting assured me that he would not misunderstand me in the future and would not conceal any doubts he might have as to whether I was treating him properly. I agreed to continue collaborating with him and I assured him that I had no intention of robbing him of any of the glory that was his due. I agreed further to have the names of those who participated in the researches, then underway in my department, in which the physiological action of pancreatic extracts was being investigated, published with the names in alphabetical order. This placed his name first and Best's second.

Banting's sour 1940 comment on the meeting was that "Macleod thought I had been working overhard, advised a holiday and smoothed everything over with a sticky candy."[51]

V

While all of this was going on, Collip was working in his lab trying to produce a purified extract. Later in his career J.B. Collip's skill at extracting hormones made him something of a legend in Canadian medical research. He was part chef, part brewer, part wizard, and, to his critics, part "messer," as he mixed and filtered, distilled and evaporated, concentrated and diluted, centrifuged and blended. A restless man by temperament, endlessly criss-crossing North America on marathon automobile trips, talking so quickly and disconnectedly people had trouble following him, Collip would finish with one batch of extract and go on to another and another, never making them the same way twice, sometimes working so quickly he had trouble recalling what he had done. It was laboratory research, but the most practical kind of tinkering – a touch of this, a dash of that, what Collip later referred to as "bathtub chemistry."

Rough and ready as Collip's methods were, they were just what the Toronto group needed in January 1922. There are no records of Collip's trials and progress as he mixed up batch after batch of pancreatic extract, testing each one for potency, perhaps several times at different stages, on his rabbits. We know his starting point was fresh whole beef pancreas ground up in alcohol. Then the permutations and combinations of possible treatments seemed practically endless. How long should it stand before the first filtration? At what temperature? Should acid be added to the alcohol? At what concentration? How should the alcohol be evaporated? How much evaporation? How many more filtrations? What about using other solvents? How do you get the fats and salts out? And so on, and on.

Actually the chemistry was fairly comprehensible, especially after the fact. The pancreatic tissue consisted of fats and proteins, water, salts, smaller quantities of other organic materials, and the mysterious active principle. Different kinds of proteins were soluble in alcohol at different concentrations and different degrees of acidity. The active principle was soluble in alcohol at the approximately 50 per cent concentration Banting and Best had first hit upon. Would it be possible to find a concentration of alcohol at which the active principle would be still soluble and most of the non-insulin protein contaminants insoluble? Or, for that matter, vice versa - the proteins still dissolved, the active principle precipitated out? The fats could be dealt with fairly easily by known chemical methods, the salts with a little more finesse. While the alcohol was critical to the whole operation, it was a constant problem to get rid of it without also somehow destroying the active principle. Banting and Best's results seemed to show that heat destroyed the active principle in aqueous solution.

Collip's method involved gradually increasing the concentration of alcohol in the mixtures, finding that the active principle stayed in solution at higher and higher concentrations, while most of the proteins precipitated out and the lipids and salts could eventually be extracted by centrifuging and washing. It was late on a January night, probably the evening of the 19th,[52] when Collip discovered a limit. (He may have been looking for it because of his observations in late December about the potency of the precipitate in one of his early batches.) At a certain concentration of alcohol, somewhere over 90 per cent, the active principle itself was precipitated out. There it was. You could "trap" the active principle (as Collip put it), or isolate it, by first producing the concentration of alcohol in which it was soluble but most of its protein contaminants were not, and then moving to the concentration that would precipitate it. The night he discovered this, Collip wrote in 1949, "I experienced then and there all alone in the top story of the old Pathology Building perhaps the greatest thrill which has ever been given me to realize."[53] Describing the chemical procedure at a dinner for Collip in 1957, Dr. R.F. Farquharson, Professor of Medicine at Toronto, ended the account of the purification by saying, "As Walter Campbell used to say, Collip then actually saw insulin."[54]

Actually that was an exaggeration, for the powder Collip produced was eventually found to consist of a little active principle in a lot of impurities. But it was far purer than any previous extract. Collip tested its potency on rabbits, waited a few days to check for abscesses, and knew he had an extract that could go back to the clinic. The treatment of Leonard Thompson with injections of pancreatic extract, Collip's extract this time, resumed on January 23.

VI

One of the more remarkable personal confrontations in the history of science occurred sometime between January 17 and January 24. There are no contemporary accounts of it, no references whatever by Collip, and only the two following accounts, neither of which should be considered totally reliable.

Banting wrote in 1940 as follows:

> The worst blow fell one evening toward the end of January. Collip had become less and less communicative and finally after a week's absence he came into our little room about five thirty one evening. He stopped inside the door and said "Well fellows I've got it."
>
> I turned and said, "Fine, congratulations. How did you do it?"
>
> Collip replied, "I have decided not to tell you."
>
> His face was white as a sheet. He made as if to go. I grabbed him with one hand by the overcoat where it met in front and almost lifting him I sat him down hard on the chair. I do not remember all that was said but I remember telling him that it was a good job he was so much smaller - otherwise I would "knock hell out of him." He told us that he had talked it over with Macleod and that Macleod agreed with him that he should not tell us by what means he had purified the extract.[55]

Best, not having read Banting's account, gave his version of the incident in a letter to Sir Henry Dale, written in 1954 and intended for the historical record:

> One evening in January or February, 1922, while I was working alone in the Medical Building, Dr. J.B. Collip came into the small room where Banting and I had a dog cage and some chemical apparatus. He announced to me that he was leaving our group and that he intended to take out a patent in his own name on the improvement of our pancreatic extract. This seemed an extraordinary move to me, so I requested him to wait until Fred Banting appeared, and to make quite sure that he did I closed the door and sat in a chair which I placed against it. Before very long Banting returned to the Medical Building and came along the corridor to this little room. I explained to him what Collip had told me and Banting appeared to take it very quietly. I could, however, feel his temper rising and I will pass over the subsequent events. Banting was thoroughly angry and Collip was fortunate not to be seriously hurt. I was disturbed for fear Banting would do something which we would both tremendously regret later and I can remember restraining Banting with all the force at my command.[56]

Except for a veiled but important reference in Banting's 1922 account, there are no other useful written records of this incident. Clark Noble once drew a cartoon, unfortunately now lost, of Banting sitting on Collip, choking him; he captioned it "The Discovery of Insulin."[57]

The one surviving artifact of the fight is an agreement signed by

Banting, Best, Collip, and Macleod, dated January 25, 1922, and entitled, "Memorandum in Reference to the Co-operation of the Connaught Anti-Toxin Laboratories in the Researches of Dr. Banting, Mr. Best and Dr. Collip - Under the General Direction of Professor J.J.R. Macleod to obtain an Extract of Pancreas Having a Specific Effect on the Blood Sugar Concentration." The two key conditions of Connaught's co-operation with the team were:

> 1. Dr. Banting, Mr. Best and Dr. Collip each agrees not to take any steps which will result in the process of obtaining an extract or extracts of pancreas, being patented, prepared by any commercial firm with aid of any of the above or otherwise exploited during the period of co-operation with the Connaught Anti-Toxin laboratories.
> 2. That no step involving any modification in policy concerning these researches be taken without preliminary joint conference between Dr. Banting, Mr. Best and Dr. Collip, and Professor Macleod and Professor Fitzgerald be held.[58]

The rest of the document spelled out technical and financial details.

What had happened? What had Collip said to Banting to cause the attack? Why had he said it? There seems little doubt that Collip said three things that night in the lab to Banting and Best: first, he would not tell them how he had made his breakthrough; second, he had told Macleod, who had agreed that Collip did not have to tell Banting and Best; and third, he might go ahead and take out a patent on his process.

What was going on in Collip's mind and what Banting and Best said to him in the course of the conversation can only be speculated upon. He was probably tired - they were all probably tired - after days of hard work and extreme pressure. I presume that Collip and Macleod had little use for Banting's conduct in the past several weeks, particularly Banting's breaking of the spirit of the collaboration by himself and Best making the extract for the first clinical test. And, it appeared, Banting had appropriated some of Collip's improvements in making that extract. Banting had shown his distrust of them; now they had no reason to trust him. It was Collip's job to purify the extract, not Banting and Best's. Collip and Macleod may have decided that Banting was trying to take credit away from Collip - that if he knew the process for making the extract he would claim it as his own. They may have believed, after the misadventure of January 11, that Banting could not be trusted not to try to forestall the rest of the team by applying for a patent. Paranoia begat paranoia. So Collip and Macleod decided not to tell Banting and Best the secret of making an effective anti-diabetic extract.

Speculating further, the kind of things Collip likely said that night are these: "Why should I tell you?...It's my job, not yours, to get it ready... What do you want to know for? So you can run your own test again?...Stick to your job, I'll do mine... You'll know in good time when *we* see how it works... Don't worry, you'll get your share of credit for the work you've done... I'm not going to let you take credit for my work... you've already tried to do it once... I don't have to put up with your kind of nonsense... maybe I'll just go back to Alberta and patent my method...." Collip could not have known Banting very well, could not have known how much of his life revolved around the work, how terribly insecure Banting was at the best of times, how desperately unhappy, suspicious, and frightened he had become as the awful pattern of recent weeks had unfolded, and how the only final outlet this blunt, unsophisticated veteran had for all his frustration and rage was to fight back. Literally.

The Connaught agreement of January 25 was probably the result of meetings in the day or so after the fight involving the principals, Velyien Henderson as the professor Banting trusted, and J.G. Fitzgerald, the director of the Connaught Laboratories. Andrew Hunter, the professor of pathological chemistry, may also have been involved. Again, except for the written agreement, there is no record of these discussions. Banting's 1922 account suggests Hunter and Henderson supported his view that Collip wanted to patent his process. Henderson's behaviour and motives in all his dealings with Banting are obscure. A number of fairly detached observers, also Best, thought Henderson deliberately fanned the flames of Banting's suspicion, perhaps because he intensely disliked Macleod. On the other hand, there may have been some concern on the part of some of the Toronto people, including Fitzgerald of the Connaught,[59] that Collip did have a purification process which might be patentable separately from anything anyone else at Toronto had done. Collip was a visitor to the university, free to go back to Alberta at any time, scheduled to leave when the term ended.[60] It would be a disaster if he left town taking his knowledge with him, as Banting was insistently and angrily claiming he intended to do. So it was time to tie Collip down, tie Banting and Best down too, and try to settle the whole mess once and for all, by putting the principles of the collaboration down on paper and getting them all to sign it. Then there would be no more need to refer to the unfortunate incident in the lab.[61]

VII

At 11 o'clock on the morning of Monday, January 23, Walter Campbell gave Leonard Thompson five cc. of the new extract made by Collip. At 5:00 that afternoon the boy was given twenty cc. The next day there were two injections of ten cc. each. Thompson's glycosuria almost disappeared. His ketonuria did disappear. His blood sugar early on the 23rd had been .520. On the 24th it dropped to .120. No extract was given on the 25th and 26th, perhaps while Collip made a new batch. It seems to have been more concentrated, with two four cc. injections becoming the normal daily dose.

The urine tests continued to be favourable, "the boy became brighter, more active, looked better and said he felt stronger."[62] This was the first unambiguously successful clinical test of the internal secretion of the pancreas on a human diabetic. Collip's process worked. Not being a medical doctor, Collip was probably not present at these tests.

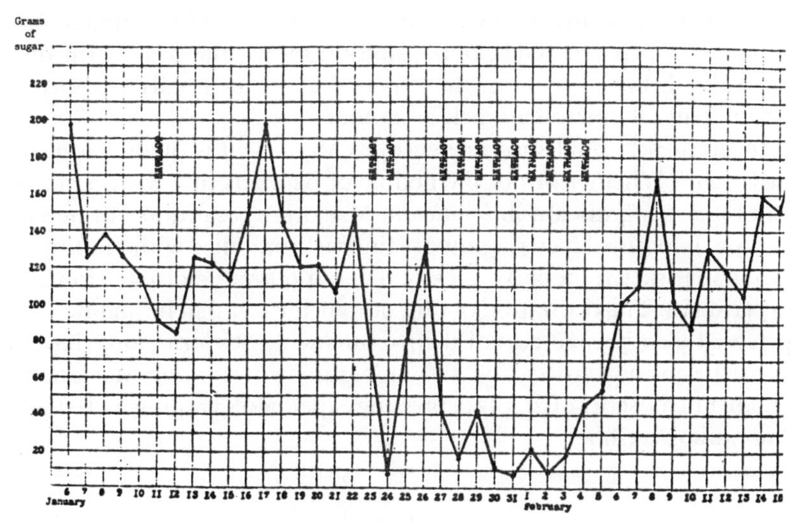

Chart 5: The effect of extract on the sugar in Leonard Thompson's urine. Taken from the 1922 published report.

The creature who had received more pancreatic extract than Leonard Thompson or any other animal was Banting and Best's dog Marjorie, depancreatized on November 18 and still receiving her daily extract late in January. There are hardly any published or unpublished records of Marjorie's condition. From January 21 to 23 Banting and Best discontinued their extract, and reported that the dog became so weak she could hardly stand. There was "decided improvement" when extract was given again on January 24 and 25.[63]

Marjorie had now lived for seventy days after her pancreatectomy, far, far longer than untreated dogs normally survived. On January 27 she was chloroformed. Perhaps it was a logical time to end the experiment, but Banting also remembered that the dog had abscesses from its injections and that the scarce supplies of extract were needed for "more acute experiments."[64] The confrontations and quarrels over credit may also have brought on the decision to kill the dog. As the dispute raged over who had been able to produce how effective an extract, Marjorie's longevity became increasingly important to Banting and Best as clinching proof that they could make potent, non-toxic extracts. It was, and is, unfortunate that they had such poor records of Marjorie, a state of affairs which may have contributed to Macleod's doubting whether the dog's pancreas had been completely removed.[65] It was apparently to resolve such doubts that a careful autopsy was done on Marjorie by an "independent and impartial" (Banting's words) pathologist at Toronto General Hospital. Dr. W.L. Robinson found that Marjorie still had a small nodule of pancreatic tissue, about three millimetres in diameter, in the submucosa of her duodenum. He could not find any islet cells in it. "It does not seem likely that so small a piece of pancreas could be responsible for the maintenance of the life of the animal, but, of course, the experiment is not finally conclusive," Banting and Best wrote in their account.[66]

In most early published references to Marjorie this qualification was included.[67] Whether or not the nodule of pancreas that Banting missed had made a significant difference in Marjorie's condition can never be known. In point of fact, the autopsy finding, combined with Banting and Best's very sketchy reports, greatly reduced the value of their longevity experiment. The realization of this, despite the fact that he and Best were personally entirely certain of their results with Marjorie, must have increased Banting's insecurity. The ambiguity surrounding Marjorie was another in the series of deeply depressing experiences Banting had endured since mid-December.

VIII

By February 1922, testing of the extract was going ahead in several directions. Six more patients in Campbell's and A.A. Fletcher's clinic at Toronto General Hospital were treated, all with favourable results. The respiratory quotient experiments were begun on dogs in January. A promising result on a diabetic dog led to a test on Banting's classmate, Joe Gilchrist, in mid-February.[68] Gilchrist came into the lab and breathed into the elaborate apparatus while the researchers measured his normal respiratory quotient, his respiratory quotient after an injection of sugar, and then his respiratory quotient after an injection of sugar plus extract. It shot up, from an initial 0.74 to 0.90, clear evidence to them that Gilchrist's system was burning carbohydrates. That one injection also cleared the sugar from his urine and helped him shake off his mental and physical lethargy.[69] Further respiratory quotient tests on two of Campbell and Fletcher's patients were also successful, as were ongoing studies of the extract's effect on ketonuria and glycogen formation in dogs. To help standardize the injections a rough test of potency was worked out in which one "rabbit dose" was defined as the amount of extract necessary to lower the blood sugar of a normal rabbit by 50 per cent (to the point where it became convulsive) in one to three hours.[70]

Banting and Best spent some time in February writing a paper on their animal experiments since mid-November. They gave a paper entitled

"The Internal Secretion of the Pancreas" to local doctors at Toronto's Academy of Medicine on February 7; it was later published in the Academy's bulletin. It should not be confused with the first major paper, under the same title, which was published that month in the *Journal of Laboratory and Clinical Medicine.* In the same month a three-paragraph summary or abstract of the New Haven presentation appeared in the Proceedings of the American Physiological Society, published in the *American Journal of Physiology.* It had been written by Macleod and was also headed "The internal secretion of the pancreas" (by F.G. Banting, C.H. Best, and J.J.R. Macleod). Banting and Best's second major paper, entitled "Pancreatic Extracts," appeared in the *Journal of Laboratory and Clinical Medicine* in May. Like the first paper, it is a straightforward description of the experiments on depancreatized dogs. The paper contains no surprises, except when it is compared with the original notebooks and the surprising number of factual errors - eighteen - are noticed. In this final research report of their work together, Banting and Best conclude that they had made "highly potent extracts" which were "however somewhat toxic, and they are apt to cause local abscesses at the point of injection."[71]

Other people were ready to write about the work, too, notably Roy Greenaway of the Toronto *Star.* "The newspapers got wind of what we are doing and through some agents of their own had enough information of a haphazard type from which they could at any time piece together a garbled account of the work," Macleod wrote to a friend. "They kept constantly prodding us for more information until at last we were compelled to publish...."[72] By the end of the third week in February there was enough clinical and experimental evidence to support a preliminary publication. The paper was entitled "Pancreatic Extracts in the Treatment of Diabetes Mellitus." Its authors were listed as Banting, Best, Collip, Campbell, and Fletcher.

After background material and a description of Banting and Best's experiments leading to the obtaining of active whole beef extract, the paper dealt with the grey area between Collip and the rest as follows:

> As the results obtained by Banting and Best led us to expect that potent extracts, suitable for administration to the human diabetic subject, could be prepared, one of us (J.B.C.) took up the problem of the isolation of the active principle of the gland. As a result of this latter investigation, an extract has been prepared from the whole gland, which is sterile and highly potent, and which can be administered subcutaneously to the human subject. The preparation of such an extract made possible at once the study of its effects upon the human diabetic, the preliminary results of which study are herein reported.

The results of the clinical tests were described, with special emphasis on

"L.T." (Leonard Thompson). The conclusions were carefully qualified, but the paper's key sentence was clear enough: "These results taken together have been such as to leave no doubt that in these extracts we have a therapeutic measure of unquestionable value in the treatment of certain phases of the disease in man."[73] The paper was sent to the *Canadian Medical Association Journal* so that it would receive quick publication.

IX

Fred Banting had little to do with the writing of this paper or the clinical work it reported.[74] He was doing very little work of any kind in the lab, and seemed to have no role in the ongoing research, either the experimental or the clinical. It had all passed into the hands of the experts - Macleod, Collip, Duncan Graham, and Campbell. "Best and I became technicians under Macleod like the others," Banting wrote bitterly in 1940. "We were asked for the extract as it was required for their experiments. We were asked to provide depancreatized dogs and other surgical work. Neither plans for experiments nor results were discussed with us."[75] He began to think about moving on, perhaps to other kinds of research. On February 4 he made a note to himself about the cure for cancer lying in the discovery of some solution, "chemical or internal secretion," that would stop the multiplication of cells. Through February and into March he read and jotted down notes about cancer research.[76]

Banting had been living and working under intense emotional pressure for the past several months - in fact, all things considered, for the past year and a half of uncertainty about his work and his future. In a sense, despite the triumph of the research, his future was just as uncertain as ever. Would he get credit for his work? What would become of him? What about his personal life? This last question seems to have been constantly on his mind, and his 1922 desk calendar indicates that in addition to all of his professional worries there were more crises in the ongoing relationship with Edith Roach:

March 11: Edith came
March 12: Edith went
March 14: Edith phoned
...

March 17: The most human letter E ever wrote
Letter of farewell
...

March 19: Phoned Edith.*[77]

Banting's attendance at the lab fell off more. The only way he could

*There is no certain knowledge, but the memory of those who were on the spot is that Edith was insisting that Banting give up the work, resume practice, and settle down.

overcome his black despair at night was to drink himself to sleep. When he could not get alcohol any other way he stole the 95 per cent pure alcohol being used in the production of pancreatic extracts at the lab. "I do not think there was one night during the month of March, 1922, when I went to bed sober."[78]

Banting's friends knew of the situation, particularly his sense of having started the work, against all odds, and then seen it taken over by others just when the good results came in. One of these friends, Dr. G.W. "Billy" Ross, a former teacher of Banting's, had probably alerted the *Star* reporter, Greenaway, who was one of his patients, to the work on the pancreatic extract. Through Ross, Greenaway met Banting and Best. He prepared a long article for the *Star* to coincide with the publication of the March issue of the *C.M.A. Journal.* "Toronto Doctors on Track of Diabetes Cure" was the *Star*'s headline on March 22, the day the *C.M.A. Journal,* containing the scholarly article, was mailed to its subscribers.

Greenaway quoted extensively from the journal article, had interviewed Macleod, and his story contained pictures of the four men - Banting, Best, Collip, and Macleod ("Have They Robbed Diabetes of its Terror?"). But Greenaway had also interviewed Banting at length, and presented the story as very much the work of Banting and Best. The article, especially one of its sub-headlines, "Banting stakes his all on the results," was the first to tell the story of the discovery from Banting's point of view.

While it created a minor flurry of interest in the Toronto press, and a short Canadian Press wire story was printed in many cities, the March announcement and publication did not capture much public attention outside Toronto. The press was always announcing miracle cures that never amounted to much; in the professional world of medicine the *Canadian Medical Association Journal* was an obscure publication with little circulation outside Canada. A much more important presentation of the results of the Toronto work was scheduled to take place seven weeks later at the conference of the Association of American Physicians in Washington, D.C.

X

In Bucharest, Romania, a few weeks after the Toronto group's first successful clinical tests, N.C. Paulesco decided he was ready to try his pancreatic extracts on humans. As yet, Paulesco knew nothing about the work in Toronto. He, too, was using extracts of whole beef pancreas procured from a local abattoir, but he was still using saline (slightly acidified and then neutralized) as his extractive. His first clinical test was on a forty-three-year-old teacher, Frère M.H., given extract on February 25 and several times afterwards. On March 3 a fifty-two-year-old woman, Madame S.G., became Paulesco's second test case.

Paulesco still prepared his extract by the methods reported in his 1921 papers. Because this aqueous extract had caused toxic side-effects in his dogs when injected intravenously (effects, he wrote, "qui la rendent inapplicable dans la pratique médicale"), Paulesco decided to administer it to humans by the safer method of rectal insertion. His patients took their extract, in quantities ranging from 125 to 1,000 cc., forced in through a 90-centimetre-long red rubber tube. Paulesco measured their urinary sugar and took occasional blood sugar readings.

The extract had no immediate effect on the patients that could not be duplicated by doses of saline alone. Paulesco thought that after several days' treatment there was a slight diminution of glycosuria and some clinical improvement. He found this somewhat encouraging. Working with depancreatized dogs, he continued through 1922 to experiment with ways of making and administering his extract. His most remarkable result came on March 24 when, according to his publication a year later, an intravenous injection of extract reduced a diabetic's blood sugar from .260 on injection to .040 eighty minutes later, and in another hour to .000! Paulesco's articles, which continue into 1924, make no mention at all of hypoglycemic symptoms, or any symptoms at all produced by this condition of "véritable AGLYCÉMIE."[79]

In April 1922 Paulesco applied for a Romanian patent on pancréine and his method of making it. He received his patent, but during the next year, in which he was apparently handicapped by lack of money, made no further progress with his work.[80] There is no evidence that pancréine was ever used successfully to treat humans.

XI

Sometime in April the Toronto group prepared a paper summarizing all the work so far: Banting's idea, Banting and Best's early experiments, Collip's purification, the clinical results, the hypoglycemic effects, the respiratory quotient, liver/glycogen, and ketonuria findings. For the first time a name was given to "this extract which we propose to call insulin." There are no records of the discussion leading to the use of this name. It was apparently suggested by Macleod that a term based on the Latin root for "island" would be more useful internationally than Banting and Best's "Isletin." It is not clear whether the naming provoked more controversy. If "insulin" replaced "Isletin" partly because the purified extract of whole pancreas was very different from Banting and Best's crude extract of degenerated pancreas, Banting and Best did not record their objections or make an issue of the naming. No one made an issue of whether or not Toronto could prove that insulin came from the islets of Langerhans. Nor did anyone realize at the time of naming that in a neat piece of scientific one-upsmanship, the word "insuline" had been proposed to describe the hypothetical internal secretion of the pancreas by E.A. Schafer in 1916. Schafer, at that time, did not know that J. de Meyer had made the same suggestion in 1909.[81]

The paper was another cautious presentation – ("While these observations demonstrate conclusively that the pancreatic extracts, which we employed, contain some substances of great potency in controlling carbohydrate and fat metabolism in normal and diabetic animals as well as in patients suffering from diabetes mellitus, we cannot as yet state their exact value in clinical practice") – and it ended with an ominous reference to "serious difficulties" encountered in attempting to prepare the extract on a large scale. The paper was entitled "The Effect Produced on Diabetes by Extracts of Pancreas," and its authors were Banting, Best, Collip, Campbell, Fletcher, Macleod, and E.C. Noble.

All of the authors had agreed that Macleod would present the paper at the meeting of the Association of American Physicians.[82] He gave it during the noon-hour session on May 3, 1922. A transcript of the ensuing discussion was published. Dr. S. Solis-Cohen briefly described a pancreatic extract that was being used, with mixed results, on patients at the Jewish Hospital in Philadelphia. Allen mentioned that his animal experiments with a pancreaticoduodenal serum were producing reductions in glycosuria and hyperglycemia. But neither wanted to detract from the Toronto presentation. "This study so careful and comprehensive, this work so thorough in its execution and so clear in its presentation, may justly be called epoch-making," Solis-Cohen said. "I am glad that I have been privileged to hear the paper." Allen, the world's leading diabetologist, said, "If, as seems to be the case, the Toronto workers have the internal secretion of the pancreas fairly free from the toxic material, they hold unquestionable priority for one of the greatest achievements of modern medicine, and no one has a right to divide the credit with them."

Rollin T. Woodyatt, who had talked to Macleod before the meeting, announced that he was convinced that Macleod "and his associates" had actually been able to extract the internal secretion of the pancreas. "I think that this work marks the beginning of a new phase in the study and treatment of diabetes. It would be difficult to overestimate the ultimate significance of such a step." Woodyatt moved that the association tender a standing vote of appreciation to Macleod and his associates. Joslin, who was also there, could not remember such an action in the twenty years he had been involved in the Society.[83]

It is probably impossible to specify one time in all the events described in the last three chapters when it could be said that insulin had been discovered. Nor was there a single definite first announcement of the results of the Toronto work begun by Banting and Best on May 17, 1921. This meeting, fifty weeks later, came close. On May 3, 1922, the Toronto group, speaking through Macleod, announced to the medical world that they had discovered insulin. They presented a complete summary of their work. Their presentation convinced their listeners that Toronto had made an epoch-making medical discovery. It was a great triumph.

They knew it. "We had it made," Walter Campbell remembered fifty years later.[84] He had been there to glory in the moment. Fred Banting and Charley Best were not in Washington for the triumph. They had decided not to go, excusing themselves by saying the trip was too expensive.[85]

Oskar Minkowski: who discovered the relationship of diabetes and the pancreas.

N.C. Paulesco: the Romanian who almost discovered insulin.

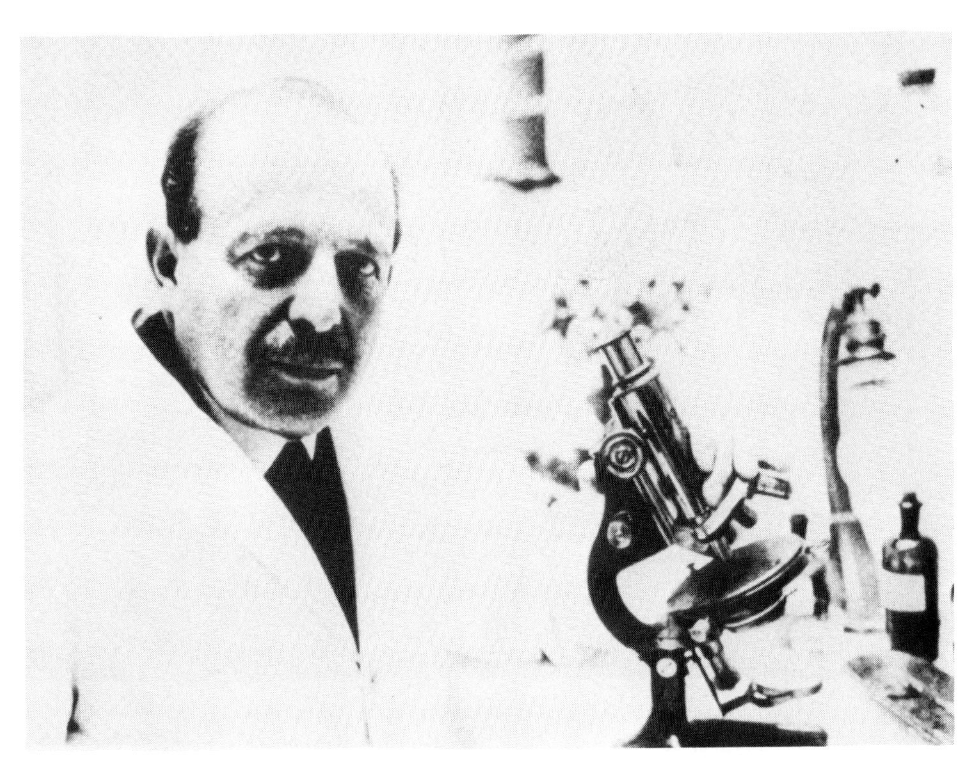

Georg L. Zuelzer: the German who almost discovered insulin.

E.L. Scott (left), and Israel Kleiner (right): the Americans who came closest to discovering insulin.

Frederick M. Allen: father of the "starvation" therapy for diabetes.

Elliott P. Joslin: the master clinician of diabetes.

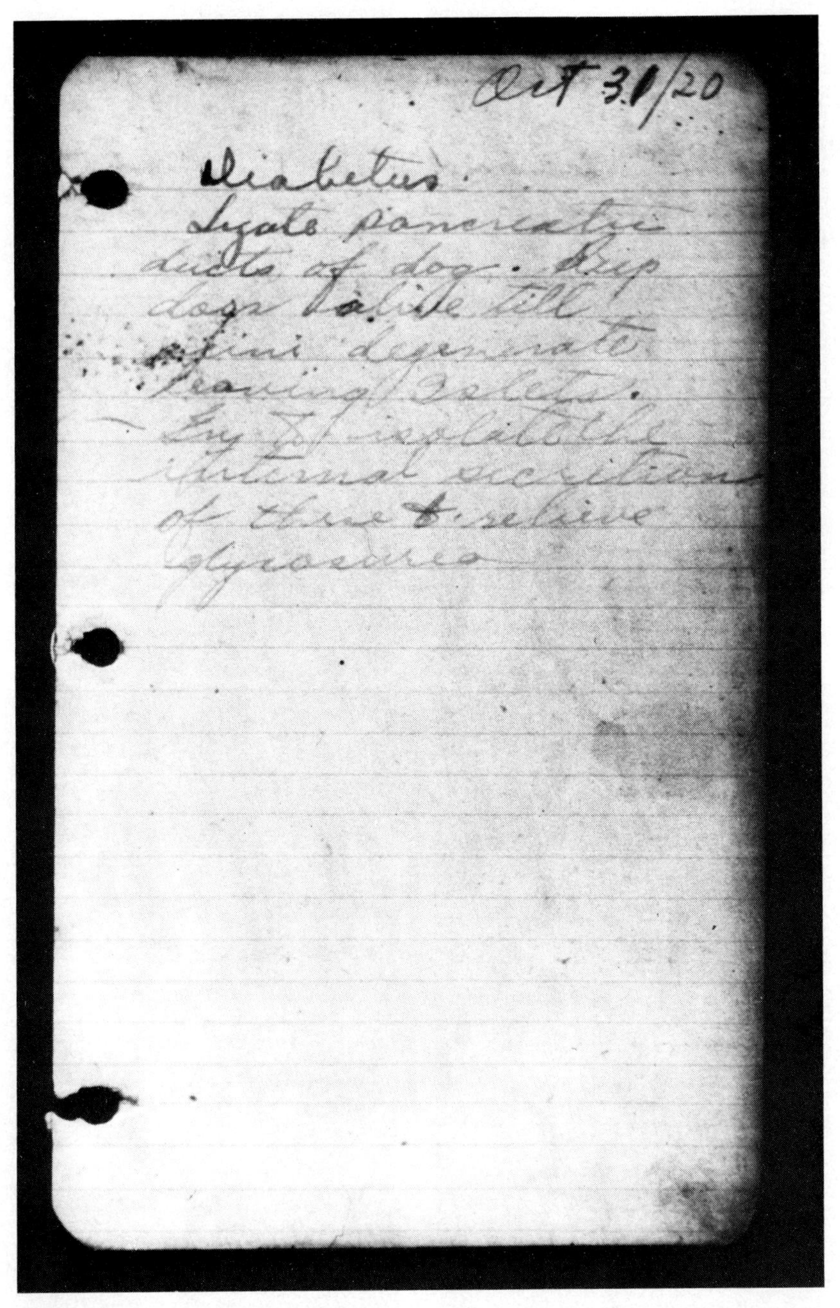

Oct 31/20

Diabetus
Ligate pancreatic
ducts of dog. Keep
dogs alive till
acini degenerate
leaving Islets.
Try to isolate the
internal secretion
of these + relieve
glycosurea

Banting's idea, as written in his notebook (October 31, 1920).
Later he always misquoted himself.

Frederick Grant Banting (1891-1941).

John James Rickard Macleod (1876-1935).

James Bertram Collip (1892-1965).

Charles Herbert Best (1899-1978).

Edith Roach: Banting's sometimes fiancée. They never married.

Margaret Mahon: Best's fiancée in 1921. They married in 1924.

The Medical Building at the University of Toronto in which Banting and Best worked.

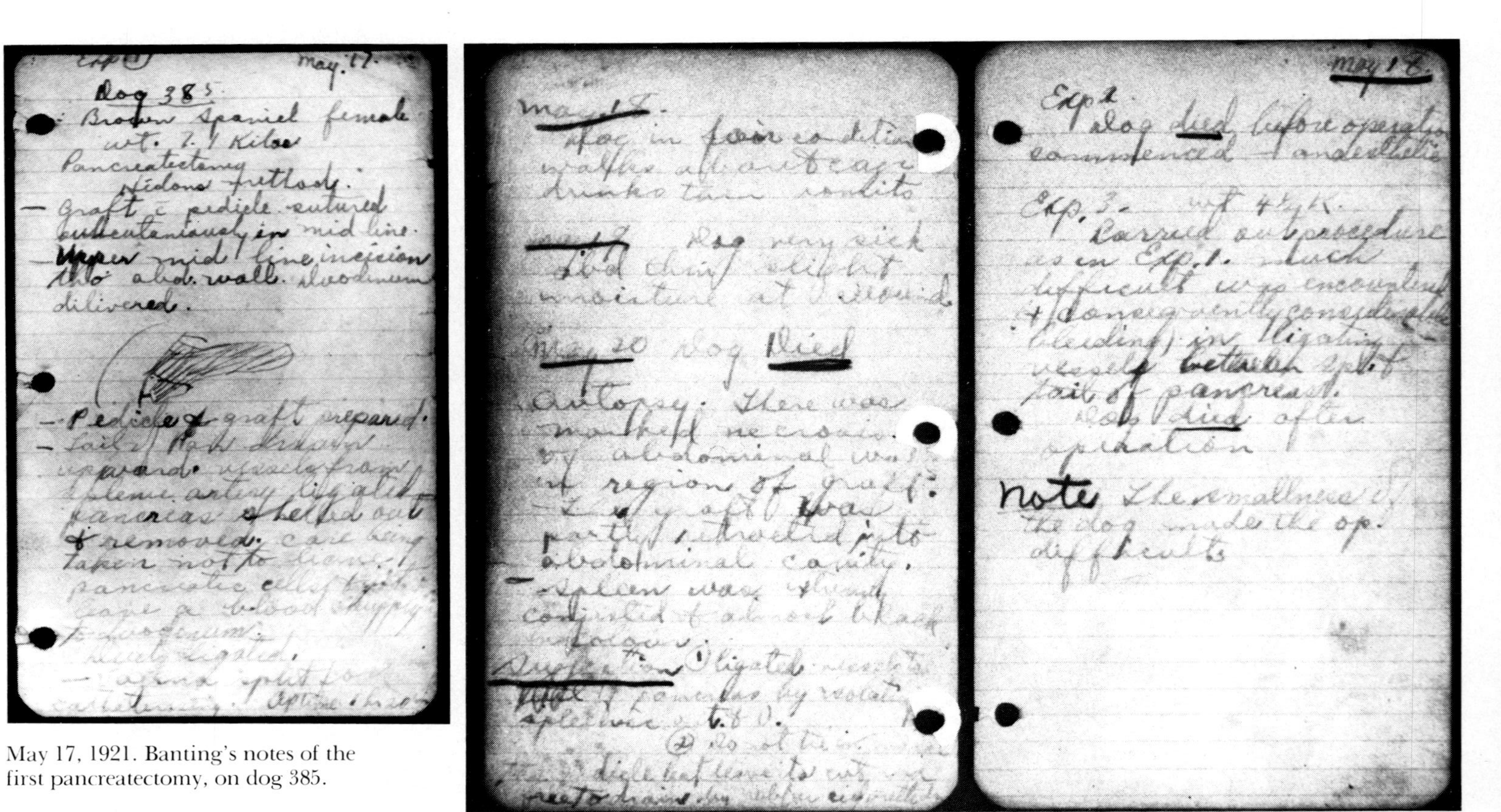

Exp (1) May 17.

Dog 385

Brown Spaniel female
wt. 7.7 Kilos
Pancreatectomy
Hedon's method.

— Graft c pedicle sutured subcutaneously in mid line.

— Upper mid line incision thro abd. wall. Duodenum delivered.

— Pedicle & graft prepared.

— Tail of pancreas drawn upward. vessels from splenic artery ligated. pancreas shelled out & removed. care being taken not to damage [illegible]

[illegible]

May 20 Dog Died

Autopsy: There was marked necrosis of abdominal wall in region of graft. [illegible] was partly retracted into abdominal cavity.

— Spleen was [illegible] congested & almost black in colour.

[illegible]

May 18

Exp 2

Dog died before operation commenced. Anaesthetic.

Exp. 3 — wt 4½ K.

Carried out procedure as in Exp. 1. Much difficulty was encountered & consequently considerable bleeding in ligating vessels between spl. & tail of pancreas.

Dog died after operation

note The smallness of the dog made the op. difficult.

May 17, 1921. Banting's notes of the first pancreatectomy, on dog 385.

The experiments continue, with heavy loss of dogs.

Banting's notes on his plan of work and discussions with Macleod.

July 30, 1921. The first administration of extract; Banting and Best's joint notebook.

Best, Banting and a dog on the roof of the Medical Building. Thought to be in the summer of 1921, but dated April 1922 by Banting in his scrapbook.

Velyien Henderson, the professor of Pharmacology who befriended Banting.

G.H.A. Clowes, research director of Eli Lilly and Company, whose interest in the Toronto work paid off handsomely.

Dr. Joseph Gilchrist, Banting's diabetic classmate; human rabbit in the early testing.

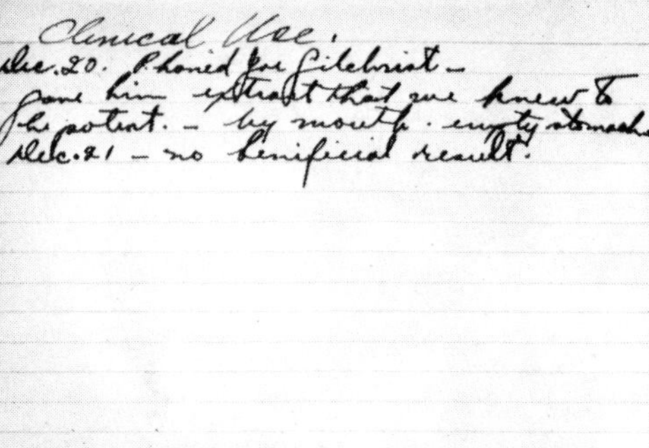

Clinical Use.
Dec. 20. Phoned Joe Gilchrist –
gave him extract that we knew to
be potent. – by mouth. empty stomach
Dec. 21 – no benificial result.

Oral administration meant that this first test on Gilchrist was bound to fail.

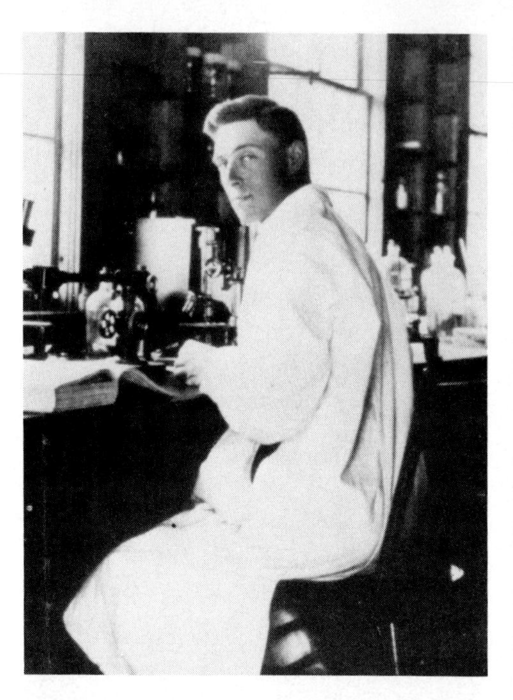

Collip in his lab, about 1922. His purification made clinical testing possible.

Leonard Thompson. A later picture of the first patient treated successfully with insulin.

Elsie Needham
In Oct. 1922. she entered H.S.C. in diabetic coma. She was the first child to recover from coma by the use of insulin

Elsie Needham. Revived from coma at the Hospital for Sick Children. The caption is in Banting's handwriting.

Elizabeth Hughes 1923
1918 wt. 75 lbs
Aug 16., 1922 wt. 45 lbs
Jan. 1., 1923 wt. 105 lbs.

Elizabeth Evans Hughes (1907-1981). Banting's prize patient, who found insulin "unspeakably wonderful." The photograph is from Banting's scrapbook.

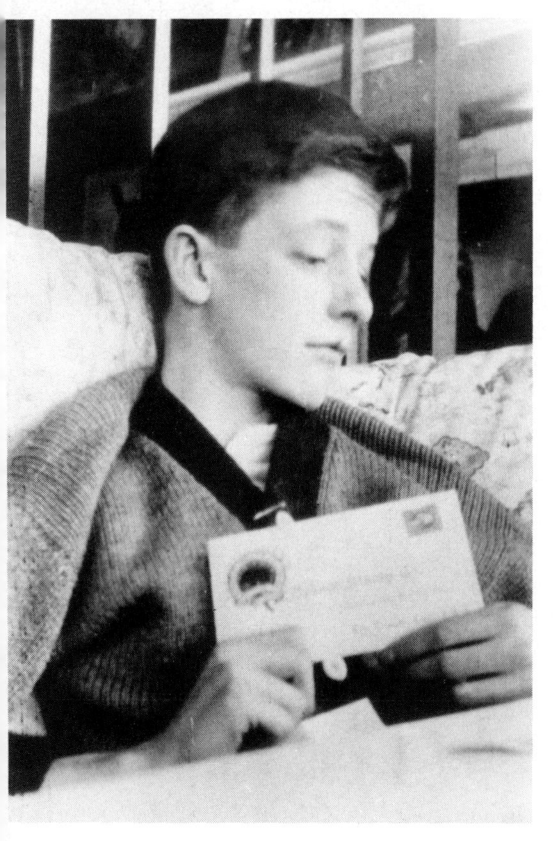

James Havens: the first diabetic to receive insulin in the United States. A snapshot taken during the first months of treatment.

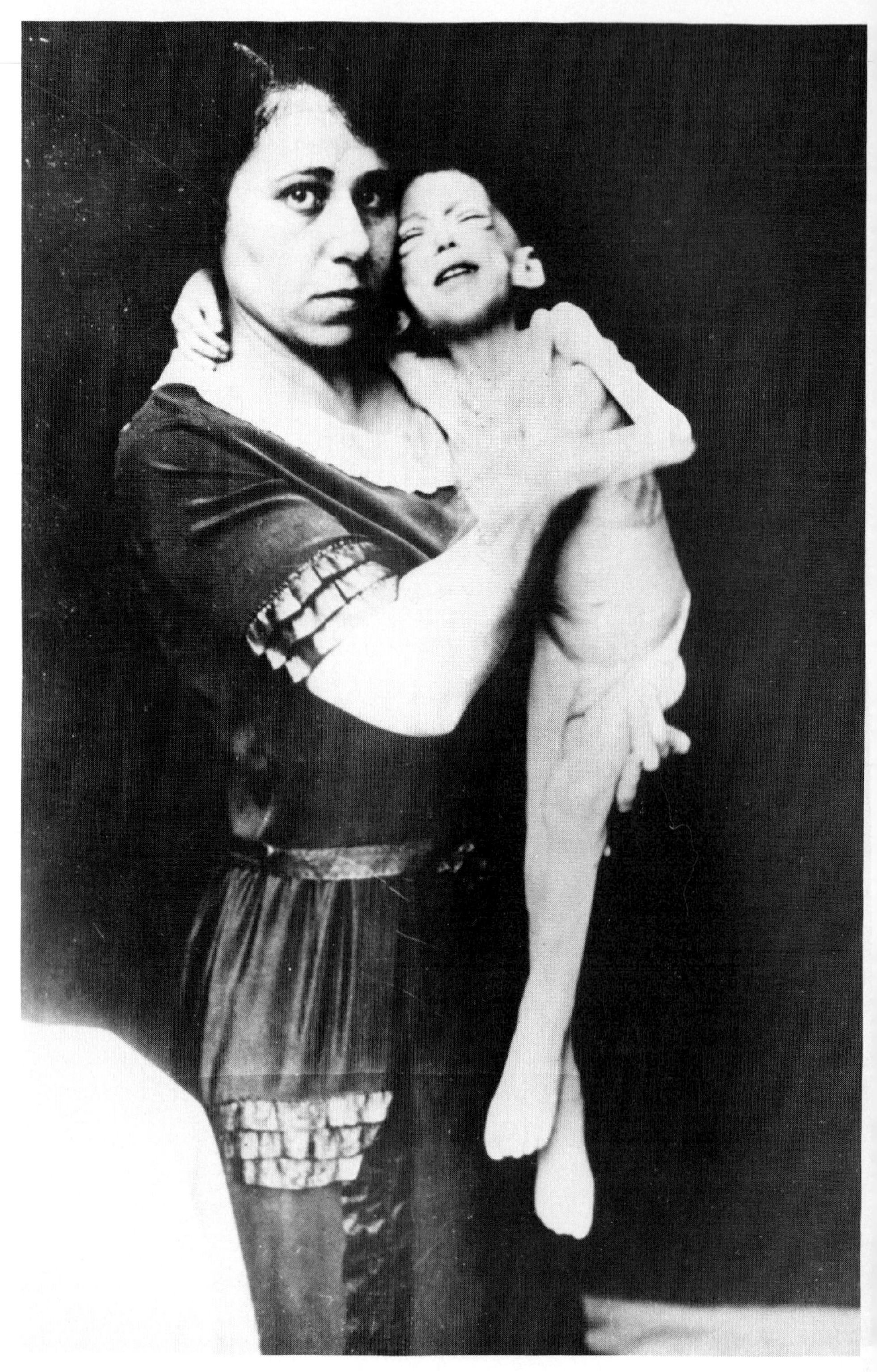

Before insulin. "J.L." Age 3 years, weight 15 lbs., December 15, 1922. *Eli Lilly and Company Ltd*

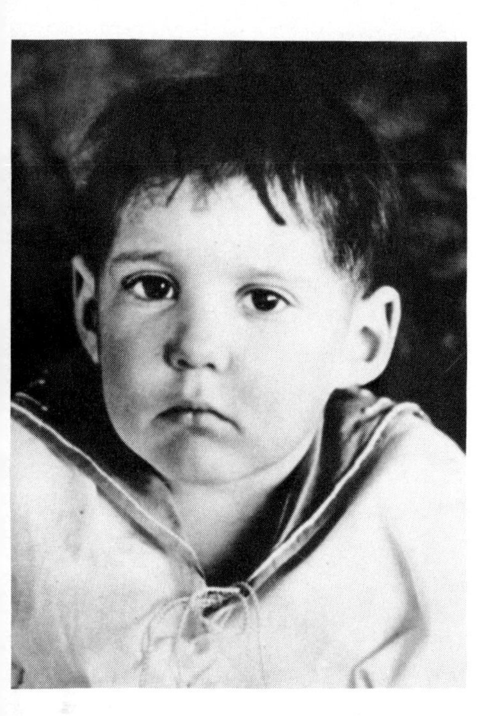

After insulin. "J.L." February 15, 1923, weight 29 lbs. These spectacular pictures first appeared in the issue of the *Journal of the American Medical Association* that introduced insulin to the profession. *Eli Lilly and Company Ltd.*

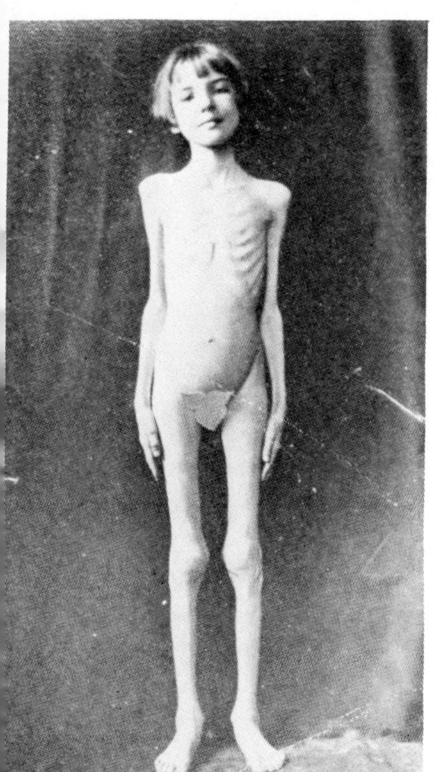

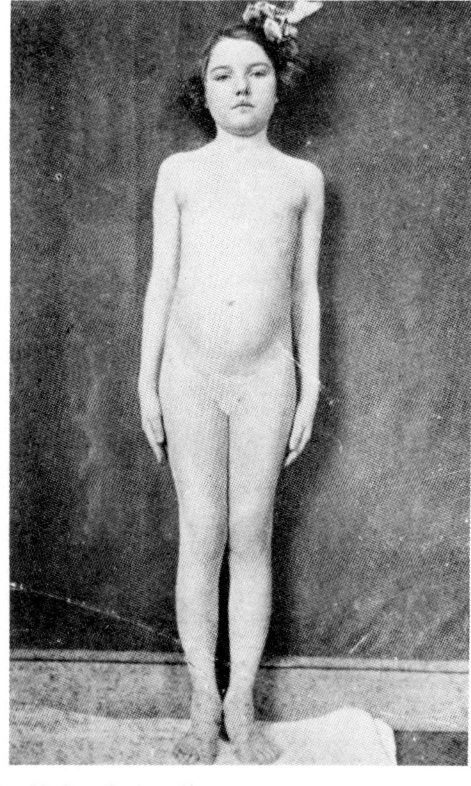

Before and after pictures of a 1922 patient of Dr. H. Rawle Geyelin. Thought to be too indelicate for lay viewing in the 1920s.

BANTING CAN GRATIFY HIS TASTE IN PICTURES

PROF. J. J. R. MACLEOD.

DIVIDES HONOR WITH CO-WORKER

DR. BANTING, who says Mr. Best contributed essentially to insulin discovery.

MR. CHARLES BEST, with whom Dr. Banting will share the Nobel Prize award.

DIVIDES HONORS WITH BEST NOBEL PRIZE FOR RESEARCH

Discoverer of Insulin Expresses Appreciation of Honor and Praise for His Colleagues

Dr. Frederick G. Banting, discoverer of insulin, who, along with Prof. J. R. McLeod, has been awarded the Nobel prize, expressed his deep appreciation this morning for the assistance which he has received from his co-workers on the discovery of insulin.

"I desire to share my portion of the award with Mr. Best," said Dr. Banting this morning. "Both in the honor and the financial aspect, and, furthermore, the fiancial end will be devoted to the advancement of medical research in some line.

"I also would like the medical sciences of Canada to share with the honors which have come. I would like to mention again with grateful appreciation the assistance which I have received from Prof. V. E. Henderson, in whose laboratories the experiments were conducted, to Dr. Fitzgerald, of the Connaught Laboratories, and to Dr. R. D. Defries for the magnificent way in which they have placed their laboratories at the disposal of Mr. Best and myself, and for their suggestions in the advancement of the work. I also desire to express my appreciation to the other members of the University, who have extended their sympathetic aid to the work.

CREDIT IS GIVEN.

"To Charles Best without doubt credit is due for his work in making large scale production possible during the past summer. He has been in charge of production and has been able to greatly reduce the cost of insulin, thereby increasing the benefit to humanity."

From the beginning Dr. Banting has given unstinted credit to Mr. Best.

"I never would have been able to do anything had it not been for him," he declared, after the success of insulin was assured. "Mr. Best was with me from the first. The idea (of insulin) first came to me while I was on the staff of Western University. Mr. Best and I made the first experiment, and we have worked side by side, sharing ideas and developing them together, and but for his unflagging devotion and enthusiasm and his patient and meticulous work we would never have made the progress we have."

Chas. Best is a young Canada, 24 years of age, holding his M.A. from the University of Toronto, and with two more years of study before him in medicine.

CONGRATULATIONS.

At its meeting last night, in addition to electing a new Chancellor, the Senate of Toronto University passed a resolution of congratulation to Dr. F. G. Banting and Dr. J. J. R. McLeod on their receiving the award of the Nobel Prize.

BANTING SHARES PRIZE WITH FELLOW-WORKER

Mr. Chas. Best Recipient of Portion of the Nobel Prize Fund

MR. CHARLES H. BEST

CITY BUSINESS MEN RISE EN MASSE TO CHEER DISCOVERER OF INSULIN

Big Audience Endorses Suggestion of Government Aid for Dr. Banting.

DR. BEST PRESENT

Canon Cody Says Province Should Double Scientific Research Grant.

Child of Nine Pays Tribute To Discoverer of Insulin

So great was the crowd that the overflow from the big ballroom had to be accommodated in the Pompeian room.

Amid great cheering President Gundy presented Dr. Banting with a life membership in the club.

Canon H. J. Cody was loudly applauded when he declared that the provincial government should make it possible for Dr. Banting to carry on his work without any financial worry, and that the $75,000 annual provincial grant to Varsity for research work should be doubled.

Seated at the head table with President Gundy and Dr. Banting were C. H. Best, Sir Robt. Falconer, Sir Edmund Walker, Canon Cody, Prof. Velyien Henderson, Prof. Anderson Hunter, Prof. R. M. McIver, Prof. D. R. Keys, Dr. R. B. Noble.

Dr. Best Introduced.

Previous to his address, his partner, C. H. Best, was introduced to the lunchers by President S. B. Gundy, and he rose for a moment and modestly resumed his seat without saying a word.

But the audience called for a speech. Mr. Best further displayed his modesty by paying tribute to the workers at the University who had not been in the limelight, but whose work had greatly aided the efforts of Dr. Banting and himself.

"While we worked on Dr. Banting's hypothesis, the work of a number of trained investigators at the University was of great value in regard to the treatment. Many of these men were working on little points which are very important, and Dr. Banting and I have taken advantage of these points, and thus made a discovery of immediate practical application." (Applause.)

In opening, Dr. Banting said it was a thing to be appreciated when so many business men would leave their desks at noon hour to hear a talk on a scientific subject.

Then he give the club a brief historical sketch of the disease of diabetes. Back in the first century it had been diagnosed by Aretus. The main symptoms were an increased hunger and thirst. Then the chemist began to work on the problem. In the 16th century Brunner removed the pancreas of a dog.

Provokes Amusement.

Dissolved in Alcohol.

Benefit to Children.

Tribute to Colleagues.

Ranks With Jenner.

CONQUEROR OF DIAB[ETES] may be the title which will ma[ke] F. G. Banting forever famou[s in the] medical world. This Toronto ph[ysician] has discovered an insulin treatm[ent for] the dread disease.

BANTING TO SHARE AWARD WITH BEST

Insulin's Discoverer Will Devote Nobel Prize to Research.

DISTINGUISHED M.D.'S TO HONOR INSULIN MEN

Many Noted Medicos Gather for Banting-MacLeod Dinner To-night

NEW DISCOVERY MADE BY INSULIN'S FINDE[R]

Dr. Banting to Announce Even Greater Thing, Declares Dr. Godfrey

Clippings from Banting's scrapbook, preserved in his papers at the Fisher Library, University of Toronto.

Collip, Best, (Mrs. F.N.G. Starr), Banting, about 1936. The only photo of more than two of the discoverers of insulin together.

Grinding pancreas to make insulin at Eli Lilly's Indianapolis plant, 1923.
Eli Lilly and Company Ltd.

CHAPTER SIX

"Unspeakably Wonderful"

A totally unexpected, almost incredible disaster in insulin production took place sometime between late February and the end of March. Certain of the fact of their discovery and of its therapeutic benefit for human diabetics, the Toronto group had gone ahead with plans to manufacture insulin in large quantities. The Connaught Anti-Toxin Laboratories was to finance and administer production. Collip was to direct insulin manufacture.[1] Special equipment was installed in the basement of the medical building. Everything seemed set for smooth progress. All the problems with purification, the fights about credit, and the rest of the strains, were surely in the past.

Then, to his and everyone's surprise, Collip found that he could not make insulin. First he could not make it in large batches using the apparatus set up in the special manufacturing area. Then he started to have trouble making it by any method, even in his own lab, apparently being unable to duplicate his own successful procedures of January and February. The result of Collip's failure was an insulin famine in Toronto during the spring of 1922, a frantic struggle by everyone on the team to find some way of regaining the knack of making insulin, and fundamental changes in policy regarding the handling and development of the frighteningly elusive discovery.[2]

I

It was one of the most trying periods in Collip's life: brilliant success in the winter; failure after failure, with more and more serious consequences, all through the spring; endless hours in the lab trying to make insulin. Everything was complicated by a serious attack of flu in the Collip household, while at the lab the breakdown in relations with Banting was total, apparently not having been restored since the fight in January. Banting and Collip probably did not speak.[3] Collip may not have been physically safe in Banting's presence: so many of the stories about the

Banting-Collip fight have it taking place in public, or centre it on the loss of insulin production, that there is a reasonable possibility of a second violent incident of some kind having taken place. He was unquestionably vulnerable to Banting's angry scorn. How could Collip have possibly lost the secret? How the hell could he have done it? Obviously, according to Banting, by being so secretive. If not secretive, or as well as being secretive, by being inexcusably sloppy.[4] Collip had known the pure joy of discovery in January. Now he knew dark nights of despair.

People who understand biochemistry tend to be more charitable than Banting was in understanding Collip's situation. Failures like these were not uncommon in primitive extractions working with unknown substances. Before and during Toronto's agony with insulin, for example, years of effort and hundreds of thousands of dollars were going into the still unsatisfactory effort to purify thyroxin, insulin's predecessor as a hormone with great therapeutic possibilities.[5] Pioneering chemists were working with delicate procedures, crude and unreliable equipment, and such frustrating unknowns as the chemical composition of the substance they were trying to produce. In Collip's case, as well, Banting's belief that his records left much to be desired was probably right. Collip's only surviving comment on the problem is a laconic statement that "great difficulties were encountered chiefly because the conditions of time and temperature which were adhered to in the original method could not be obtained in a large scale process with the facilities then at hand."[6]

A few humans had been given insulin, all at Toronto General Hospital. There are few records of how they were dealt with when the supply failed. Some of them, having regained enough weight and strength to carry on starving for a few more months, were put back on their diets. Leonard Thompson, for example, was sent home in May without insulin. The most needy cases received whatever small amounts of insulin Collip could produce. The neediest of these was a young girl, a friend of Best's from a Toronto suburb, who was admitted in February suffering from emaciation, dehydration, and severe acidosis. She was given insulin as supplies permitted. The injections eliminated the acidosis. There was no more insulin to inject. The acidosis returned. The girl gradually slipped into a coma. The doctors gave her massive doses of weak, only partially prepared extract, and were able to bring her back to consciousness. This was the first "recovery" from coma at Toronto. It was only temporary. "Collip gave us the last bit of partially completed extract at two o'clock one morning," Campbell recalled, "and then no more could be completed for days. It was not enough." The little girl's death in April 1922 was the one time in Toronto that a patient who had been treated with the extract died for lack of it.[7] Some years later in England, the first patient to receive penicillin suffered a similar fate.

II

It was a season for real-life melodrama. Banting, it will be remembered, spent most evenings in March drinking himself comatose to get his mind off his troubles. Charley Best came to his room on the night of March 31, Banting wrote later. The young man found the boarding-house room blue with smoke and Dr. Banting half drunk. Best proceeded to give Banting a bawling-out. In passing, he mentioned the situation at the lab and the opportunity they had to go back to work together trying to make an effective extract.

Banting said he wasn't interested. They could have the whole damn thing. He was going to finish the teaching term with Henderson and then get out of Toronto and find a place where there were decent people to live with.

"Then Best said probably the only thing that would have changed my attitude, 'What will happen to me?'"

"'Your friend Macleod will look after you', I said."

"Best replied, 'If you get out I get out'."

"There was silence for some moments. I thought of all the joy of the early experiments which we had known together. Here was loyalty. I emptied my glass. 'That is the last drink which I will ever take until insulin circulates in diabetic veins. Shake on it, Charley. We start in tomorrow morning at nine o'clock where we left off.'"

"Best was pleased. We sat down and as we had done hundreds of times, planned experiments."[8]

III

While these larger-than-life events were taking place among his associates, J.J.R. Macleod was worrying about the future of their work. Toronto had announced to the world its discovery that certain extracts of pancreas were effective in the treatment of diabetes. Toronto knew how to make these extracts... in theory. In reality Toronto could not make effective extracts in large quantities, sometimes not in any quantities. The researchers were sure it could be done, but they had no idea when they themselves would be able to do it again.

Suppose someone else set to work and learned how to make effective pancreatic extracts. The ugly question of patenting had already been raised within the Toronto group. Surely it was a much more pressing question when outsiders were considered. Suppose some enterprising drug company, or even an enterprising chemist, took up the pancreatic extract problem now, either found out the basic details of the Toronto people's methods, or, knowing success was possible, worked out some successful variation, and then took out a patent on the discovery?

Drug companies were certainly interested. Late in March, Clowes of Eli Lilly and Company wrote Macleod about his firm's continuing interest in developing the new extract. He urged a reconsideration of the decision not to work with a major manufacturing firm:

> Public interest in this work will naturally be very great and the demand for the product will be such as to lead to attempts on the part of unprincipled individuals to victimize the public unless some steps are taken to arrange for the manufacture of the product by the procedures recommended by Dr. Collip and the control of the product by means of such tests as you and your associates would consider necessary.

If Clowes knew about the researchers' collaboration with the very small Connaught Laboratories, he dismissed it as inconsequential. The Lilly company would be delighted to work with Toronto, Clowes wrote, and hinted, perhaps intentionally, perhaps not, that Toronto could be bypassed: "I have thus far refrained from starting work in our laboratories on this question as I was anxious to avoid in any way intruding on the field of yourself and your associates until you had published your results. I feel, however, that the matter is now one of such immediate importance that we should take up the experimental end of the question without delay, preferably cooperating with you and your associates...."[9]

Macleod replied that Clowes' firm would have first consideration if Toronto decided it needed help, but that for the next month or two the group would continue on its own. Toronto hoped to publish its method for everyone to use, and would try to protect the public by publishing specifications for the determination of insulin's toxicity.[10] Actually, Macleod was not so sure of his course and in early April began seeking other advice. He approached the deputy minister of health for Canada, who consulted with the commissioner of patents and confirmed the unhappy possibility that a competitor's patent could interfere with Toronto's work, even bring it to a complete halt. At best, the litigation necessary to frustrate such competition, on the ground of Toronto's announced priority, would be lengthy and expensive.[11]

Macleod also wrote to at least one other discoverer, E.C. Kendall, who had isolated thyroxin at the Mayo Clinic in 1914. Kendall had patented his process of isolating thyroxin, and enthusiastically recommended that the Toronto group do the same with their pancreatic extract. He explained to Macleod the arrangement between himself, the brothers Mayo, and the University of Minnesota, by which the patent had been given to the university. It had then established a special committee to license manufacturers of the product.[12]

Macleod was more cautious than Kendall about patenting. Chemists and drug companies had few qualms about taking out patents on their processes or products (Kendall wrote of the Toronto situation, for example, "I can see no more reason why the man that separates the active constituent of the pancreas should not share financially as much as the man that makes a new wireless telephone"). But medical men, such as Macleod and Banting, were bound by their profession's code to make all advances in health care freely available to humanity. If nothing else, it would violate a physician's Hippocratic oath to engage in the profiting from a discovery that patenting normally implied. During preliminary discussions of this problem in Toronto, Banting was apparently particularly reluctant to be in any way associated with patenting.[13]

The possibility of losing the discovery seemed so real, however, that the group decided Toronto had to have the insurance patenting offered. On April 12, Banting, Best, Collip, Macleod, and Fitzgerald wrote jointly to the president of the University of Toronto, Sir Robert Falconer, explaining the situation. They proposed that a patent on the process be taken out in the names of the two "lay members" of the group, Best and Collip, and then immediately assigned to the Board of Governors of the University of Toronto. It was to be a purely defensive manoeuvre, one which would never stop anyone else from making the extract. In fact the point was to stop anyone from ever being in a position to stop anyone else:

> The patent would not be used for any other purpose than to prevent the taking out of a patent by other persons. When the details of the method of preparation are published anyone would be free to prepare the extract, but no one could secure a profitable monopoly.[14]

The Board of Governors of the university agreed to the arrangement. An application was filed for a Canadian patent in the names of Collip and Best.

IV

All four of the principal researchers worked long hours in April and May trying to regain the secret of making insulin. Although their later accounts tend to disagree on credit for important suggestions, it seems that the research was more than ever effectively a team effort, with at least three of the four making vital contributions.[15]

They gradually became convinced that the crux of the problem was in the heating that the extract experienced as part of the process of evaporating off the alcohol. Best discovered significant variations in the pressure of the water being supplied to the crude vacuum pumps they were using. These caused significant variations in temperature and distilling time. (A similar problem twenty years later frustrated early attempts to purify penicillin.)[16] Macleod, who had been investigating different grades of alcohol, as well as the influence of different degrees of acidity, then turned his attention to what was happening in the evaporation. He found that the high temperature was causing some of the proteins in the solution to break down, an observation which seemed to reinforce previous experience that heat somehow neutralized the active principle. Macleod suggested abandoning the use of vacuum stills, and going back to the warm-air current method of evaporation that Banting and Best had used earlier at his urging. Collip, too, had decided the temperature had to be kept down, and to do this had experimented with acetone rather than alcohol as the principal extractive.[17]

By mid-May the group had recovered the ability to make insulin. The method involved using acetone with slight acidification. (The degree of acidity was the other variable that was constantly tinkered with; the solubility of elements in the mixtures varied according to the degree of acidity as measured by pH determinations. As was realized later, adjustments in the pH range of the solutions were in fact far more important than the temperature of distillation.) The pancreas-acetone mixture was filtered and then set out in enamel-lined trays placed in a make-shift wooden tunnel. A big old exhaust fan, formerly used in the medical building's heating system, supplied the wind. Coils in the roof of the tunnel heated the air as it passed over the trays. After an hour in the tunnel, five hundred cc. of solution in a tray would be reduced to fifty cc., the temperature never exceeding 35C. The rest of the process, involving Collip's method of "trapping" the active principle in various percentages of alcohol, was fairly straightforward, though it took several days before the final product emerged.

The method produced a few cubic centimetres of insulin solution. It was expensive, mainly because of the cost of alcohol, and hazardous. "You can't imagine a more dangerous set-up," Peter Moloney told me. He was the first chemist added to the production facility to work on insulin in that spring of 1922. In 1980, when we talked in his room in St. Michael's College, the distinguished, white-haired, chuckling old man, still an active chemist as he approached his ninetieth birthday, brought back vividly the reek of acetone that spring and summer, the rattling of the motor driving the big fan, and his horror when a bottle of picturic acid was shaken off its shelf, fell to the floor, and shattered. Only the placing of its cork stopper, Moloney thought, saved an explosion that would have ignited the acetone, causing a dreadful fire. Toronto would have sacrificed its medical building and several chemists in its haste to make insulin.[18]

V

The rediscovery of a way to make insulin made it possible to consider resuming clinical tests. Banting, as we have seen, had played little part in the clinical work at Toronto General Hospital, for he had been denied an appointment to the hospital's staff. In February and March it had seemed to Banting as though he had no further role to play in the development of the discovery. As he pulled himself together that spring, however, probably relying heavily on such friends as Velyien Henderson for advice, Banting must have realized that he had an unchallengeable claim to use the extract. A very large number of people, including Banting himself, believed that he had discovered it. It would be unthinkable to deny Dr. Frederick Banting, a licensed physician in good standing, priority in the clinical use of insulin. If Banting did not get his way the amount of trouble and bad publicity he and his friends could cause was practically unlimited.

The fact that Duncan Graham would not give Banting an appointment at Toronto General Hospital was not the barrier it had first seemed. Why should TGH have a monopoly of the clinical tests of insulin simply by virtue of being the university's chief teaching hospital? In the spring of 1922, Dr. F.G. Banting established an office at 160 Bloor Street West in Toronto and began the private practice of medicine. This one step instantly gave him the right to use the facilities of TGH's private patients' pavilion for his private patients. Then, early in April, Banting was interviewed about the discovery by the Director of Medical Services for the Canadian Department of Soldiers Civil Re-Establishment, which handled the affairs of war veterans; several weeks later Banting was appointed head of a new diabetes clinic at Toronto's Christie Street Military Hospital, where he had worked briefly in 1918-1919. Now he had all the facilities he needed.[19] About the same time, an agreement was reached with the Connaught Laboratories on the distribution of insulin for clinical use. One-third of the production was to go to Banting for his private practice, one-third was to be used in Banting's Christie Street clinic, and one-third would be available for work at Toronto General and the Hospital for Sick Children.[20] Macleod began referring all the inquiries he received from diabetics to Dr. Banting, "my clinical associate."

By mid-May enough insulin was being produced by the new method to permit resumption of limited clinical testing. Dr. Joe Gilchrist received his second injection on May 15. Gilchrist had agreed to work at the Christie Street clinic under Banting, and so served as both physician and patient. In the early months of sporadic production and frequent impurities, Gilchrist became Toronto's self-proclaimed "human rabbit," testing each new batch on himself after it had been tried on the rabbits.[21]

There was also enough insulin in mid-May to allow Banting to meet the urgent request of Dr. John R. Williams, who had come to Toronto from Rochester, New York, some miles away on the other side of Lake Ontario, to see if he could get some insulin to try on his most desperately ill patient. Jim Havens, son of a vice-president of Eastman Kodak, had been diagnosed as diabetic seven years earlier at age fifteen. He did fairly well on an Allen diet until 1920 when his capacity began a sharp decline. The boy was treated by Allen, and his father supported Allen's "heroic efforts" to get research going at the Physiatric Institute. By early 1921, however, James Havens, Sr., had given up hope that Allen or anyone could help young Jim.[22] When news came of the discovery in Toronto a year later, James Havens, Jr., was a 73½-pound skeleton, living on 820 calories a day, barely able to lift his head from his pillow, crying most of the time from pain, hunger, and despair. According to Williams, he was

> a most pitiable spectacle. Blood sugar 450 mgs. Plasma bicarbonate [a measure of acidosis] 24.9 volumes per cent. For weeks the patient had suffered severely from pains in his legs, which made the constant use of codeine necessary. The edema and profound weakness confined him to bed and he was rapidly approaching death, when through the great kindness of Doctors Banting and Macleod, extract was supplied for his treatment.[23]

Havens got his first insulin on the evening of May 21, 1922. He was the first person treated with it in the United States.[24] The first injections of one or two cubic centimetres (throughout this period one rabbit "unit" of insulin, roughly defined as the amount necessary to send a rabbit into hypoglycemic convulsions, was usually about one cubic centimetre in volume) were very painful and had no effect, confirming, it is said, Williams' hesitancy about trying the new cure in the first place.[25] On May 26 Banting went to Rochester to examine Havens. He advised doubling and then tripling the dosage. Within a day or so Havens' urine was sugar-free, his blood sugar was down to normal, and his clinical condition greatly improved. Banting agreed to have fresh supplies of insulin sent by train from Toronto.* Two weeks after first receiving insulin, Jim Havens was able to rise from his bed and walk. "The patient is very much better," Williams wrote Banting. "His appearance indicates much greater improvement than the laboratory studies suggest. Dr. Joslin made the statement to me one time, that of all the ways of measuring the condition of a diabetic, he thought the clinical appearance to be one of the best and that is my experience...The greatest advance noted is in his state of mind."[27]

Williams had come to Toronto personally to plead for insulin. Others, alerted by the May 3 paper in Washington, were beginning to do the same or to write Banting or Macleod asking when the new treatment would be available. "I have some really heart-breaking cases under my care at the present time," the chief pediatrician from the Johns Hopkins Hospital in Baltimore wrote in a typical appeal, "two of them lone children of different families whose carbohydrate tolerance is gradually going down. They know of your work and are pestering me to get some of the material if I can.

*Small bribes appear to have been necessary to forestall customs' inquiries about the strange substances.[26]

I do not wish to pester you, but only to let you know how anxious I am to use some of the 'insulin' if I can get it."[28] But Toronto had no extra insulin to give to him or anyone else. For several weeks Jim Havens was the only diabetic outside of Toronto who was being treated with insulin.

VI

The pressure to produce more insulin was mounting daily. All attempts to produce the hormone in large quantities continued to fail. Toronto's recovery of a method in mid-May was accompanied by a realization that the group had to have help. The ubiquitous George Clowes had spoken to Macleod again at the Washington meeting at the beginning of May, had written offering advice on the American patent situation, and continued to urge Toronto to collaborate with his firm. The Torontonians finally came around and invited Clowes to come to Toronto on May 22. (They also invited Rollin Woodyatt of Chicago, who was offering an informal collaboration that would put his expert staff, financial backing, and the pancreas resources of the Chicago stockyards at Toronto's disposal. Woodyatt, however, was unable to come to Toronto at the time specified.)[29] Clowes brought with him a chemist, a patent attorney, and the vice-president of Eli Lilly and Company, Mr. Eli Lilly. In two or three days of meetings at the King Edward Hotel, the Americans and Canadians worked out an agreement for the development of insulin.

The proposed collaboration was explained in another formal letter, written on May 25 to Falconer from the research team plus Fitzgerald of the Connaught. They now recommended that the University of Toronto Board of Governors accept from Collip and Best a United States patent on the process, for which they were applying. The Board would eventually license North American firms making insulin for sale and collect royalties from them to support research in the university. For now, however, the group recommended that a temporary exclusive licence be given to Eli Lilly and Company of Indianapolis. The explanation of the recommendation was as follows:

> Experience on production of "insulin" on a moderately large scale in the Connaught Laboratories has shown that this is fraught with many difficulties not encountered in the small laboratory scale, and we do not believe that production in amounts that are adequate to supply the demands for it can be accomplished without further experimentation in its preparation on a much larger scale than is possible here. To make this further step possible it will be necessary for us to collaborate with some well equipped and properly staffed commercial house engaged in this work. After careful consideration we have decided that it is much better to arrange to deal with one

firm rather than several, partly because concentrated effort is likely to be more efficient than divided effort, and partly because we could not act as consultants to several establishments at the same time....

We have chosen to collaborate with [Eli Lilly and Company]. We recognize this firm will be placed at an advantage over its competitors through this collaboration with us, but we believe that it is much less serious than there should be further delay in proceeding intensively with production on a large scale, and, moreover, we propose to give other firms, as well as hospitals and other non-commercial concerns, every chance to do the best they can by publishing the details of the method as at present used by us in the Connaught Laboratories in full at an early date (within three months). By this step, our proposed co-operation with the Lilly Co. cannot be criticized as unethical or unfair or as in any way prejudicial to the free manufacture of "insulin."[30]

Toronto's decision to collaborate with his company was a triumph for and testimony to the persistence of G.H.A. Clowes. It also reflected a vote of confidence in both Clowes and his company by the Toronto researchers, who had not acted impulsively or without consideration of their several alternatives.

In 1922 Eli Lilly and Company had been making and selling pharmaceuticals for forty-six years from their base in Indianapolis, Indiana. It was a family-owned "ethical" drug company (no patent medicines, no extravagant claims, and advertising and sales to doctors and pharmacists only), which had grown to become a major, though not dominant factor in the industry. In 1921 Lilly employed about eleven hundred people and did just over $5 million worth of business. The firm was managed, according to the founder's son and president, J.K. Lilly (whose own son, Eli, had been the family man in Toronto), with the aim of being "conservatively progressive." Part of the house's progressiveness in the early 1900s had been the creation of a substantial research facility. At the end of the First World War the Lilly family had decided to strengthen further the firm's links with the scientific community, even though the short-term returns from such ventures might be minimal. As part of this continuing policy G.H.A. Clowes was hired as a special research chemist in 1919 and appointed director of research the next year.

George Clowes (pronounced *clews*) was an Englishman in his mid-forties, a minister's son who had taken a Ph.D. in chemistry in Germany, done post-doctoral studies in England and France, and then emigrated to America as a land of greater opportunities than Britain. Before moving to Lilly, Clowes had spent many years at a state research institute in Buffalo, New York, ninety miles from Toronto. He was a serious researcher in his own right, most interested in problems relating to cancer, but with a restless curiosity about all kinds of knowledge. His job with Lilly was almost unprecedented in a commercial organization, Clowes wrote, in giving him virtually free reign. During the summers, for example, he worked at the Marine Biological Laboratories at Woods Hole, Massachusetts, where he had his own lab, assistants, and no commerical responsibilities. His research reports emphasized the absolute necessity of continuing to strengthen the firm's links with the scientific community, including the universities, and raising its prestige and authority among researchers. Few broad corporate strategies have ever paid off as quickly and magnificently as this did for Eli Lilly and Company.[31]

Macleod had known Clowes for some years, was impressed by his stature as a scientist, and by his company's enlightened support of research. He and the other Torontonians were probably also impressed by the plans Clowes outlined to them for the development of insulin. The firm had recently been very active in work on glandular products, and had a good team of chemists ready to work on insulin. It wanted an exclusive licence for an "experimental period" of one year, during which there would be a complete pooling of knowledge between Toronto and Indianapolis. There would be a several-stage development of the product involving large-scale clinical tests in Toronto and the United States, with Lilly supplying extract free of charge in the initial stages and then selling it at cost. Lilly would share any improvements it made in the manufacturing process with Toronto, and if any improvements were patentable would pool the patent rights for all territory outside of the United States. At the end of the experimental period, Lilly wanted a licence to manufacture insulin on the same terms as Toronto would license other manufacturers. As Clowes had proposed in earlier letters, the firm thought it would be appropriate for insulin licensees to pay Toronto royalties on all insulin sold.

The collaboration was formally established in an "Indenture," dated May 30, 1922, between the Board of Governors of the University of Toronto and Eli Lilly and Company. It was intended to be a close, but not necessarily exclusive, relationship. The Lilly company was prohibited from divulging details of the process to other parties, for example, but Toronto was not. As they had told Falconer, the Toronto team still intended to publish their method in order to make sure others would know how to make insulin when Lilly's exclusive rights expired and to protect themselves from charges of unethical secrecy. As well, the Lilly agreement limited the company's territorial rights to the United States, Central and South America. To handle insulin in Britain and the rest of the Empire, perhaps Europe too, the Torontonians had decided to offer the patent rights to the British Medical Research Council, for administration in a way parallel to the University of Toronto's handling of the Americas. At the end of May, Macleod wrote the Medical Research Council conveying the offer. On its part, the Board of Governors appointed a small committee to work with the discovery group in carrying out the licensing and development arrangements. These bodies soon evolved into Toronto's Insulin Committee.[32]

The Toronto group was anxious to get policies for developing insulin in place during May, not only because of the demand from doctors and diabetics, but also because of the imminent break-up of the group. Collip's appointment at Toronto expired on May 31. Such negotiations as there may have been about his staying on seem to have dissolved in the quarrels with Banting and then the difficulties making insulin. Whether or not Collip wanted to stay on but was not wanted, or was wanted but was fed up with the fighting, is not known.

At the end of May the Toronto researchers read six short papers on their work at a session of the annual meeting of the Royal Society of Canada. The first two of these, scheduled to be published several months later, contained methods for making insulin as developed by Banting, Best, and Collip. There were two recipes - "The Preparation of the Earlier Extracts," by Banting and Best, and "The Preparation of the Extracts as used in the first Clinical Cases," by Collip - an indication for the record of who had done what by members of a team that had fallen apart.[33] Best and Collip then travelled to Indianapolis. On June 2 and 3 they told the Lilly chemists all they knew about making insulin and helped with the first attempt to extract it. The process worked.[34] His time in Toronto over, J.B. Collip went back to his job at the University of Alberta.

VII

Beginning work immediately, the Lilly company poured men and money into insulin production. But they were not the first to make insulin in the United States. Dr. W.D. Sansum of the Potter Metabolic Clinic in Santa Barbara, California, had noticed Banting and Best's first publication and in April had written Banting to ask about progress. When Banting told him of the delays, Sansum decided to try making pancreatic extract himself. He and his associates tried various methods. As soon as they learned to use alcohol as an extractive and normal rabbits for testing (from Banting and Best's May article, combined with a letter Joslin had published immediately after the May 3 meeting), the Potter group found they could make potent extracts. On May 31 they began administering insulin to an adult male patient, and soon succeeded in making him sugar free. They tried to increase their supply of the extract early in June, collecting the pancreases from sixteen hundred sheep. Just as had happened in Toronto, they found that the attempt to scale up production failed completely. Macleod learned of the California work in mid-June when Sansum wrote to him asking for advice.[35]

Macleod had expected some such development. It was only through professional courtesy that Rollin Woodyatt was delaying trying to make extracts and other diabetologists, including Allen, were starting to become impatient. They all had dying diabetics on their hands, patients they had encouraged to carry on in the faint hope that some treatment would be discovered. Now the announcement of the treatment had come out of Toronto, but no treatment. Some of the patients' life expectancy was a matter of weeks.[36]

To meet the clinicians' demand, while at the same time usefully spreading out the research job, Toronto and Lilly had agreed that a select group of physicians and institutions would be given the extract for testing purposes as soon as it became available. Until then, the Torontonians saw no reason why other researchers should not be able to make insulin. Macleod sent both Sansum and Woodyatt details of the method. To honour the Lilly agreement, he required them not to divulge the method to anyone likely to produce the extract commercially.[37]

Of course experimental and clinical work would continue in Toronto, with the Connaught Laboratories, small and makeshift as its facility was, doing everything possible to increase insulin production for the city and for Canada. After Collip left, Best was placed in charge of Connaught's insulin manufacture. Banting handled the clinical work through his private practice and his Christie Street patients. Macleod was to carry on experimental development. To finance the research, Macleod applied for and was awarded an $8,000 grant from the Carnegie Corporation. Part of it was to be shared with Collip, who fully intended to carry on research into insulin at the University of Alberta. Macleod himself was travelling east, to spend the summer of 1922 at the Marine Biological Station in St. Andrew's, New Brunswick. He was intrigued by the thought that insulin ought to be easily procurable from those species of fish in which the islets of Langerhans were anatomically separate from the rest of the pancreas. Perhaps fish insulin would be easier and cheaper to produce than the dribs and drabs of semi-pure beef insulin they were struggling so hard to make in Toronto.

VIII

The University of Toronto was awakening to the importance of the discovery made in its Physiology Department and first tested at its teaching hospital. Part of the institution's consciousness involved realizing how curious it must appear to outsiders that Banting had no university appointment (his job in the Pharmacology Department had expired) and no position at Toronto General Hospital. How strange, too, that no further clinical testing was going on at Toronto General. (According to Banting, this fact was driven home to the chairman of the hospital's Board of Trustees, C.W. Blackwell, when on a trip visiting American hospitals he was asked everywhere about insulin and had to admit nothing was happening at his hospital and he knew nothing about insulin. Blackwell broke off his trip, came back to Toronto, and began discussing with Falconer how to get Banting at work treating diabetics at the hospital.)[38]

Unless something was done quickly, the university and hospital faced the prospect of losing much of the prestige attaching to what was starting to look like a very great discovery. Banting was completely outside the university. Suppose he went even further outside, and, as he had threatened at least once in the past, left Toronto entirely. He was beginning to get offers, some of them princely.*[39] What an embarrassment if the principal

*Some of the concrete offers are cited in the endnote, but one bizarre encounter, described by Banting in his 1940 account, deserves special mention:

> I was sitting in my room one evening. It was raining hard. I was called to the phone and a man asked if he could see me. He said he would come up immediately. A few minutes later he arrived. He was a big man and there seemed hardly room for him. He looked about and his first words were, "Well for God's sake".
>
> I sat at my desk and he sat on the only other chair, which was of the hard stiff back kitchen variety. "So this is where you live." "Well you are a damned fool." "Now listen to me" and he proceeded to tell me how the wife of a friend of his had been under the care of the best Doctors in New York and despite the diet treatment she got worse and worse. She had been given insulin and she felt entirely different. He knew the woman. She had tried everything and "knew them all", you could not fool her. She was all for insulin. That was good enough for him. All I had to do was to hand over the patents to a group on Wall Street and he would guarantee me $1,000,000 cash. Insulin would be patented in every country in the world. Ten percent royalty would be collected. The company would keep five & I would be given 5% royalty in addition to the $1,000,000 cash. A chain of clinics would be established across the United States and Canada one or two in every large City. I would be Medical Director and would have all the clinical cases for study and could have all the laboratories I desired. I would thus be relieved of financial responsibilities, being independently wealthy, and could devote myself to scientific research. I need not even see patients if I did not wish - "except a few very wealthy ones by appointment". My time would for the most part be my own. They would look after the whole organization.
>
> To me as I sat in my little room filled with smoke, it was not even a temptation. It meant that the suffering diabetic would be exploited.
>
> When he had finished I only asked him one question. "What would you do for the poor diabetic who could not pay?" He replied, "that can be easily worked out, they can be treated in an out patient clinic". "Not good enough", I replied. "I am afraid you do not understand the situation. The indigent diabetic is our greatest problem. Every effort must be made to reduce the cost of insulin and remove the necessity of expensive diets so that they can look after themselves."
>
> He said that there was one more thing that he would like to point out. "If such a scheme as I have outlined is not followed, then you will have every doctor in the USA making money out of the diabetics, and they will form clinics and make the money out of diabetes that you should be making."
>
> "I am afraid this scheme is not very attractive to me." He then gave me proof of his financial backing. He gave me his card and asked me to think it over and get in touch with him. He phoned long distance from New York a couple of nights later. He could not understand why I still declined.

This incident took place in the summer of 1922, probably late in August. It is impossible to know how accurately Banting remembered the encounter and whether the American visitor was a reputable businessman or a fly-by-night promoter. I suspect the latter. Nonetheless, the incident was important in reinforcing Banting's belief that he had turned down incredible fortunes to keep insulin in Canada.

discoverer of insulin, as most people saw him, left Canada. Even if Banting stayed in Toronto, his non-relationship with the university would be embarrassing. On the other hand, those who knew Banting's limitations might also have realized that he was not in fact competent to direct major clinical experiments on diabetics. The clinical reports Banting was likely to produce on his own would very likely pale in comparison with those of the first-class American diabetologists. This, too, would be embarrassing. Banting may have realized it himself. As his friend Dr. D.E. Robertson put the situation to Duncan Graham, "Campbell knows all about diabetes but can not treat it and Banting knows nothing about diabetes and can treat it." Finally, if the enterprising American clinicians got ahead of Toronto, making it possible for American diabetics to get insulin in preference to Canadians, the result might be a national outcry. Altogether it was a very delicate situation.[40]

Even after the insulin famine eased in May, Walter Campbell appears not to have been getting supplies of insulin for his patients. Perhaps there was too little available. Perhaps Banting, as he himself implied in 1940, was conspiring to withhold insulin from the hospital until he was given a clinical position.[41] In any case, an *ad hoc* university committee, chaired by Falconer, met in mid-June to resolve the doctors' conflicts. Agreement was reached on the conditions by which Banting, collaborating with Duncan Graham, Walter Campbell, and A.A. Fletcher, would have clinical facilities at Toronto General Hospital. These would also entail a university appointment for Banting. The slow grinding of the university and hospital bureaucracies, combined with Banting's heavy schedule that summer, made it impossible, however, to get the clinic started before the latter part of August.[42]

IX

During July Macleod was working in New Brunswick. Best was spending a few weeks with his family in Maine. Banting was alone in Toronto. He was being deluged with requests for insulin from physicians, diabetics, diabetics' families, people who had come to Toronto, people wondering if they should come to Toronto, people wondering if insulin could come to them. July heat in Toronto was oppressive again this year, Fred wrote Charley, "and even worse than the heat as a disturbance is that diabetics swarm around from all over and think that we can conjure the extract from the ground." Diabetics were literally camping at the doors of the lab trying to get insulin.[43]

The standard reply to all inquiries was that insulin was still in the experimental stage, supplies were severely limited, and the inquirer would be informed when the situation changed. All available production was going to Jim Havens in Rochester and to Gilchrist and a handful of diabetic soldiers at Christie Street Hospital. Thinking the supply situation was improving, Banting gave in to some of the most desperate pleas; in mid-June and early July he agreed to treat a few private patients who were otherwise about to die. Four living skeletons, three children and one adult, were brought to Toronto from points in the United States and Canada.

Elizabeth Hughes was not among them. The fourteen-year-old diabetic had clung to life through the winter of 1921-22, a pathetically starved little girl, five feet tall but weighing no more than 52 to 54 pounds. In the spring of 1922 she was taken to Bermuda with her nurse to enjoy the climate. She contracted the diarrhoea epidemic on the island. Both her weight and her carbohydrate tolerance slipped further. From May 19 to June 2, 1922, Elizabeth received less than 300 calories of nourishment a day. Her weight, fully clothed, fell below 50 pounds. As indomitable a girl as ever existed, a kind of real-life duplicate of the heroines of girls' literature, Elizabeth fought off the lassitude and despair that overtook most diabetics in the final stages of their sickness. She continued to exercise every day and made it a personal triumph to walk up the ramp to the ship that brought her home from Bermuda.

Elizabeth's mother, Antoinette Hughes, had learned about the discovery in Toronto. Allen and other doctors told her that in this case the newspapers were right; there was something to it. On July 3 she wrote Banting to ask whether anything could be done for her "pitifully depleted and reduced" daughter. Banting's answer on July 10 was the standard discouragement. All the Hughes family could do was try to keep Elizabeth going, hoping she would last until insulin was beyond the experimental stage. In fact it was impossible to build up her tolerance, and Elizabeth continued her drift towards death from starvation. A friend of J.J.R. Macleod's, whose moving appeal on behalf of a poor fisherman on Prince Edward Island had met with the same response, wrote, "It is pitiful that so great a boon should be in sight, yet not in reach."[44]

X

The insulin situation was a nightmare. Every attempt to increase the quantity of extract being produced in Toronto failed. When Best left on holidays, there were problems procuring pancreas. Then there was a shortage of acetone. Worse still, the quality of extract that was being produced was not good. Protein impurities caused abscesses in many of the patients; salts still in the solution made many injections excruciatingly painful.[45] Strong extract seemed to have the worst side-effects, but weak extract had to be injected in painfully large doses to do any good. Banting resorted to rectal administration of extract to try to minimize the pain. There was fleeting optimism, then realization that the insulin was having no effect.[46]

By the end of June it seemed that the optimism after the first success with Jim Havens had also been premature. Dr. Williams was still enthusiastic about Havens' subjective improvement and his weight gain, but wrote Banting that there was little laboratory evidence of progress. On a steady regime of eight cc. of extract a day, Havens was still excreting 200 grams of sugar and showing a blood sugar averaging .350. The boy was beginning to complain about the severe pain caused by the injections, suffered from an abscess, and his morale was starting to weaken. From time to time he had to be given a day of "rest" from his suffering.

As he pondered Havens' case, Williams became more and more interested in the possibility that a Rochester colleague, Dr. John R. Murlin, might have an alternative worth trying. Murlin was noticed in chapter one as one of the researchers who had continued work on pancreatic extracts despite the disparagement of them in the years before the war. Before being ended by the war, Murlin and Kramer's research, which had started well, had led them into a long blind alley.[47] Murlin left the pancreatic extract problem until October 1921, when Paulesco's results encouraged him to start up again. He was making interesting progress with respiratory quotient experiments on animals when Banting came to Rochester at the end of May to give insulin to Havens. Murlin met Banting, learned more about Toronto's methods, and, with his colleagues at the University of Rochester, launched a feverish program of extract preparation and testing. He found that extract made by Banting and Best's methods had fatally toxic side-effects. So he worked on a wide range of alternatives, and towards the end of June told Williams he believed he had an extract that could be taken by mouth through a duodenal tube.

Williams and James Havens Sr. hesitated to have yet another experimental remedy tried on poor Jim, but the Toronto extract finally became so painful they saw little to lose in Murlin's alternative. On July 9 and 10 Jim Havens was given massive doses of Murlin's pancreatic potion (a hydrochloric acid perfusate). Its only effect was to make the boy violently ill. Later in the day on the 10th, Williams thought Havens was heading for coma. He quickly went back to Toronto's less unsatisfactory extract:

> I injected 8 cc. of the extract into the buttocks. He immediately complained of a sensation all over his body as though he had been poisoned, and of a profound burning in the stomach. I at once gave him by mouth a dram of soda bicarb in 12 ounces of water. This did not relieve the burning or apparently ease the symptoms, but in a few minutes he vomited more than 2 quarts of undigested food and fluid. Shortly after that intensely itching wheels [*sic*] broke out on his body. I thought he would die but he came out of it all right.

Havens' father wrote to Banting that they had now backslid to just about where they had started with Jim almost two months earlier.[48]

Progress or not, the pressure on Banting to take more patients continued to grow. Early in July he was phoned by Dr. L.C. Palmer, a local surgeon who had been a fellow medical officer at Cambrai in 1918. Palmer had a fifty-seven-year-old, severely diabetic patient, Mrs. Charlotte Clarke, who was suffering from a gangrenous infection in her right ankle. She seemed to be under a death sentence, for only amputation could stop the spread of the infection. Severe diabetics rarely survived amputations, and in a case like this most surgeons would not even try.

Banting could not turn down a fellow soldier's request for consultation. He decided that they should go ahead and try the amputation, using insulin. What the hell, why not? On July 10, Charlotte Clarke, who was nearly comatose, was given her first insulin. On the 11th, Palmer amputated her right leg above the knee, using a general anesthetic which he had not thought she would have been able to stand without insulin. After the operation he was still skeptical: "I did not feel that wound would heal and looked for the worst possible results," he wrote in his summary of the case. Mrs. Clarke came out of the operation showing large quantities of acetone in her urine. Banting injected insulin to control it. "It did not seem possible that she could get better," Palmer wrote. This was the first major operation performed on a diabetic with the help of insulin.[49]

It may have been responsible for precipitating yet another round in Toronto's continuing insulin crisis. Banting wrote afterwards that he had taken five other patients off insulin to supply Mrs. Clarke. It was poor quality insulin, in any case, and there was very little of it. By mid-July, production at the Connaught Laboratories was apparently at the point of failing completely once again. Williams, who had come to Toronto in desperation to get something pure enough to use on Havens, later wrote that "Toronto insulin had become intolerable." Banting was beside himself, Peter Moloney remembered, to get insulin to keep his patients alive.[50]

Could Eli Lilly and Company come to the rescue? When the firm's work on insulin began early in June, Clowes planned to run ongoing small-scale experimental programs in tandem with a series of factory-scale attempts at mass production. A team of chemists, headed by George Walden, devoted their full time to the insulin work. The schedule called for fairly large quantities of insulin to be on hand by October.[51]

Lilly's preparations, made from pork pancreas, were potent from the beginning. As always, however, it proved painfully difficult to increase the yield. The first shipment of Lilly insulin, ten five-cc. bottles labelled "Iletin," had arrived at the physiology department in Toronto on July 3. Best, who was about to leave for holidays, immediately took four bottles of it for Banting's clinic. George Eadie, who was doing rabbit tests on the extract in the department, reported to Macleod that he later gave Banting two more bottles because the clinic was so short of insulin.

(In New Brunswick, Macleod was distressed to learn from Eadie that

R.D. Defries, acting director of the Connaught Laboratories, had written Lilly asking that future shipments be sent directly to Banting. Macleod wrote Defries to make sure the physiology department got some of the extract for testing. "Please do not misunderstand my attitude in this matter," he told Defries, "but you must know how disagreeable and upsetting things were last winter and to avoid this I am trying in the future to have every thing work strictly according to prearranged agreements." On his part, Banting was upset that Macleod had in the first place, "on his own initiative," instructed Lilly to send its extract to Eadie. He also found that Macleod's technicians had taken over the little room he and Best had used, leaving him with no lab space - or, after yet another quarrel, research money - in the medical building.)[52]

Clowes came to Toronto on July 16 to go over the results of Banting's first clinical tests of "Iletin" and plan the future testing program. He was surprised to learn of Connaught's production problems and the severe shortage in Toronto. Clowes wired Indianapolis to ship more insulin. He suggested to the Connaught chemists that they try evaporating the alcohol at still lower temperatures as well as getting a more complete separation of fats. The Lilly people had been skeptical of Toronto's makeshift wind tunnel evaporation method from the beginning, Clowes wrote Banting after his visit. Lilly had always used vacuum distillation, and Clowes thought Toronto would be wise to scrap its system and get new vacuum equipment.[53]

Banting decided to go to Indianapolis to study Lilly's method for himself. "I have a hunch that Clowes is holding out on us since he would not tell us how that [first Lilly] batch was made," Banting wrote to Best. "And furthermore since the extract we're making here is 'pretty rough', I think they might supply us with some for the patients are needing it very badly."[54] On the 23rd he went to Indianapolis with D.A. Scott, Connaught's latest addition to its insulin team.

Banting's suspicions about Clowes were groundless, for the Lilly group went out of their way to help the Canadians. In fact Clowes and the Lilly family took an instant liking to Fred Banting and decided to support him every way they could. Banting and Scott were shown complete details of the production facility, and the insulin supplies Clowes had promised were waiting for Banting. J.K. Lilly described the visit in a letter to his son, Eli:

> When they left Toronto, there was not a single unit left in the city. Banting...has a large number of patients, and he certainly was in trouble. We had 150 units ready for him, and when I told him he could take it back with him, he fell on my shoulder and wept, and when I told him that the next evening we would send him 150 units, he was transported into the realms of bliss. Banting is really a fine

chap and we must back him to the limit.[55]

Probably because supplies were so limited, Charlotte Clarke, the diabetic amputee, had been given no insulin after the seventh day of her post-operative period. Initially it seemed as though the insulin had done its job, for Palmer was able to remove the stitches from her wound and report nearly perfect healing. On July 25, however, the wound broke completely open. "The outlook was most discouraging," Palmer wrote. "It did not seem possible to ever get the wound to heal again." Two days later Banting was back in Toronto with his fresh supply of Lilly insulin. The wound immediately began healing again.[56]

XI

Banting came home convinced that Toronto had to have vacuum stills like those being used to make insulin in Indianapolis. It was expensive apparatus, costing several thousands of dollars that the Connaught Laboratories did not have. Banting decided to get the money. Most of the university's senior administrators were out of town, so he went directly to the chairman of the Board of Governors, Sir Edmund Walker. Walker was past president of the Canadian Bank of Commerce and a commanding patron of the arts and sciences in Toronto. He agreed to see Banting in his splendid downtown office. Banting explained to Walker that $10,000 was needed immediately for better equipment in the insulin plant. Walker replied that it was quite impossible to get that amount so quickly. Such an expenditure would have to be approved by the Board of Governors. The Board would not be meeting again until the university year began in September.

Banting was furious. Several offers of financial help from wealthy Americans had been transmitted to Toronto through American doctors. One of these doctors, H. Rawle Geyelin of New York City, had particularly impressed Banting during a visit to Toronto earlier in the month. "Now Banting, I am going home to put my most severe diabetics to bed so that they will live long enough to get insulin," Geyelin told him as they parted at the station. "Let me have some as soon as possible...and by-the-way if you need money for your research I might be able to help you." Listening to Walker's explanation of why the University of Toronto could not meet his request, Banting thought of Geyelin, got to his feet, and, as he remembered it in 1940, said to Walker:

> "Mr. Chairman, we got to get this still and I want to know if you damned Board of Governors will or will not accept the money if I get it for them."
>
> The dignified old gentleman was amazed and completely nonplussed at my boldness. He gasped and stared and stammered. "I could see no objection."

"Thanks," said I, turned on my heel, and left without further word.

Banting took the train to New York a day or two later to ask Geyelin for the $10,000. Geyelin phoned one Robert Bacon, who had a very sick diabetic child. After a few minutes' conversation Geyelin turned to Banting and asked how the cheque was to be made out. Banting wired Defries in Toronto to go ahead and order the first small vacuum still. He also wired Indianapolis to add Geyelin's name to the list of clinicians who would receive insulin when it was ready.[57]

XII

Banting went on from New York to visit Allen at his Physiatric Institute in Morristown, and Clowes at Woods Hole; he and Clowes then saw Joslin in Boston. Lilly was ready to begin supplying Joslin, Allen, and other leading diabetologists with insulin. (Sansum in California, it will be remembered, was using his own insulin clinically; Woodyatt in Chicago had also begun making insulin and first used it on patients in June.) Clowes and Banting discussed how to go about this in a way that would protect Toronto's - and Banting's - priority in the work. Clowes' idea, consistent with his original plan, was to form a small co-ordinating committee to plan the course of the testing, with a view to the results being published in a special issue of Allen's *Journal of Metabolic Research.* Banting would be a member of the committee, an editor of the journal, and at the head of the list of authors in the special issue. Clowes thought the issue could be published by the end of 1922; Lilly would bear the expense of distributing it throughout the United States. "And if this were done," Clowes wrote Banting on August 11, 1922, "you would not only get full credit for your work but it would be the first step toward securing the Nobel Prize in medicine for you and your associates."[58]

Clowes and the Lilly family had begun to grasp the full significance of what they were doing. "I am almost overwhelmed with this tremendous situation," J.K. Lilly wrote Clowes, "and experience some difficulty in keeping my feet on the ground and my brain in normal operation.... Macdonald [one of the Indianapolis clinicians] says it looks to him like the biggest thing that ever happened in medicine, and that is saying a good deal for some very big things have happened in medicine." "You have certainly entered the holy of holies," he added a few days later, "and are sitting on the throne with the elect. It is a marvellous development and I rejoice in it."[59]

But insulin did not come easily to the Lilly company either. Just as the Americans thought they had mastered the process and were proceeding in a straight line towards commercial production, unforeseen problems started to develop. Every lot was not coming out at full strength. In early August several lots were not successful at all, apparently because the United States government had forced a change in the kind of alcohol the company was allowed to use. Having just made a commitment to supply the clinicians with more than seven hundred and fifty units of insulin a week (of which Banting was to get five hundred), Lilly found itself "right on the ragged edge" of a serious supply problem.[60] The experimental program, aimed at developing better manufacturing processes, had to be suspended to meet the promise to the clinicians. This greatly distressed Clowes, who persuaded Banting to cut back his allotment from five hundred to three hundred and fifty. Joslin and Allen had agreed to give Banting advice on his patients' dietary regime, and Clowes hoped this would enable him to stretch his insulin just as far. Banting's cutback enabled Williams in Rochester to begin receiving Lilly "Iletin" to use on Jim Havens instead of the painful Toronto stuff.[61]

Clinical tests began at the Methodist Hospital in Indianapolis on August 3.[62] In Boston, Elliott Joslin received his first insulin on August 6. Thinking about the trials he would begin the next day, Joslin was too excited to sleep that night. It is said he was too nervous to make the first injection himself, so it was given by his associate, Dr. Howard Root. The patient was a forty-two-year-old former nurse, Miss Mudge, who in five years of diabetes had starved herself down to 69 pounds - "just about the weight of her bones and a human soul," Joslin put it.[63] Miss Mudge was an invalid from her diabetes; only once in the past nine months had she found strength to go out on the street. The immediate effect of her first injection of insulin was not that dramatic, Joslin remembered. But six weeks later Miss Mudge was walking four miles daily.[64]

Dr. Frederick Allen made his planned visit to Toronto on August 8 (partly to give Banting help) before beginning to use insulin in Morristown. While he was away, rumours spread among the patients at his Physiatric Institute that something momentous was about to happen. One of the nurses, Margate Kienast, later described their reaction:

> ...the mere illusion of new hope cajoled patient after patient into new life. Diabetics who had not been out of bed for weeks began to trail weakly about, clinging to walls and furniture. Big stomachs, skin-and-bone necks, skull-like faces, feeble movements, all ages, both sexes - they looked like an old Flemish painter's depiction of a resurrection after famine. It was a resurrection, a crawling stirring, as of some vague springtime.

She remembered the scene when the patients heard that Dr. Allen had come back:

> Bed immediately after dinner was the rule for our patients. But not

that evening. My office opened on the big center hallways. I could see them drifting in, silent as the bloated ghosts they looked like. Even to look at one another would have painfully betrayed some of the intolerable hope that had brought them. So they just sat and waited, eyes on the ground.

It was growing dark outside. Nobody had yet seen Doctor Allen. His first appearance would be at his dinner, which followed the patients' dinner hour. We all heard his step coming along the covered walk, past the entrance to the main hallways. His wife was with him, her quick tapping pace making a queer rhythm with his. The patients' silence concentrated on that sound. When he appeared through the open doorway, he caught the full beseeching of a hundred pairs of eyes. It stopped him dead. Even now I am sure it was minutes before he spoke to them, his voice curiously mingling concern for his patients with an excitement that he tried his best not to betray.

"I think," he said, - "I think we have something for you."[65]

On August 10 Allen began administering insulin to six of his most critically ill patients. Their first doses were minuscule, half a unit or less per injection, partly to spread out the supply, partly for fear of hypoglycemic reaction. Even so, the effects were striking. "Our first results with your pancreatic extract have been marvellously good," Allen wrote Banting on the 16th. "We have cleared up both sugar and acetone in some of the most hopelessly severe cases of diabetes I have ever seen. No bad results have been encountered either generally or locally. We have been able to increase diets, and already an effect seems evident in the form of increased strength.... I only wish that we could have several times as much extract as is available just now."[66]

XIII

Allen's most desperately ill patient had been Elizabeth Hughes. She still clung to life, but her only progress was downwards. It may have been while talking with Allen in Toronto that Banting agreed to add the child to his list of patients. Or there may have been one final appeal to him from the family, possibly through Dr. Lewellys Barker, who was once again on hand and was in touch with the Hughes family. By August 12 it had been decided to bring Elizabeth to Toronto. Allen told Banting he would find her a model patient for treatment. "There could not be a child, who for her own self deserved your care more than Elizabeth, in addition to any consideration due on account of her family." Allen must have offered to take Elizabeth as one of his first insulin patients at Morristown; the family seems to have been influenced by the prospect of going right to "the fountainhead" for insulin (where, it should be added, supplies of the fluid were far more plentiful). Elizabeth, her mother, and her nurse came to Toronto on August 15.[67]

When Banting examined Elizabeth Hughes on the 16th, he must have marvelled that she was still alive. The pathetic child would turn fifteen in three days. Banting's handwritten notes of his examination of Elizabeth survive in his papers:

> wt 45 lbs. height 5 ft. patient extremely emaciated, slight aedema of ankles, skin dry & scaly, hair brittle & thin, abdomen prommt, shoulders drooped, muscles extremely wasted, subcutaneous tissues almost completely absorbed. She was scarcely able to walk on account of weakness. Respiratory, digestive & cardio-vascular systems normal.[68]

He began insulin treatment at once. The first injections, one cc. twice a day, cleared the sugar from Elizabeth's urine. Banting immediately began increasing her diet. It had been 889 calories (actually 789 through July, but on the 29th Allen had allowed an extra 100 calories of fat daily, probably to hold off death from starvation). At the end of the first week's treatment Banting had Elizabeth up to 1,220 calories; another week and she was on a normal girl's diet of 2,200 to 2,400 calories.

Antoinette Hughes had had to go back to Washington. In long, chatty letters Elizabeth kept her mother informed of her progress. Elizabeth realized from the beginning that insulin was going to bring her back to health. She was an extraordinarily fluent writer, but had trouble finding words to describe what this experience meant to her. "To think that I'll be leading a normal, healthy existence is beyond all comprehension," she told her mother in her first letter. "Oh, it is simply too wonderful for words this stuff," she burst out a few weeks later.[69]

The day Fred Banting examined Elizabeth Hughes he met his friend Dr. D.E. Robertson at lunch at the university's faculty club. The newspapers had found out about Elizabeth; her trip to Toronto to get insulin from Dr. Banting was reported throughout North America. During lunch Robertson looked carefully at Banting and then asked whether he had worn the suit he had on now when he had met the Hugheses. Yes, Banting said, it was the only suit he owned.

"You are coming with me," Robertson told Banting at the end of their meal.

"I asked no questions. He took me to the most expensive tailor in Toronto and said, 'make this man a suit,' 'let me see your blues'. I was measured while he selected the cloth. Then he said 'better make him an overcoat' & he selected the cloth & directed how it was to be made. 'I don't know when he'll have enough money to pay you, but I vouch for him.' "[70]

It was a time for new suits and celebrations. On August 19 Clowes wired that the process was now working splendidly and Toronto's quota could be restored. On the 21st the diabetes clinic at Toronto General Hospital finally opened for business. As an attending physician under Duncan Graham's direction, Banting was to be paid the then princely salary of $6,000 annually (much more than his associates, Campbell and Fletcher, were allowed).* His clinic at Christie Street had not gone well, for the first patients had been plagued by pain and abscesses, reactions which discouraged other diabetic veterans from volunteering. But the situation suddenly changed dramatically. One of the "faithful" asked for a weekend's leave and permission to take his insulin supplies with him, Banting remembered. The doctors consented. The soldier returned to the hospital on Monday all smiles. "For the first time in three years I am a man again." Insulin had restored his sexual desire and potency. "By night," Banting wrote, "every diabetic in the hospital was asking for insulin."[71] It was mostly Lilly insulin they were getting, but by the 22nd the Connaught facility, newly equipped with the special vacuum apparatus, was about to produce its first substantial batch of truly potent insulin.[72]

In Rochester Jim Havens had already been switched to the American product. In Toronto Mrs. Charlotte Clarke was learning how to use her new artificial leg. In her rooms at the Athelma Apartments, on Grosvenor Street just next to Toronto General Hospital, little Elizabeth Hughes found herself slowly awakening from her nightmare of diabetes, diet, and starvation. "Isn't that unspeakably wonderful?" she exclaimed to her mother.[73]

*Banting also collected fees from his private patients, but these appear to have been very modest, much less than fashionable physicians traditionally charged those who could afford to pay.

CHAPTER SEVEN

Resurrection

Elizabeth Hughes' case was particularly wonderful. Clinicians were very cautious in their early use of insulin; most were so used to diet therapy that they only gradually, even reluctantly, increased the food allotment of diabetics receiving insulin. Banting was different. Having never treated any diabetics, he had no preconceptions about the relation between insulin and diet. When he started treating Elizabeth Hughes, he took the common sense view that the main thing a young girl who weighed 45 pounds needed to do was to gain weight. So he prescribed a very liberal diet for Elizabeth, moving her rapidly up to 2,500-2,700 calories daily. "When I said she was to have bread and potato, both patient and nurse thought that I was joking and breaking faith with the gods, Joslin and Allen. The child was so delighted that she could hardly wait until the next day. The nurse thought she was breaking faith with her profession in obeying such an order and tested every single specimen separately for 24 hours and could scarcely believe it herself when there was no sugar."[1]

On her daily chart Elizabeth meticulously recorded the first piece of white bread she had eaten for three and a half years (August 25), the first time she had corn for supper (August 29), the reintroduction of macaroni and cheese to her life (September 7). Her diet was still unbalanced, with the intake of carbohydrates kept down and the extra calories supplied by fats, mostly through a daily pint of thick cream. Five weeks after coming to Toronto, Elizabeth had gained a little over ten pounds. "I declare you'd think it was a fairy tale...," she wrote her mother,

> I look entirely different everybody says...gaining every hour it seems to me in strength and weight...it is truly miraculous...Dr. Banting considers my progress simply miraculous, none of his other patients coming near me in diet etc., and so I consider myself especially lucky. He brings all these emminent Doctors in from all over the world who come to Toronto to see for themselves the workings of this wonderful discovery, and I wish you could see the expression on

their faces as they read my charts, they are so astounded in my unheard of progress....[2]

Elizabeth was one of Banting's private patients. She lived outside the hospital, and only Banting, her nurse, and Elizabeth knew the details of her diet ("I know if they did know they'd nearly roll off their seats. It's our great big secret!... Wouldn't Dr. Allen have ten fits if he knew what I was on now?")[3] Through the autumn of 1922 Elizabeth Hughes came back to life and health. Every week she gained two or two and a half pounds, and in October realized she was starting to grow taller. Her invalid's routine of reading and sewing, sewing and reading, was interrupted more and more often by movies, concerts, plans to go back to school, trips to Niagara Falls, and very special nights out:

> Last night I had the loveliest time! Dr. Banting came over about quarter past seven and asked us to go out and take a drive with him. Blanche couldn't go...so I went with him alone. Gracious I felt so grown up going out with a man alone at night!
>
> Well first he took me to his office where he showed me no end of interesting things about his work, clippings etc. and then he took me up to his room to see his favorite books and paintings.... Well then we drove around a bit and ended up by going into the Connaught laboratory which is in the basement of the Medical Building over at the University, and seeing extract made from the first stages to the last. It was the most interesting thing I've seen in a long time and you must prevail on him to let you see it done when you come up. They are putting out such large quantities now that that enormous plant is running night and day with the men working in relays. Oh it certainly was one glorious evening and I shall never forget it.[4]

The wonderful progress was not without complications. Every new batch of insulin was of different potency; while adjusting to it Elizabeth often had hypoglycemic reactions. They tended to be mild, though, and not very frightening, in fact led to the glorious treat of being allowed candy, for Banting armed her with molasses kisses to bring her blood sugar back up from hypoglycemia. The impurities in her insulin also caused pain and swelling, and the injection routine became particularly unpleasant when weak insulin had to be given in larger doses: "Imagine, I have to take 5cc. at a time. Isn't that awful?" Elizabeth wrote on October 25,

> but it seems they have had no extract for the last few days and I suppose we were lucky to have even that poor stuff. We only have a two cc. syringe you know and so Blanche fills that and gives it to me and then unscrews it from the needle which is left sticking in to me (I feel like a pincushion) fills it again, and gives me that (am left a pincushion once more), and then have the fifth cc. It really is quite a

process, and altogether takes about twenty minutes for the whole performance. My hip feels as if it would burst too, but it doesn't, although my whole leg is numb until I walk on it a bit, then it recovers rapidly, and within an hour I would hardly know anything had been given.[5]

Elizabeth's hips, the fleshiest part of her body, were nothing more than a mass of swollen lumps from her insulin injections, she told her mother. She always added that she could endure anything for the sake of her new diet and what it was doing for her.

> I want if you can possibly find them, the links that were taken out of my little silver watch and my gold bracelet Mrs. Crozier gave me. My arm is fattening out so much you will be glad to hear that my watch is really becoming quite uncomfortable, so I need another link put in and I saved them just for this special immergency, although I must say I didn't ever expect it to come....[6]

I

Elizabeth Hughes was one of several hundred North American diabetics receiving insulin that autumn. Lilly was supplying a dozen or so leading American clinicians, Sansum was still making his own insulin in California, the clinics at Toronto General and Christie Street hospitals were in full operation, and in September the university's Insulin Committee, uneasy at the overwhelming United States presence in insulin development, decided that insulin should be made available for clinical testing in Montreal, Winnipeg, Kingston, and London, Ontario.[7] Taking time from his laboratory research, Collip made the first insulin used clinically in Edmonton, Alberta. The United States list was also steadily expanded. In mid-February Clowes estimated that upwards of one thousand diabetics were receiving Iletin from more than two hundred and fifty physicians in sixty clinics in the United States and Canada.[8] Clowes and the Americans were very careful to maintain Toronto's priority in the publication of results, and generally to honour Canadian sensibilities. Nevertheless, by far the largest trials were those run by Joslin, Allen, and Woodyatt in the United States.

The early use of insulin was both a clinician's delight and a baffling challenge. A delight because it worked so well, a challenge because so little was known about insulin and its effects, so much had to be tried. What was the proper dosage, for example? A simple question which actually begged a host of other questions. Should the dose be high enough to keep a diabetic's urine sugar free, or should a little glycosuria be permitted? What levels of blood sugar should be aimed at? Was it necessary to worry about blood sugar at all if the urine was sugar free? How could and should the effects of insulin be spread out over twenty-four hours? Should the dosage be kept low to avoid hypoglycemic reactions? What kinds of doses were appropriate in the emergency situation of coma? And on and on and on. Relating the daily intake of insulin to that of fats, proteins, and carbohydrates posed an endless series of problems. Measuring how insulin actually worked raised yet another category of issues, some of which would be wrestled with back in the physiologists' and biochemists' labs. Laboratories where, it had to be admitted, the scientists still did not know exactly what insulin was. Its chemical composition was a mystery.

That mystery was solved in part in the mid-1920s, when it was realized that insulin was a protein, and later the unravelling of its exact structure in the 1950s would be a milestone in molecular biology. But in the generations since the discovery of insulin, most of the questions about its action and proper use, and others arising from them, have been and still are being investigated by thousands of researchers around the world. In the autumn of 1922 the handful of physicians who had insulin were like explorers who had found an unknown continent and were trying to map it all in a few months. Governments did not offer guidelines, help, or hindrance. To make this job just a little more challenging, the basic tool, insulin, remained in limited supply to the end of 1922, and neither the Toronto nor the Lilly product had as yet come close to being purified or standardized; the best Lilly could obtain was a potency variation from lot to lot of plus or minus 25 per cent.

In their animal research the Toronto group had discovered the potentially lethal effects of too much insulin. So the clinicians were constantly on the watch, orange juice or candy at hand, for the typical symptoms of hypoglycemia - anxiety, restlessness, sweating, trembling, sudden hunger, behaviour analogous to drunkenness - and for that reason there were perhaps fewer severe hypoglycemic reactions than there would have been otherwise. There were virtually none, for example, in Walter Campbell's clinics at Toronto General Hospital. One night at Christie Street, however, the doctors found a hypoglycemic patient trying aimlessly to clamber up the wall. Dr. Joe Gilchrist himself had a hypoglycemic reaction on the street one day; it became more dramatic and severe every time it was described during the next thirty years. One version had Gilchrist arrested for drunkenness and his Christie Street patients trooping down to the courthouse to get their doctor out.[9] Banting maintained that Gilchrist was the first human to experience the effects of an overdose of insulin,[10] but Williams in Rochester may have witnessed the first truly severe hypoglycemic reaction in a human when, during the first days of treating his second patient, Lyman Bushman, about June 1922, "we threw him into profound insulin shock. He was so lifeless that the chief of our surgical staff pronounced him dead. We immediately restored him by the injection of some glucose, and it was looked upon as a miracle in the hospital."[11]

Both Williams and Woodyatt had patients die from what was later diagnosed as hypoglycemic reaction.[12] Allen, on the other hand, and probably many others, lost patients in coma from being too conservative, afraid to give the enormous dosages necessary in some of the most severe cases. In the period of rationing there sometimes was not enough insulin on hand to try heroic doses. And there were hard ethical dilemmas, as when Allen had to decide whether to keep giving vital supplies of insulin to a totally insane diabetic. This patient would die without insulin; rational patients badly needed it; Allen thought the large doses necessary for the patient's survival might be contributing to his insanity. He slid around the problem by transferring the patient to a psychiatric ward at Johns Hopkins which had some insulin available.

One of the great hopes in the early days was that insulin might actually cure diabetes. Perhaps it allowed a diabetic's pancreas to rest and the islet cells to regenerate.[13] Enough of the early patients were so responsive to minute doses of insulin, or seemed able to get along with greatly reduced doses, or seemed able to do without it completely after initial recovery, that the issue was in doubt for the first year or more of experimentation. On balance, however, the clinicians were realizing that while insulin often gave their despairing patients enormous psychic regeneration, it did not seem to bring about any lasting change in a diabetic's impaired metabolism. Banting tried a tolerance test on Elizabeth Hughes in November, for example, taking her off insulin to see how many calories she could handle without showing sugar. She reported the results to her mother:

> With regard to my old tolerance test, it doesn't seem to be panning out very well and that is why I guess we will be able to go [home] sooner than we thought. Of course nothing is the matter, only I don't seem to be able to stand any more than my old diet of 933 calories. We have tried to raise twice both times with only two grams of fat more, but each time I either showed sugar or showed that I was very much on the edge. We are trying it out one more day, and if I show again this time I can go back on my old wonderful big diet tomorrow, thank goodness! Well this proves the marvel of insulin all right, for it shows that its that stuff alone thats carrying all of my extra calories. Isn't it wonderful to think that just that liquid stuff does all that work for my poor old tired out pancreas? Dr. Banting thinks that on newly-developed cases their tolerance would be much more likely to be raised under this treatment, but for an old case like mine where this pancreas has had so much strain and where my tolerance has been much lowered by various setbacks he thinks the pancreas is just about done its duty, and can't take care of even a gram more. Thank goodness for the insulin![14]

As insulin was gradually purified, the incidence of pain, abscesses, and swellings caused by the injections dropped to insignificant levels. Almost all patients easily learned to give their own injections. Even so, everyone would have preferred an easier method of administration. Clinicians experimented with every other possible route - oral, rectal, vaginal, intranasal, intravenous, and by inunction (rubbing into the skin). Subcutaneous injection of insulin turned out to be the only practical method. Intravenous injection was occasionally used to get quick results in coma cases.

Quite apart from hypoglycemic problems, some diabetics sometimes barely survived taking insulin. In Woodyatt's clinic at Presbyterian Hospital in Chicago, for example, a 78-pound boy, Randall Sprague, started receiving insulin on September 21, 1922, a day before his sixteenth birthday. For two days he received what Woodyatt referred to as "Macleod's insulin." It had no effect. Then he was started on insulin Woodyatt was making himself. It was effective, but the limited supplies of it were exhausted by October 16. So, according to Sprague,

> ...on October 21 I was started on the Lilly extract in a dose of 5 cc. daily. Urine sugar promptly cleared.
>
> However, after being on Iletin for one week, I experienced a severe anaphylactic reaction with symptoms persisting for two days, including generalized skin eruption, nausea, vomiting, fall in blood pressure and profound weakness. I was very ill and thought I was going to die. To put it mildly Woodyatt was irritated by this episode and sent a telegram to Eli Lilly and Company in which I believe he advised them to discontinue distribution of the extract until its content of foreign protein was reduced. Since no other insulin was available I was off insulin for six days and urine sugar increased to as much as 30.39 grams daily and tests for acetone and diacetic acid became positive...insulin was urgently needed.
>
> On November 4, 1922, Woodyatt insulin from a new batch was started and the dose gradually increased to 3 cc. daily. Glycosuria and ketonuria cleared. [15]

The next spring Sprague found he could take Lilly's insulin without difficulty. Nearly sixty years and some 45,000 injections of Iletin later, the distinguished diabetologist and endocrinologist, Dr. Randall Sprague, wrote that account in a letter to me.

Another victim of anaphylactic reaction was Jim Havens. Poor Havens, who had so much trouble with Toronto's early impure insulin, found out in August and September that Lilly's preparation caused him even worse side-effects. Havens, it turned out, was one of the small group of diabetics allergic to pork insulin. Connaught's beef insulin was still impure and painful. Havens had regressed almost back to where he had started when, in October, 1922, he started finally getting special beef insulin from Lilly.

Then there was patient No. 24 from Allen's Rockefeller study, a fifty-three-year-old manufacturer whose life had been saved in 1914 by under-nutrition. In possibly his most successful case, Allen had brought the man back from the point of death, stabilized him at 1,500 calories a day and a body weight of 97 pounds, and restored his strength so he could resume work. In December 1922 an attack of influenza shattered the man's carbohydrate tolerance. In January 1923 he began getting insulin at the Physiatric Institute, one of only three patients in Allen's Rockefeller Institute group who survived to receive insulin. The same discipline and courage and vitality that kept Elizabeth Hughes alive shines between the lines of Allen's clinical reports of this case. And insulin seemed to have had the same transforming effects on patient 24 that it did on Elizabeth and so many others. In the early spring of 1923 he could walk ten miles a day.

On April 27 he suddenly developed a mysterious cough and fever. On April 28 he fell into a coma. On April 29 he died. Insulin proved useless. Allen could not explain what caused his death.[16]

Insulin did not free diabetics from careful dietary control and self-discipline. Some of the more reckless diabetics, or those who believed they had been cured by insulin, thought they could abandon diet, abandon insulin, or abandon both. This was a suicidal course. One of Joslin's early patients, Thomas D., omitted insulin for five days, kept his diet, developed a mild infection, and died in coma despite the doctors' emergency efforts with insulin. Less serious but still annoying and harmful was the disruption caused in the Toronto General Hospital clinic by a food-smuggling operation[17] carried out by one of the patients who arranged for an enterprising newsboy to tie cakes and candies to the end of a string let down from a third-floor window.*

All the failures, mistakes, and problems faded into near insignificance when compared with the success the clinicians were having. It was as simple as this: insulin worked wonders, near-miracles, time after time. An eight-year-old boy was carried into Joslin's office by his parents. He had been so hungry on his diet that he would burn his hands stealing food from a hot oven. "Dr. Joslin, do anything you want with Frederick, you can't make him any worse", his parents told the doctors. Two months of insulin treatment later, the mother wrote Joslin that Frederick was feeling fine and wouldn't touch a particle of food other than his diet. He walked down town every afternoon, the neighbourhood children staring in amazement at a little boy who had not been able to go out for the past two years. "Nothing we can say based upon laboratory results can equal in importance statements of this character," Joslin and his associates con-

*It was his unwillingness to control his own and his patients' diet properly that caused the infamous Canadian quack, Egerton Y. Davis, to disbelieve in insulin until the effects of rationing were observed during the Second World War. See Harvey Cushing's biography of Sir William Osler.

cluded in the account of this case in their first paper on insulin. They followed it with the case of Dorothy Z., a five-year-old girl who could not climb stairs when she started getting insulin, but soon could walk up and down stairs and dance with her brothers. And later in the article the case of a little Finnish child, Annie N., who came into the hospital, sank towards coma, was given insulin, and within thirty-six hours sat up in bed, played at the window, and threw a kiss to the doctor as he left her room. "It still remains a wonder," Joslin wrote, "that this limpid liquid injected under the skin twice a day can metamorphose a frail baby, child, adult, or old man or woman to their nearly normal counterparts."[18]

The most spectacular of insulin's triumphs came when comatose diabetics were virtually resurrected by the injections. In September at the Hospital for Sick Children in Toronto, Banting and Dr. Gladys Boyd brought an eleven-year-old girl, Elsie Needham, back to life after many hours of deep unconsciousness. By January she was back at school, to all appearances a healthy, normal child, with years of life ahead of her.[19] Another child restored from near coma in Toronto was Leonard Thompson, who had been readmitted to Toronto General in October "in a state of severe acidosis bordering on coma with marked dehydration." This time the boy was put on insulin permanently.[20]

Allen gave insulin to the largest number of patients, 161, in the first year of clinical trials, and his 181-page report is a gold mine of clinical detail. Because of his methods he had an unusually large number of living skeletons waiting for insulin to put flesh on their bones. Also because of his methods he had a large number of former patients who had failed or refused to accept his therapy, but came back to the Physiatric Institute for insulin and found they could tolerate the reasonable diet that went with it. Frederick M. Allen was the most scientific, strictest, sternest, least emotional of the clinicians. Nevertheless, the language of transcendence crept into even his accounts.

> Though the patient was an extremely poor and uneducated tenement dweller, she followed treatment with religious scrupulousness,... though she lived the life of an emaciated invalid and death from inanition seemed to be the ultimate prospect, this treatment was the only possible means whereby a patient with diabetes of this severity could have been kept alive to receive salvation through insulin....
>
> The child has become the picture of health, and pictures of her condition before and after insulin treatment would show a miraculous contrast. [21]

Actually it was Dr. Rawle Geyelin who supplied the most striking of the "before and after" pictures, some of which are reprinted in this book, to accompany the classic papers in the special insulin issue of Allen's *Journal of Metabolic Research.* And it was J.R. Williams who came closest to breaking the professional shell of the case-hardened clinician when he wrote of his first case, Jim Havens, "The restoration of this patient to his present state of health is an achievement difficult to record in temperate language. Certainly few recoveries from impending death more dramatic than this have ever been witnessed by a physician."[22] Rollin Woodyatt, possibly the most dedicated, certainly the most terse (his first paper on insulin was only nine pages long) of the remarkable group of diabetologists, wrote Macleod in October that "this Insulin effect is as striking and the results as brilliant as anything I have ever seen in medicine or surgery."[23]

Late in November, following Clowes' plan, the leading clinicians came to Toronto for a conference to discuss their results, co-ordinate their publications, and advise the manufacturers on dosage and standardization. The group included Allen, Joslin, Woodyatt, Geyelin, Williams, a team from Indianapolis, and Russell Wilder of the Mayo Clinic. Between meetings, which would have included a good deal of "best case" story-telling and comparisons, Banting took several of the doctors to see *his* prize case, Elizabeth Hughes. She had been dreading their visit, especially having to see the humourless Dr. Allen.

> Well all the doctors came at last just as we were about to sit down to lunch the way I just had a feeling they would....There were six of them and they all stood in the door and just stared at me until I got so nervous I didn't know what to do. It seems to me that every time I looked up I met the eye of one of theirs fixed on me. It was terrible.
>
> There were Dr. Joslin, Dr. Allen, Dr. Banting, Dr. Fletcher, Dr. Woodchat and Mr. Best. Dr. Allen acted nicer than I've ever seen him and Dr. Joslin was simply adorable. No wonder everybody is crazy over him. Of course these two were ushered in first and Dr. Allen said with his mouth wide open - Oh! - and thats all he did. He just kept saying over and over again that he had never seen such a great change in anyone and he actually cracked a joke as he was leaving saying he was glad to have been introduced to me or he wouldn't have known who it was. Now I think thats very good for him. He's grown very fat but his nose hasn't filled out any unfortunately and its as flat as ever. Of course as he was going he gave us an invitation to come to Morristown when I came home and of course I accepted the invitation with alacrity. On the whole he conducted himself so much better than I ever thought he would that everything went off beautifully.
>
> And Dr. Joslin is the sweetest man, all he could do was to look over at me and smile and say that he never saw anybody with Diabetes look so well. Dr. Banting...said that they had had the honor of

hearing Dr. Joslin say at one of their meetings up here after he had seen me that I was the most wonderful case of Diabetes he had ever seen treated. Now think of that coming from a man like that and think of Dr. Allen probably sitting there and hearing him say it. I asked Dr. Banting if the latter had anything to say and he said no that Dr. Allen had spoken a few words during the whole time he was here. A man of few words is the word. Its a joke among the doctors though that what Dr. Allen misses in saying he always takes out in writing.[24]

Four days later, on November 30, 1922, Elizabeth Hughes went home to Washington. It was Thanksgiving Day in the United States.

I have described these early cases in such detail to emphasize just how striking the effects of insulin were - much more visibly and dramatically so than insulin's impact today, when few diabetics decline anywhere near coma before obtaining treatment. One more case study deserves to be published exactly as it was written. This is not merely a description of a splendidly successful use of insulin, but is also one of the best prose passages Fred Banting ever wrote. In later life Banting wanted, perhaps above all else, to be a writer. Here are the last few pages of the account of the discovery of insulin that Banting wrote out in longhand in 1940, and never got around to revising or polishing:

Another striking example...was the case that came to Toronto early in June 1922. I was in my office on this afternoon and a man carried his wife into the office under his arm. He was a very handsome man of 30 & he deposited in the easy chair 76 lbs of the worst looking specimen of a wife that I have ever seen. She snarled and growled and ordered him about. I felt sorry for him. I placed her in hospital more in pity for him than in regard for her.

She was one of the most uncooperative patients with which I have ever delt. It was in those early days when insulin was very scarce & precious and we endeavoured to get as much experimental knowledge as possible from each dose. She would steal candy or any kind of food she could lay hands on. She demanded that her poor husband come to the hospital early in the morning and every night he must not leave until she was asleep, yet she scolded him cursed him and treated him like nothing all day long. I could never understand why he took it all patiently, unruffled and even cheerfully. She was a terrible looking specimen of humanity with eyes almost closed with aedema, a pale and pasty skin, red hair that was so thin that it showed her scalp, and what there was was straight and straggling. Her ankles were thicker than the calves of her legs and her body had sores where the skin was stretched thin over the bones. Above all she had the foulest disposition that I have ever known. I could not understand and I marvelled at & sympathised with the poor husband.

She was in hospital some weeks and improved considerably and then he took her home. I was frankly glad to see the last of her. For his sake I had been kind. As a case to follow she seemed hopeless. I did not write to them nor did I hear from them.

A year later I was at my desk early one morning when the phone rang. A cheerful chuckling voice asked if I would be there for ten minutes. I said I would. The receiver was hung up. I went on with my correspondence.

In a few minutes I heard the outer door open and a moment later my office door was thrown wide open as in rushed one of the most beautiful women I have ever seen. She was a stranger. I had never seen her before yet she threw her arms around my neck and kissed me before I could move from where I stood. Over her beautiful head I saw the laughing face of the patient husband. I stood back. The three of us stood hand in hand. I looked at them. The husband said "Doctor I wanted you to see her now. This is the girl I married - before she had diabetes." We laughed and talked. She was a devoted wife. He was no longer the slave but did most of the talking. I asked them many questions. As they went out he whispered "I'll have to take some insulin myself doctor."

Months later I received a tiny envelope with the name and a pink ribbon. A daughter. And I wondered if the little one had red hair, and I prayed she would never have diabetes.

It was difficult to find words and images to describe the transformation insulin wrought. Metaphors of salvation and resurrection were never far from writers' and diabetics' consciousness. Elliott Joslin, a man close to his Puritan heritage, felt the parallels between the sacred and the secular most strongly. In later life he often talked of how insulin reminded him of what he called the "Banting Chapter" of the Bible. "By Christmas of 1922 I had witnessed so many near resurrections that I realized I was seeing enacted before my very eyes Ezekiel's vision of the valley of dry bones. Ezekiel XXXVII, 2-10:

...and behold, there were very many in the open valley; and, lo, they were very dry.

And he said unto me, Son of Man, can these bones live?

And...lo, the sinews and the flesh came upon them and the skin covered them above: but there was no breath in them.

Then said He unto me, "Prophesy unto the wind, prophesy, Son of Man, and say to the wind, Thus saith the Lord God: 'Come from the four winds, O breath, and breathe upon these slain, that they may live.'"

So I prophesied as he commanded me, and the breath came into them, and they lived, and stood up upon their feet, an exceeding great army.[25]

II

Insulin was one of North America's first great contributions to medical science and practice. Its use only gradually spread to Europe and the rest of the world. If we exclude Zuelzer and his acomatol, the first European to use insulin was Dr. R. Carrasco-Formiguera, a young Spaniard who was spending the 1921-22 year studying at Harvard. He happened to be present when Banting gave the first presentation at New Haven. In June Carrasco-Formiguera wrote Macleod asking for details so he could try insulin on a desperately ill patient in Barcelona whom he had been keeping alive in the bare hope of something like this being discovered. In September Carrasco-Formiguera and an associate, Dr. Pere González, managed to make up a brown fluid containing insulin. It was very impure. Carrasco-Formiguera had to test each batch on himself: "sometimes marked and persistent pain made me decide not to use a particular batch." On October 3, 1922, he gave ten cc. of his extract to Francesc Pons in Barcelona. The results were promising, but the patient later died when the doctors temporarily ran out of insulin. Carrasco-Formiguera was soon treating other patients, though, and later undertook to supervise the manufacture and distribution of insulin in Spain.[26] Nobody else on the continent appears to have used insulin clinically until 1923.

No one in Britain seems to have paid much attention to the reports of Banting and Best's researches published in North America in early 1922. (The first inquiry from Britain was by a Canadian at the Royal Infirmary in Edinburgh, Jonathan Meakins, who wrote Macleod on June 17, 1922, asking for more information so he could treat a diabetic colleague. Macleod sent directions on July 8, but Meakins did not use insulin, it appears, until January 1923.)[27] Word of insulin reached the highest medical circles in Great Britain virtually out of nowhere - or almost nowhere: an obscure medical school in a far-off colony - in June 1922, when Macleod wrote the secretary of the Medical Research Council conveying the University of Toronto's desire to give the council complete British patent rights to the anti-diabetic extract. Fitzgerald of the Connaught Laboratories was in England the next month and discussed the situation with the officers of the council. It was a young organization, created by the British government just before the war as an offshoot of developments in British health insurance, and was still feeling its way. The MRC scientists were interested, but skeptical.[28] Miracle cures were always being announced in medicine, even by people who should know better. Dr. Henry H. Dale, director of the biochemistry and pharmacology department of the council's National

Institute for Medical Research, suggested that he and a biochemist colleague, Harold Dudley, visit Toronto to reconnoitre the alleged discovery.[29]

Dale and Dudley were in Toronto in late September and early October. They also studied insulin production in Indianapolis and visited several of the American clinics. Their reaction was immediately enthusiastic. "The thing is undoubtedly a true story," Dale wrote Walter Fletcher, the secretary of the MRC,

> and the progress reflects enormous credit on all concerned with getting things to their present stage, with the very poor equipment they have had hitherto. Banting & Best are fine fellows & the whole story at this end, is perfectly straight, & there is no sign of anything but unselfish enthusiasm.[30]

Dale and Dudley's report to the MRC at the end of October was a long, detailed discussion of insulin and its effects. The Englishmen were particularly impressed by the "striking case" of Elizabeth Hughes, whom they had seen in Toronto. They were less impressed with the situation involving the applications for patents on insulin manufacture, doubting that the patents would be sustained if challenged, worrying, as we will see, about the relationship with Lilly. They rather grudgingly admitted that patenting was probably necessary and that the MRC ought to go ahead and accept control of the British patent. At the least, Dale and Dudley wrote, holding the patent would enable the council to "exercise a moral control over the manufacturers, and would induce the latter to submit to a system of supervision, as regards this product, which the law does not enable the Council at present to enforce." They recommended, and the council accepted, a two-streamed program under which the council would supervise experimental manufacture and clinical testing on the one hand, while working with selected manufacturers toward large-scale commercial production on the other hand.[31]

These arrangements were completed in mid-November. Clinical testing in Britain began in December and January, well after hundreds of lives had been saved and immense publicity generated in Canada and the United States. A leading British diabetologist, P.J. Cammidge, published a remarkably wrong-headed letter in the *British Medical Journal* in November, doubting that insulin would ever amount to much.[32] Despite that, the news from America produced a considerable clamour for insulin in Britain, where it was estimated that at least ten diabetics died every day.[33] "I think about one hundred of my patients pray every night that you should develop diabetes mellitus," one of the more offensive of the Harley Street practitioners wrote Fletcher in early December, implying that the MRC lacked an appropriate sense of urgency.[34]

In fact the council was following Dale's advice not to get caught in Toronto's trap of having to sacrifice crucial experimental work, which would be invaluable in manufacturing insulin, for the sake of keeping a few premature clinical users going. Its policy was designed to save the maximum number of lives in the long run, and is difficult to criticize. These were excruciatingly difficult choices.

The British parallel to the Elizabeth Hughes case was that of Paula Inge, the beautiful eleven-year old daughter of the noted theologian and dean of St. Paul's Cathedral, W.R. Inge. Paula's diabetes was diagnosed in November 1921. The doctor's first estimate was that she might last only three weeks. A year later she was still alive, completely faithful to her starvation diet. In December her father began making inquiries about the possibility of insulin being available to treat Paula. When Dale informed him that it was still in the experimental stage and not available, Inge accepted the situation, writing back, "Godspeed the MRC."[35]

Fewer than fifty diabetics in eight hospitals in Britain received insulin in the winter of 1922-23.[36] Although Dale and Dudley made important improvements in the manufacturing process, British researchers and manufacturers experienced many of the same difficulties familiar to North Americans (as well as such extra problems as a slaughterman's strike, combined with a Canadian embargo on the shipment of live cattle to Britain, which caused a raw material crisis).[37] By the end of March, Eli Lilly and Company, which had made its production breakthrough, was offering to supply Britain with American insulin. "Plainly, the American supply cannot be kept out simply in the interests of British manufacturers while people are literally dying for want of it," Fletcher wrote the Minister of Health, explaining why the MRC was ordering large supplies of American insulin. [38] In mid-April, when the Lilly supplies arrived to supplement the small quantities produced by Burroughs Wellcome and the joint venture between Allen & Hanburys and British Drug Houses, it was possible to make insulin available to most of the seriously ill British diabetics.

Paula Inge was not among them. The story told by those who knew Dale is that she never received insulin at all. Her father's accounts suggest that she was finally given the new treatment. But something must have been wrong with the product or the dosage, for Paula Inge fell into coma and died on Maundy Thursday in March 1923. Her parents consoled themselves with the belief that God had given them a whole year's grace before taking their daughter. Medical science, Dean Inge thought, had done all it could for her.[39]

III

One of the most distinguished foreign scientists who visited North America in 1922 was Professor August Krogh of the University of Copenhagen. Krogh was the most recent winner of the Nobel Prize in physiology and medicine; for his work on capillary action during exercise he had been given the 1920 prize, one of only two awarded since 1914. Brought to the United States to lecture on his capillary work, Krogh found American medical men talking insulin everywhere he went. So he decided to come to Toronto to study insulin at first hand and consider the possibility of undertaking its manufacture in Denmark.

J.J.R. Macleod was delighted at the prospect of a visit from Krogh, hoped the Dane would be his house guest while in Toronto, and arranged special dinners and lectures. Krogh was in the city on November 23 and 24. He spent much of his time with Banting and Macleod, gave a guest lecture on the capillaries, and left for home with authorization from the University of Toronto to introduce insulin into Scandinavia.[40] During the winter of 1922-23 Krogh and his associate, Dr. H.C. Hagedorn, began the organization of Danish insulin manufacture, establishing a special non-profit Nordisk Insulin Laboratory. There were vast supplies of pork pancreas available from Denmark's bacon factories. By the end of 1923 Danish "Insulin-Leo" had joined Lilly Iletin and the British insulins in mass production.

Inquiries were coming to Toronto from as far away as Peking (where American medical missionaries hoped to make insulin) by the fall of 1922, but outside of Britain and Denmark there was no other rapid progress towards large-scale manufacture. In normal times the fiercest commercial competition might have come from Germany, with its world leadership in chemistry and pharmaceuticals. In the early 1920s, though, Germany was in chaos from the effects of the war and then its uncontrolled inflation. It would be well into 1924-25 before extensive manufacture of insulin started in Germany. In the meantime a handful of German researchers and clinicians learned about insulin, not always accurately. Carl von Noorden apparently tried insulin in late November 1922, decided it had only transitory effects, and gave up. Eighty-year-old Bernard Naunyn wrote to his former student, Minkowski, saying he did not believe the reports on insulin; they were only another case of American exaggeration.[41] Minkowski, the greatest of all the pre-insulin researchers, was more careful. Having read an article by Macleod in a November 1922 issue of the *British Medical Journal* (his clinic apparently could no longer afford the North American journals), Minkowski wrote Toronto for off-prints and advice on where he could obtain insulin. "With the greatest impatience I am looking forward to the moment," he wrote in January 1923, "in which the utilisation of your discovery in the interest of the patients committed to my charge shall be possible for me too."[42]

A young medical student, Martin Goldner, was among those present at the moment Minkowski had anticipated. Many years later, having come to the United States and established himself as a diabetologist, Dr. Goldner wrote this description of the scene:

It was in Breslau, and I believe during the Spring Quarter of 1923, that Professor Minkowski, at his regular morning lecture, showed us students the first vial of insulin which had come to Germany. The lecture hall was crowded as always, and the entrance of the Professor was greeted with the usual tramping of feet, followed by silence. Minkowski stood in front of us, tall, quiet, his white hair and beard blending with his white coat, and he looked at us with his unforgettable eyes, understanding and kind, yet penetrating. His appearance commanded respect and admiration, as well as confidence and devotion. He had one hand in the pocket of his coat. From this he lifted the small vial.

"This," he said, "is the first insulin to reach our country. It has been sent to me by Dr. Banting and Dr. Best, of Toronto, who have discovered it. It was once my hope that I would be the father of insulin. Now I am happy to accept the designation as its grandfather, which the Toronto scientists have conferred on me so kindly."

Loud and long tramping of the students' feet expressed their applause. Then followed the usual case presentation. That particular morning, two patients were brought into the amphitheater, one an elderly diabetic man with an ulcer on the foot; the other, a diabetic child in keto-acidosis. The Professor discussed the conditions, and then asked to whom he should give the precious hormone, since he did not have enough for both, and could not possibly look forward to an early shipment. The students suggested the child, who appeared near death. Minkowski shook his head sadly. There was good hope, he said, that the old man's diabetes could be improved, and his leg ulcer healed under the influence of insulin, but there was little chance of saving the child's life, the keto-acidosis being the final stage of the disease, from which there is practically no return. Even if insulin could have had some effect, it would have been only temporary, prolonging the agony without preventing the doom; in spite of all sympathy, he said, the physician must be realistic and use cool and prudent judgment.

The students were quiet. The lecture proceeded. Many of us felt that we had witnessed a historical moment: seeing not only the old professor's gracious acknowledgement that his life's aim had been fulfilled by the ingenious research of younger men, but also the dawn of a new era in the treatment of diabetes - an era which would prove much more successful than the professor could have imagined.[43]

The insulin situation in France was, if anything, less satisfactory than in Germany. Only a few French researchers experimented with insulin in the early months (the earliest seems to have been Dr. F.L. Blum of Strasbourg). France did not allow drugs to be patented, but had no quality control regulations, leading to many companies dabbling in poor-quality insulin while hospital labs made their own. When Banting visited France in the summer of 1923, he found the situation so "deplorable" and the insulin so bad that it was hopeless to even try to give consulting advice.[44]

Despite the slow French progress, Paris was the site of one of the most bizarre incidents in the scientific rivalry for credit for work on insulin. At the December 23, 1922, meeting of the prestigious Société de Biologie, Eugène Gley, one of France's most noted endocrinologists, asked that a sealed envelope he had deposited with the society in 1905 be opened. It contained a short statement, "Sur la Sécrétion Interne du Pancréas et son Utilisation Thérapeutique," in which Gley, who had worked with pancreatic extracts back in the early 1890s, described experiments he had carried out in 1900-1901 with extracts of pancreas he had caused to degenerate by occluding the ducts. Intravenous injections of the extract had considerably diminished glycosuria in diabetic dogs and caused subjective improvement in their condition. Gley proposed more extensive research along the same lines and the need to isolate the active principle of his extracts, sealed his "cachet", and never took up the work again. He had not had the resources to maintain the animal facilities required for such extensive work, he explained in 1922. Gley offered no explanation for his quixotic, irresponsible gesture (he ought to have at least published the idea, so that better-equipped researchers could pursue it), made no claim to be the discoverer of insulin, and congratulated Macleod as having achieved "une grande simplification" of his method.[45]

Another researcher in a war-torn country who was also so short of resources that his lab could not afford the North American journals, learned about insulin from notices in a Paris medical publication. Early in February 1923, Nicolas Paulesco wrote to Banting from Bucharest asking for offprints and enclosing one of his papers.[46] Banting was never a very good correspondent, was being deluged with mail and other obligations, would not have bothered to read letters in French carefully, and in any case had no more offprints. He did not answer Paulesco's letter.

IV

August Krogh's first experiments in insulin manufacture included making it from fish as well as hogs. This was because he had visited Toronto just at the peak of Macleod's belief that fish were the wave of the future in insulin production.

Macleod's enthusiasm stemmed from his work at St. Andrew's, New Brunswick, in the summer of 1922. He went there having known ever since the early publications of his fellow Scotsmen, Rennie and Fraser, that in the bony fishes (teleosts) the islet tissues are distinct from other pancreatic tissue. It was a fairly simple matter for Macleod to prepare separate extracts from the different tissues of teleosts; the extracts of islet tissue were potent, those of the zymogenous pancreatic tissue were not. Thus Macleod supplied the first experimental verification of the hypothesis that insulin was the secretion of the islets of Langerhans. He rushed his results into print just as purists, such as Cammidge, were questioning the propriety of his lab's use of the term "insulin" for its extracts which were, after all, made from whole pancreases.[47]

"I do not think I ever enjoyed two months work so much," Macleod wrote about that summer by the sea.[48] Even more exciting than the neat proof that insulin had been rightly named, was his finding that certain common varieties of teleosts, especially sculpin and angler- or monk-fish, seemed to contain large quantities of insulin which could be easily extracted. At a time when the Connaught Labs were producing only about five units of insulin from a kilogram of beef pancreas, Macleod found he could get three to four units from less than a gram of angler-fish pancreas.[49] In September 1922, fish seemed to Macleod to be the answer to all the production problems Lilly and Connaught had been having with beef and pork insulin. The sea creatures would be all the more practical as a source of supply in such seafaring countries as Britain and Denmark, so Macleod had strongly suggested to both Dale and Krogh that they do extensive investigative work on fish.

This search for an alternative source of supply underlines the fact that even as the volumes increased, North American production remained difficult, erratic, and expensive throughout 1922. Even with their new vacuum equipment, the Connaught Laboratories group at Toronto still could not make insulin in large quantities. "Like any of our previous experiments the first large scale trial turned out to be an almost complete failure," Best wrote in mid-November, referring to Connaught's use of yet another manufacturing process. During one of the recurring periods of optimism at Connaught in September, the Insulin Committee had made its decision to expand Canadian clinical testing outside of Toronto. But suddenly the Torontonians had to ask Lilly to ship the necessary insulin to Canada, for Connaught could not even meet the needs of Banting and other clinicians in the city. Canada did not stop relying on American insulin until the early summer of 1923.[50]

The Americans were able to produce a supply for the Canadian clinics only because they were working so hard with vast amounts of money, pancreas, alcohol, skilled manpower, and rabbits. And they, too, had their discouragements, such as the constant complaining in the autumn of 1922 that Iletin lacked potency. Part of the difficulty was explained when it was realized that Toronto and Indianapolis had drifted into different rabbit tests for potency (Lilly began using fasting one-kilogram rabbits as opposed to the well-fed two-kilogram Toronto rabbit, which gave them a much weaker basic unit), but even then the Lilly product seemed to be suffering badly from rapid deterioration.[51]

Chief chemist George Walden's attempts to prevent the deterioration led to the company's great advance in insulin production and purification. Comparing stable batches and those that deteriorated, Walden found that the degree of acidity, as measured by pH, varied from batch to batch. Marked deterioration took place within a pH range from about 4.0 to about 6.5. The British researchers in Dale's laboratory came up with a similar realization early in their work, but were content simply to adjust the pH levels so that the dangerous range was avoided.[52] Toronto had not made the crucial observation because it had stopped using tricresol in manufacture. The tricresol, it was later realized, had been throwing the solutions into the 4.0 to 6.5 pH range - the "preservative" was thus the catalyst causing deterioration. It was one of the information breakdowns between Toronto and Indianapolis that the Lilly chemists did not know of Toronto's abandonment of tricresol.[53]

Whereas the other labs had, by design or luck, found ways of avoiding the deterioration, Walden at Lilly took pains to study the process itself. He realized that the weakening of the insulin solution actually involved the gradual formation of a precipitate which contained the active principle, thereby reducing the activity of the remaining solution. At the "wrong" pH, insulin was being precipitated out of insulin solutions. Walden's advance came in the discovery that this precipitate was a far purer, far more potent hormone than anything they had seen before. To get it, all you had to do was learn to profit from adversity: instead of avoiding the "isoelectric point" in the pH range at which insulin precipitated out, the thing to do was to go for it. Deliberately adjust pH's to the isoelectric point so as to cause the maximum precipitation. Collect the precipitate, fiddle with it a bit more, and you had by far the best insulin yet, in Walden's words, "a product having a stability many times as great and a purity ranging from ten to one hundred times as great as the best product hitherto obtainable."[54]

Walden gradually evolved his isoelectric precipitation method between October and December 1922. Once its possibilities were fully appreciated, Lilly had solved the production problem. This took place in the winter of 1922-23. By February 1923 the Americans were building up huge reserves of insulin. In his March correspondence with the British, Clowes offered to supply insulin in any quantities: "We can produce in Indianapolis a sufficient amount of Iletin to supply the entire needs of the civilized world." Clowes' one remaining problem was standardization. The company was spending $2,500 to $3,000 a week on standardization - running through thousands of rabbits, over a hundred thousand of them in the first six months* - to get a consistency which, though not at all bad in rabbit

*The great Lilly rabbit hunt, undertaken at a time before the animal supply industry had developed, was one of the firm's more herculean efforts. So was the job of getting the

testing, still varied by up to 10 per cent from batch to batch.[55]

In accordance with the Lilly-University of Toronto agreement, the American insulin had been distributed free of charge throughout the experimental period. In mid-January Lilly entered the planned second stage, selling their insulin at cost. The company had adopted the clinicians' advice to switch to a smaller unit. Iletin was first sold at wholesale for five cents per new unit (the equivalent of twenty cents per original unit). In 1923 dollars it was an expensive treatment – the forty cents to one dollar a day that the diabetic patient was paying for insulin would be the equivalent of about $6 to $15 in 1980s money.[56]

Nor was Iletin as yet available to all diabetics. There was considerable concern, vocal in Toronto but apparently shared in Indianapolis, about the risk involved in insulin treatment. An overdose, of course, could be lethal. More time was needed, it was felt, to educate ordinary physicians about insulin and the need to use it carefully. From the beginning, Clowes' plan had included publishing the clinical results and educating the ordinary physician. Now the clinical reports on insulin had been delayed and would not be ready for publication until late in the spring. The Insulin Committee also wanted more accessible literature and training programs for general practitioners to be made available; and it would be helpful to have insulin approved by the American Medical Association's Council on Pharmacy. Lilly, who were not yet certain of the durability of their product, accepted Toronto's views. The policy for the early months of 1923 was to make insulin available only through selected hospitals and physicians while stepping up the educational work.[57]

This cautious North American policy led to the curious anomaly that the first country in which insulin became fully available on a commercial basis was the United Kingdom. There, a joint committee of the Medical Research Council and the Ministry of Health decided in mid-May that since the insulin supply far exceeded the demand, all restrictions on its sale could be removed. British drug firms were "at liberty to supply insulin for distribution in the country and for export through the ordinary commercial channels."[58] The further irony in the British situation was that imported Lilly insulin, sold by the British drug companies at a very substantial profit, at first made up about 80 per cent of the British supply. Because of laxer control, American insulin was being sold more freely in Britain than it was in the United States or Canada.

British health officials later regretted having been so precipitate, concluding by the end of 1923 that insulin was being "lavishly and wastefully used" by practitioners who needed to be more fully educated on the subject. By then it was too late to return to controls. It seems that the

---

rabbits back in captivity when they managed to break free, creating a hopping pandemonium in the hallways of the Lilly Science building.

clamour for insulin resulting from its late experimental introduction in Britain had led to a too early deregulation of its distribution.[59]

The crowning irony in the British situation developed in the early summer of 1923 when Burroughs Wellcome began selling the first insulin to reach Mexico, Cuba, and other Latin American countries. Insulin was still under clinical trial in Canada and the United States, and Lilly was still awaiting Toronto's permission to export to anywhere other than Britain. Here was a competing firm first in the field - and not with a competing product, actually, but with Lilly's own insulin! "To cap the whole situation," Clowes wrote Toronto's patent attorney, C.H. Riches, incredulously, "Burroughs Wellcome are actually using our product with which to establish for themselves the priority in entering the world market and all the advantages that accrue therefrom."[60]

V

Another complicating factor affecting insulin distribution in North America was the vexing matter of rights to its manufacture. Issues involving the patent situation and the future of the Lilly-University of Toronto relationship were confused, delicate, and carried immensely important consequences for the future handling of the magic hormone.

The patent applications had been filed in the names of Collip and Best in the spring of 1922, about the same time as Toronto gave Eli Lilly and Company its one-year exclusive licence to make insulin. The agreement with Lilly allowed the company to take out American patents on any improvements it made in the manufacturing process; it would assign patent rights for the rest of the world to Toronto. The document also contained a passing reference to trade names that Lilly might use to describe the extract. According to Clowes' later recollections, at the May 1922 meetings in Toronto Mr. Eli Lilly had made clear how important it was to the company to have a brand name for its product. One of the Torontonians had apparently suggested Banting and Best's original term, "isletin," which the Americans then modified, either because of their spelling idiosyncracies or to avoid exact duplication of a possible generic name, to "Iletin." The Canadians seem to have thought it an insignificant matter at the time.[61]

It was no longer insignificant by the end of the summer of 1922, when the Torontonians began to worry that "Iletin" might be used so widely in the literature that it would effectively become the popular and scientific name for insulin, somewhat the way Bayer's "Aspirin" and Parke Davis's "Adrenalin" were being used.[62] Then they learned that Lilly had filed a patent application in the United States, separate to the Collip and Best application, but one which covered the Toronto method. Clowes explained that the action was simply to help protect the Lilly interest; Lilly's application would probably be thrown out as conflicting with the Toronto application, but would deter possible competitors. The explanation did not impress the university's Insulin Committee, or Riches, its patent adviser. The disapproval was transmitted to Indianapolis.[63]

H.H. Dale's visit from Britain that autumn added to Toronto's alarm. Dale, who had worked for a time at Burroughs Wellcome and had an insider's alertness for machinations in the pharmaceutical industry, instantly decided that Toronto had blundered badly in its handling of Lilly. "Macleod and Best were already very suspicious of their conduct in certain directions," Dale wrote Fletcher in late September,

> Unfortunately they missed the point of real importance and I am afraid they have given the whole game into their hands, by not only allowing but almost inviting them to register a new trade name "Iletin" for their version of "Insulin". The Lilly game, which these people could not understand, seems to me perfectly obvious. If they can make use of their start to get the name "Iletin" used by all the clinicians as the name for the hormone, they will easily upset the patent, get clear of control & snap their fingers at the competition. They are already pursuing a policy which can only have the effect of rendering the Toronto patent doubtful, & this policy had reduced Macleod and Company to a state of complete bewilderment.[64]

It is not clear that Dale, in doing all he could to sow distrust of Lilly in Toronto, fully understood the complexity of the situation. Quite apart from a clash of personalities between the Britishers, Clowes and Dale, which clouded the issues - Dale tended to condescend to the voluble upstart, Clowes, while the bemused Americans in situations like this joked about the arguments being "our Englishman against theirs" - there were serious threats of competition on the horizon to worry the Lilly company.

Some came from quack products that fly-by-night, and not so fly-by-night, companies got out on the market remarkably quickly. In 1922 and 1923 the pharmaceutical division of the big meat-packing firm, Armour and Company, was selling "Insulase," the Digestive Ferments Company of Detroit was selling "I'Lang-Hans," the Philadelphia Capsule Company was offering "Insulans," and the Harrower Laboratories of California were peddling their "Pan-Secretin" in both Britain and North America. Most of these products were in capsule or tablet form and were advertised as easier to take than insulin. Most did contain extracts of pancreas; all, of course, were totally useless. Some of the manufacturers undoubtedly knew this; others took an attitude then common in the drug industry, which Best nicely described in a report on the Digestive Ferments Company:

> It is the policy of these companies to manufacture any biological a

few doctors may imagine is giving encouraging results in the treatment of disease. They do not inquire into the merits of the product. They are not in a position to understand if they did inquire. They distribute anything they can sell.[65]

More seriously, there was the prospect of legitimate competition and challenges to Toronto's priority. In Britain, for example, the first articles on insulin had produced immediate claims by two doctors to have discovered effective pancreatic extracts before Toronto.[66] Much closer to home, the physiologist John R. Murlin, at Rochester, was certain that as early as 1916 he had produced a pancreatic extract that could burn sugar, and that now, in 1922, he had a practical way of making his own anti-diabetic pancreatic extract. After the unfortunate experience with Havens in July, Murlin had tested his perfusate on other patients, apparently with some success. The Rochester newspapers, perhaps prodded by Murlin, proclaimed that he had conquered diabetes. These claims caused serious trouble for the Havens family, when Lilly and Toronto, believing that the Havens and Dr. Williams had co-operated with Murlin, told them to go ahead and rely on Murlin's extract. In early September Murlin was involved in discussions with another American drug company, the Wilson Laboratories, about its possible manufacture. Murlin was outraged to learn of Toronto's patent plans; he was certain that he had priority over the Toronto group, and that E.L. Scott also had priority by virtue of his use of alcohol as an extractive in 1912. Murlin wanted Lilly to agree not to interfere with Wilson making his extract, provisionally named "Glycopyren." When he learned that this was unlikely, he approached Scott, suggesting that the two of them challenge Toronto's patent application and perhaps take out a patent in their own right. Scott was lukewarm to the idea, but was surprised at Toronto's action and felt the patent examiners should be informed of the contributions of earlier workers. [67]

The United States patent examiner was already well enough informed. On November 10 the Collip and Best application was rejected on the ground of there being two prior American patents. By far the most important of these was U.S. patent number 1,027,790, awarded on May 28, 1912, to the researcher who had forestalled them all, Georg Zuelzer! There is no record of Toronto's reaction to this news. It may have been astonished surprise, for none of Toronto's publications before this date contains any reference to Zuelzer. If Macleod and company knew of Zuelzer's work they had forgotten it or not bothered with it. Their next paper about insulin began with a reference to Zuelzer.[68]

It was possible to pursue the patent application to a higher level. Late in November, Riches represented Toronto at a hearing before the chief American patent examiner, who was impressed, friendly, and helpful. The Torontonians were advised to present all the evidence they could muster regarding the clinical results of their process. Macleod went to Washington to testify. Telegrams and affidavits were solicited from Joslin, Allen, and Woodyatt. Banting asked Charles Evans Hughes to do what he could to have the patent commissioner expedite matters: Hughes obliged with a useful letter to the patent commissioner describing what insulin had done for Elizabeth. Toronto's amended patent application stressed the major differences between its method and Zuelzer's, and argued forcefully that neither Zuelzer, who had admitted as much in his patent application, nor any other researcher had been able to produce a non-toxic anti-diabetic extract. Toronto had.* Murlin may have gone to Washington himself intending to contest Toronto's claim. If so, his effectiveness was more than nullified by the fact that his fellow Rochesterian, Dr. John R. Williams, considered him to be the next thing to a quack and confidentially supplied Macleod and Lilly with inside information undermining Murlin's claims about his clinical tests.[69] As they had promised, Lilly amended their duplicate patent application to remove all conflict with Toronto's priority.

On January 23, 1923, an American patent on both insulin and Toronto's method of making it was awarded to the discoverers.[70] Before that, however, there had been an important formality. Some months earlier, Lilly's patent attorney had pointed out that one Dr. F.G. Banting seemed to have played a major role in the invention that Collip and Best were trying to patent. There was a distinct danger of a Collip-Best patent being voided on the ground of their not being the sole inventors, and even of charges of

*Zuelzer's patent specified several methods for making extract. Most of these started with allowing fresh pancreas to "self-digest," i.e., letting the activated digestive enzymes destroy the protein matter in the glands. Zuelzer did not believe the internal secretion was a protein. It is. Self-digestion was a very likely way of destroying it, thus producing inactive extracts.

But Zuelzer did not always wait for self-digestion, and sometimes proceeded quickly, apparently using first saline and then alcohol as extractives. In neither his patent application nor his publications did Zuelzer specify his methods clearly. In 1926, at a time when there was some worry that Murlin might be organizing a challenge to Toronto's patent situation, the Insulin Committee tested all of Zuelzer's methods. In twenty-two experiments, Toronto's chemists twice were able to produce an extract that lowered blood sugar. But it was very impure, causing severe shock. They could not purify it by any of Zuelzer's methods, nor could they produce the "fine, grey, feebly smelling powder" Zuelzer claimed to have obtained. They left open the possibility of his having achieved his results by some other method (IC, Zuelzer file, report enclosed in Hutchison to Riches, Oct. 17, 1926). Zuelzer's biographer, Mellinghoff, also complains about his vagueness regarding his methods. A view, expressed later by Macleod, that Zuelzer's human patients may have been suffering from hypoglycemic reactions, does not seem consistent with either their symptoms or the likely potency of the extract. It may be that some of Zuelzer's post-1911 animal tests, using extraction methods he never published, did produce hypoglycemic convulsions.

Generally, many of the precursors had produced active extracts, i.e., extracts containing insulin. But none of them had been able to purify their extracts sufficiently to eliminate the various possible toxic impurities. These included other proteins; peptides; adenosine derivatives; histamine; serotonin; prostaglandins; lysolecithin and other lipids; bile salts from adjacent tissues; and pyrogens from gram-negative bacterial contaminants.

perjury being brought against Collip and Best. Whatever he felt about his Hippocratic oath, Banting's name simply had to be on the patents. At the formal request of the Insulin Committee, and with assurances that the university would defend him against any criticism, Banting agreed to have his name added to the application.[71] The British and Canadian patents, which were non-controversial, were also finally in the names of Banting, Collip, and Best. For a consideration of one dollar each, the three men recognized by patent offices as the discoverers of insulin promptly assigned their patent rights to the Board of Governors of the University of Toronto. The Board assigned the British patent to the Medical Research Council.

There had been no continuing friction between Toronto and Indianapolis in the handling of the basic patent application. At first it seemed as though the "Iletin" question was also going to be resolved. At the end of September, Lilly agreed that the word "Insulin" would be used as the generic name of the pancreatic extract in all the clinical publications, with "Iletin" being used only to refer to the specific Lilly product.[72] Then, at the same December meeting with Allen, Joslin, and Toronto's Insulin Committee which had decided to recommend a reduced unit, Clowes apparently also committed Lilly to abandoning the use of "Iletin." The Insulin Committee's minutes record as one of the conclusions of that meeting:

> The Eli Lilly Co. will immediately take the steps necessary to discontinue using the name Iletin for their preparation. They will in future use the name Insulin (Lilly). The Toronto committee have given this question much consideration, and they are very anxious that the name Insulin be adopted for whatever product is manufactured with their approval and endorsation.[73]

Macleod sent a copy of these minutes to Clowes in Indianapolis. Later in January, when Lilly started selling insulin, the product was still named Iletin. The words "Insulin, Lilly" had been added in small print underneath "Iletin" on the label.

Clowes was apparently over-ruled or repudiated on the trade name by the Lilly family themselves. They considered retention of "Iletin" vital to the company's future handling of insulin.

The Lilly problem as a commercial organization with a major investment to protect was to develop a strategy for maintaining its lead in the marketing of insulin after its head-start period had ended. What would happen when Toronto started licensing other American manufacturers, who might have very small start-up costs compared with Lilly's very heavy research and development expenditures? Brand or trade names are used in retailing to encourage consumers to distinguish among products, and in the hope of stimulating consumer loyalty to brands they like. The drug industry has been no exception in trying to build markets for products through establishing trusted brand names; and, whether one agrees with them or not, its representatives are sincere in arguing that a brand name is a valuable guarantee of product quality.

The Lilly executives believed that Toronto had accepted this brand-name strategy from the beginning – indeed had supplied the suggestion to call it "Iletin." They thought Toronto understood the importance of a brand name to the company, and claimed that without such an understanding they never would have accepted a situation in which, after the experimental period, they would be just another licensee.[74]

The company tried to interest Toronto in alternative approaches. Admitting that it might seem "to a degree selfish," J.K. Lilly suggested to Clowes that Toronto would get the very best results by licensing only one company in each country to make insulin. That way there would be great efficiencies of scale in manufacture, no competition for raw materials, economy in testing, minimal advertising costs. And no squabble over brand names.[75]

The idea of giving a private company an American monopoly on insulin was of course unthinkable. J.K. Lilly probably threw it out as a gambit in his struggle over "Iletin," and, just as important, in the company's manoeuvring to persuade Toronto to extend its exclusive licence, which would expire on May 30, 1923. As winter changed to spring, Clowes bombarded Toronto with endless letters offering all possible reasons for renewing the licence for another year.

Most were variations on the theme that Toronto was largely responsible for the unforeseen delays in getting insulin on the market. First there had been the hold-ups in the experimental program the previous summer while supplies had to be rushed to clinicians. Now there was the deliberate restraint from distribution, which Toronto had requested, to await publication of the clinicians' papers and other literature. Lilly had not had time to enjoy or exploit its monopoly, and its whole position would be undercut if Toronto arranged for other American manufacturers to come on the market on June 1. Clowes described the situation very frankly to Defries of Connaught:

> ...I had always understood that the Toronto Committee realized our predicament and that a tacit understanding existed to the effect that we should be granted an extension of time so far as the experimental period was concerned, sufficient to permit us to expand our output to the point of supplying the entire requirements of the United States for a period of at least a few months, which would enable us to recover our initial expenditures, and, what is far more important, pull the price down to a point at which competition would be unattractive.
>
> ...If our competitors were to start level with us without any of our initial costs...we should be severely handicapped. Our only chance

to make a good recovery is to have a six months monopoly during which period we supply the entire United States.[76]

Whatever tacit understanding might have existed, there was nothing in writing to justify an extension. Toronto thought about the situation and decided not to extend Lilly's privileges. "Insistent requests from the manufacturers, the granting of licenses to several manufacturers in England, our published statements, the satisfactory evolution of manufacturing methods are among the reasons," Macleod wired Clowes.[77]

The company graciously accepted Toronto's decision. Probably because it was already playing its trump card: Lilly was applying for a patent on Walden's method of isoelectric precipitation. This was perfectly allowable within the terms of the agreement. But C.H. Riches believed that the claims in the Walden application were so broadly drawn that if it were accepted Lilly would control totally American insulin production. "In my opinion," Riches wrote Macleod, "these product claims have been drawn for the deliberate purpose of securing to the Eli Lilly Company a monopoly in the United States of the production and sale of Insulin by any method whatsoever, and conflicts with the policy of the University in doing the greatest good for the greatest number."[78]

Toronto had its own trump card to play. In the autumn of 1922 a team at Washington University in St. Louis, led by Dr. Phillip Shaffer, had discovered the isoelectric precipitation method of purifying insulin independently of Lilly. Shaffer's method might be patentable in its own right; certainly it could be used to interfere with Lilly's patent application. At a meeting of the Insulin Committee on April 2, it was decided to send Riches and Defries to St. Louis to consult with Shaffer about patenting his method ("it would leave in the hands of the University something to offer to other manufacturers which would put them in a position to compete with the Eli Lilly Co.," the minutes record rather forlornly, "and at any rate it would show these manufacturers that the University had done all in its power..."). At the same time Toronto decided to inform the American Medical Association's Council on Pharmacy, which was considering Lilly's submission of Iletin for approval, that the discoverers themselves did not approve of the company's use of its trade name. The furthest Toronto would go would be to allow "Iletin" to be used in brackets and in small print after "Insulin."[79]

The Canadians found Shaffer convinced that he and his associates, Doisy and Somogyi, were the discoverers of the isoelectric precipitation method. They would oppose the Lilly application if Toronto wanted them to.[80] Armed with this consent, Riches and Defries went on to Indianapolis for a showdown with the American manufacturers.

The surrender came in a characteristic four-hundred-word telegram from Clowes to the Insulin Committee:

While we consider ourselves legally and morally entitled under our agreement with committee to take out strongest possible patents on our discoveries and whilst we are not in the least concerned about Shaffers claims as our process is superior to and differs essentially from his and we are satisfied of our priority, nevertheless we would not consider doing anything that might embarrass Toronto University....[81]

Lilly agreed, in effect, to enter into a patent pool, allowing Toronto to make Walden's and any of their other improvements available to other licensed American manufacturers. In turn, though, they had exacted a key concession from Toronto. The Insulin Committee agreed to drop its objections to "Iletin" - which were blocking approval by the AMA - so long as the identification "Insulin, Lilly" was given equal prominence on the company's label.[82] In the bargaining in Indianapolis, the Americans had possibly reminded Toronto that flexibility was in both their interests. A breakdown in relations and a patent fight between Lilly, Shaffer, and Toronto could well remove Toronto from the American scene entirely and considerably diminish its glory for the discovery. As it worked out, everyone was satisfied except Shaffer and his colleagues, who soon came to feel that Toronto and Lilly were hogging the glory for the momentous improvements in insulin production.[83] At the end of June, a new, non-exclusive licensing agreement was concluded. During this spring of friction and negotiation, the fears of the Lilly people must have been considerably eased when they realized that it would take many more months before Toronto would be ready to license competitors. Lilly's effective monopoly actually lasted through the second year, for the first of the new American licensees, Stearns, did not start selling its insulin until June 1924.

VI

There was a distinct possibility in the spring of 1923 that all of these struggles about patents, licences, and so on, might be sheer wasted time. Toronto and Lilly were making insulin from cattle and pork pancreas. The basic patent had been amended to cover insulin made from fish à la Macleod's recent work and enthusiasm - though firms located in Indiana and Ontario would be in serious trouble if the best raw material was found in the sea. And the patents did not cover anti-diabetic substances made from other ingredients. Would it be possible to find such substances? Banting and Best had thought that the best proof of the potency of their pancreatic extracts came when they kept the depancreatized dog, Marjorie, alive for ten weeks. In the spring of 1923, in a lab several thousand miles from Toronto, another depancreatized dog was being kept alive for week after week. It was being given injections of an anti-diabetic extract made from onion greens. This extract, named "glucokinin," actually seemed to work better than insulin. Its discoverer was J.B. Collip.

The trail Collip followed to glucokinin began in Toronto in January 1922, as he reflected on the team's discovery that its pancreatic extract caused glycogen formation in diabetic animals. "It was predicted by me," he wrote fifteen months later, "that wherever glycogen occurs in nature a substance somewhat analogous to that produced by the pancreas of higher animals would be found."[84] Among the lower animals a great deal of glycogen could be found in clams (Collip had shown this in earlier work) and at a plant level it was present in yeast and other fungi. Did clams and yeast also contain insulin? Or something like it?

The idea was taken up in Toronto that winter. W.P. Warner, W.B. Dixon, and C.S. Dixon found in yeast a substance that, in several experiments, lowered the blood sugar and reduced the urinary sugar of diabetic dogs. For some reason they chose not to publish their results, and the work was dropped.[85] As soon as he got back to Alberta that spring Collip picked it up again. He had clams shipped in from the Pacific coast, made an extract of their tissue using his insulin method, and injected it into rabbits. The extract worked: his rabbits' blood sugar gradually fell until they suffered hypoglycemic convulsions and died.[86]

Perhaps because of the expense of getting a clam supply, Collip then turned back to yeast. He tried extract after extract in late 1922, failing every time. Collip was a very frustrated young scientist that winter - deeply embittered by the lack of credit coming his way for the discovery of insulin, harassed by the need to make insulin for diabetic patients in Edmonton's University Hospital, desperate to get on with his difficult research.[87]

On January 26, 1923, Collip got his first clearly positive results with yeast extracts,* and quickly began multiplying his findings, using a variety of extracts of both baker's and brewer's yeast. At the same time he was rethinking his hypothesis and beginning to wonder whether something analogous to insulin might be present in nature wherever sugar was burned.[88] After consulting botanists at the University of Alberta, Collip began experimenting on other plants, starting with the green onion, in quest of the universal hypoglycemic agent.

Collip was not the only researcher working along these lines. In No-

*Collip's exact dating of this result in his early accounts suggests that for him it was another thrilling moment of discovery, similar to the exhilarating experience the night he first realized he had purified insulin. One evening in 1981 I received a call from a seventy-nine-year-old lady who thought she should pass on to me her vivid memory of the morning in 1923 when she, a pharmacy student at the University of Alberta, saw Collip emerge from his lab, utterly dishevelled, having not had his clothes off for days, to announce that he had got it. She thought she remembered him discovering insulin, but it was undoubtedly one of the later "discoveries." Is the pure elation of discovery just as authentic an emotional experience for a scientist when it is later found to be no discovery at all?

vember 1922 the idea of an insulin-analogue in plants occurred to Best and D.A. Scott after a conversation about insulin with Woodyatt. They, too, consulted botanists, and on New Year's Day, 1923, began work with dahlia tubers and potatoes. Their results were inconclusive. In England, however, another pair of young scientists, L.B. Winter and W. Smith, had been pursuing the yeast question; in mid-February Winter and Smith publicly announced that an insulin-like substance existed in baker's yeast.

The announcement caused great consternation at the Medical Research Council. Would manufacturers stop work on animal pancreas to await leavening of the yeast situation? Reassuring letters were issued to the manufacturers. Patent applications were quickly amended. There was also consternation on the part of Winter and Smith's supervisor at Cambridge, the distinguished biochemist F.G. Hopkins. He was not worried about the possibility of unsound science and a premature announcement; instead he was upset that his own students had beaten him to the punch. Hopkins, too, had been getting good results from yeast, but had not been quite ready for publication. "The enterprise of the partners is terrific!" he wrote ruefully. Winter and Smith were already testing their yeast extract on a human diabetic, and seemed to be getting good results.[89]

Collip had started testing his yeast and onion extracts on depancreatized dogs. In Toronto the news of Winter and Smith's findings stimulated Best and Scott to try again with their vegetable substances. They made extracts of potatoes, rice, wheat, beetroot, and celery - every one of which lowered the blood sugar of normal rabbits.[90]

The most extensive of all the "beyond insulin" research had gone on at Lilly in the summer of 1922, where, according to Clowes, they had investigated yeast, fungi, clams, and other marine forms, "in fact every possible source of insulin." Nothing had worked very well, but now - determined to maintain an advantage, for they were facing a possibility of having to write off their total investment in animal insulin - the Americans were ready to start up again. Clowes wanted to know much more about yeast.[91] Macleod was still working on extracting insulin from fish, as were chemists at the Mayo Clinic in Minnesota. There would be immense rewards for the discovery of something better than the insulin now being used. An American newspaper reported in February 1923, that "the race to discover the source of quantity production of insulin...is going on in every medical centre in the U.S. and Canada."[92]

Collip thought he had won the race, or at least the experimental scientists' heat. On March 21, 1923, he announced to a New York City meeting of the Society for Experimental Biology and Medicine that he had discovered a new hormone, present in yeast, onions, barley roots, sprouted grains, green wheat leaves, bean tops, lettuce, "and probably universally present in plant tissue," a hormone which appeared to be "just as essential to the metabolism of sugar in the plant as a similar hormone, produced in the higher animal by the islets of Langerhans, is to the metabolism of sugar in the animal." Collip suggested that the new hormone, the plant equivalent of insulin, be named glucokinin. He concluded his communication with a confident prediction:

> That this hormone will be useful in the treatment of diabetes mellitus in the human subject there can be little doubt. Judging by the results obtained on diabetic animals it will in some ways be much superior to "Insulin." Its effect develops slowly and is long maintained.[93]

Collip was determined not to be done out of credit for this discovery. Learning of Winter and Smith's work on yeast, he fired off a telegram to the British journal *Nature*, claiming coincident priority with them for the yeast work and drawing attention to his glucokinin announcement.[94] In the spring of 1923, he rushed long, detailed descriptions of his experiments into print. Admissions that the work was still incomplete were counterbalanced by claims that his discovery opened up "a new field of investigation of great scope in plant physiology." In April Collip reported that he was ready to start testing glucokinin on a human diabetic. If it proved clinically serviceable, he noted, it might be better than insulin because its effect developed more slowly and lasted longer. As well, it would be readily available everywhere.[95]

The press learned of Collip's research. "Green Onion Tops to Cure Diabetes" was the heading most newspapers put on a Canadian Press dispatch late in April describing the discovery of glucokinin. The reporter predicted that in the near future it would supplant insulin, giving final victory over diabetes.[96] Collip wrote Macleod saying how annoyed he was at the "most undesirable publicity" (his students, it seemed, had talked to reporters), but was still convinced that he was on the right track. "The other day I got the first lawn grass of the season," he added in the same letter, "and from 200 g. of green grass I obtained an extract which caused convulsions in a 13 oz rabbit the day following the administration."[97]

The depancreatized dog lived for sixty-six days on just three injections, its urine being sugar-free most of the time. In his next papers Collip revealed that glucokinin had another remarkable characteristic: the blood of a rabbit made hypoglycemic by glucokinin could be used to make another rabbit hypoglycemic, and so on through several more rabbits. With such powerful animal passage of the active principle, he predicted in June, "the production of this potent serum in quantity can be very readily carried out."[98]

Back in Toronto, Best and Scott continued their vegetable research, rushing their findings into print, drawing attention to the fact that Collip's work and theirs was conducted over the same period of time. By using Connaught's new methods of insulin preparation, they reported, they had made extracts of potatoes and beets that rapidly produced convulsions in normal rabbits, just as insulin did. These convulsions could be relieved by dextrose, just as the insulin convulsions could.[99]

This was work done at the Connaught Laboratories. In the Department of Physiology, Macleod was much more skeptical about the search for the "pseudo-insulins." One of his assistants re-did Winter and Smith's experiments on yeast, and disagreed with their interpretation of their results. "My impression is that they have shot off half-cock," Macleod wrote. "There is certainly something wrong with this work.... It looks to me as if these boys had been carried away by their enthusiasm and have published unwisely." Macleod was puzzled by Collip's work on glucokinin, finding it difficult to know what to make of such strange findings by a colleague whose ability he respected so highly, but was nonetheless able to suggest various problems with Collip's findings: many substances produced hypoglycemia, some of them through damaging the liver; glucokinin might simply be stimulating insulin production from the pancreas of the rabbits; blood pressure was a complicating factor; and so on. Macleod did not respect Collip's rush to claim priority, mentioning to Dale how annoyed he was at the "foolishness" Collip's cable to *Nature* reflected. "It is incredible that he should have done such a silly thing although there may have been extenuating circumstances which I do not know of." And then he added one of his rare bitter remarks: "If every discovery entails as much squabbling over priority etc. as this one has it will put the job of trying to make them out of fashion."[100]

VII

Collip spent the summer of 1923 trying to make glucokinin work. The more he published, the more qualifications and doubts crept into his work. Had his dogs really been totally depancreatized, no matter what the gross autopsies found? Were the rabbit tests truly reliable? Any rabbit starved long enough was bound to become hypoglycemic, whether injected with extracts or not. In any case, was the very slow hypoglycemic effect, produced by large doses of these vegetable extracts, really comparable to the powerful action of insulin? Did Collip actually know enough about glucokinin to test it on a human? Nothing more was said of the impending clinical test. Indeed, Collip had nothing at all more to say about glucokinin after a paper written in June. He published very little in the rest of 1923 and in 1924. He is said to have passed his own epitaph on these experiments a year or two later when he told a colleague, "You're right; I don't think there's any insulin in potato peels."[101]

Winter and Smith's clinical tests of their yeast extract did no special harm, but led nowhere.[102] Best and Scott carried on work on a purified beetroot extract into 1924, but became more interested in a series of experiments that seemed to show the presence of substantial quantities of insulin everywhere in the mammalian body.[103] Like Collip's glucokinin work, and as the chemists working on standardizing insulin were also realizing, all of these results showed a great deal about the unreliability of blood sugar testing on rabbits.

People continued literally to fish for insulin. During the winter of 1922-23, Macleod's fish insulin had been clinically tested by Walter Campbell. It worked just as well on humans as pork or beef insulin, and for many species of fish was easier and cheaper to extract from the pancreatic tissue, gram for gram, than was the insulin coming out of slaughterhouses. The only obstacle to commercial production of insulin from the sea was the cost of collecting the raw material. It was an easy matter for meat-packers to cut out pancreases at an early stage in their animal dis-assembly lines. How could a similar procedure work in the fishing industry?

Macleod sent two of his students, Clark Noble and N.A. McCormick, to spend the summer of 1923 in Atlantic Canada working on the raw material problem. "It will be a great achievement if you can cheapen the production of insulin," Macleod wrote Noble. And went on, revealingly, "I will await your further reports with great expectations. If we could produce with our comparatively simple equipment in the laboratory as much insulin as the extravagantly furnished insulin factory, and at a much lower cost - it would of course be a great thing and might cause certain highly distended bubbles to burst."[104]

It turned out that the sculpin and and monk-fish and skate Macleod had worked with were not practical sources of supply, largely because the fishermen did not normally take them. The students discovered, however, that cod and pollock islet tissue could be easily snipped out and was just as loaded with insulin. They ran elaborate experiments measuring the cost of having small boys cut out islet tissue and store it in alcohol during the dressing of cod onshore. Then, to see if a year-round supply could be obtained, they spent a week on an Atlantic trawler as guests of the very interested National Fish Company. The students learned that it was impossible to gather islets of Langerhans as fast as the fishermen gutted one catch of cod and washed the mess overboard to be ready for the next. The summer's results were inconclusive, not living up to Macleod's great expectations.[105]

There were other strategies and other experiments. American Bureau of Fisheries experts were calculating the cost of collecting shark pancreas in Florida waters ("Sharks Join War on Diabetes"), and at least one shipment came to Lilly for testing. Krogh in Denmark and the MRC people in Britain ran experiments on a wide variety of fish. Henry Dale wished he were young enough to go to sea himself in search of insulin. Just as well he did not, for the zoologist who was sent out in mid-winter got caught in one of the worst gales in years and was lucky to survive.[106]

VIII

In the meantime the final arrangements were worked out in North America for the sale of insulin through normal channels. A several months' delay was caused when disparities in unit strength between Connaught's and Lilly's product led to another change in the basic unit. Toronto had determined to "discipline" the Americans for having adjusted their rabbit unit almost at will.[107] The change took the British completely by surprise. Having based their unit originally on the Lilly unit, they would have to make an expensive, unforeseen change. While Lilly took the matter more or less philosophically, using it to generate more appeals for Toronto to go slow on licensing competitors, Dale was outraged by this latest "arbitrary and inconsiderate" dictation from "those blundering Toronto amateurs."[108] He soon realized, however, that the inconvenience was much less than they had first thought: the British stock of experimental rabbits had gradually developed a tolerance for insulin, causing the British unit to increase in strength, unknown to the manufacturers, by just about exactly the amount Toronto was dictating. There was, as it were, hardly a hare's breadth of difference in the units.* So no change was necessary to make the adjustment.[109]

The British were also finding that they could do without imported American insulin. The MRC had decided to protect British diabetics from becoming dependent on foreign supplies. Ignoring all the Lilly arguments that the lowest possible prices were in diabetics' best interests, the MRC stopped importing the foreign product in the summer of 1923. Domestic production seemed adequate to meet the demand. Actually it was not, and with Britain facing an insulin famine in September, the MRC once again had to order emergency supplies of Lilly insulin. These were duly shipped, without a word of complaint about the restrictive British policy. Dale,

*The problem of standardizing insulin had been and continued to be far more complicated and less facetious than this cursory treatment implies. The resolution of the problem in the years 1923 to 1926, culminating in international acceptance through the League of Nations of a specific quantity of powdered insulin as the basic unit, is an important chapter in the history of bioassay. H.H. Dale made his most useful contribution to insulin research in persuading Macleod, Krogh, and others to give up the futile attempt to use rabbit or mouse units, which would inevitably be variable, as standards of measurement. (See Feldberg 1970, Dale 1959.) But that statement, too, is a gross simplification, for the international standard has always been related to animal tests. G.A. Stewart wrote in 1974 that "The rabbit blood sugar and mouse convulsion methods of assay are still the only internationally accepted methods for the determination of insulin potency, and have been used whenever a new International Standard for Insulin has been established."

who had so distrusted Lilly a year before, commented on "the willingness of the Lilly firm to be used by us, even to assist the development of British manufacture."[110]

For the time being, Lilly's attention was focused on their giant American market. Even under the restrictions in force since January, distribution had expanded to the point where twenty to twenty-five thousand American diabetics were receiving insulin from about seven thousand physicians in mid-September 1923.

By mid-October, dozens of clinical and popular articles had been published to educate the doctors and public about insulin, the unit had been adjusted to conform to Toronto's standards, new labels had been prepared, the price had been reduced to half of January's level, and "Iletin (Insulin, Lilly)" was ready to go on the market. For months the company had been patiently explaining to its salesmen why the product everyone was clamouring for could not be made freely available. "Now Gentlemen," the salesmen's newsletter said at last, "we place in your hands for development, The Greatest Advance in Medicine for Fifty Years."

CHAPTER EIGHT

Who Discovered Insulin?

Insulin may not have been the greatest, but it was certainly one of the most important medical discoveries of the modern age. To be sure, it had not come out of a vacuum. Most people were not aware of the specialized background from which insulin came: the years of work by hundreds of researchers on the pancreas and diabetes, the evolution of endocrinology as an important field of research, the improvements in chemistry leading to quick, accurate blood sugar tests, and other developments. The man on the street did know, however, that he lived in an age when many great things were being achieved in medicine. The microbe-hunters, such as Pasteur and Koch, had found the causes of dreaded infectious diseases, surgeons like Lister had made operations a way of saving rather than taking lives, and humanitarians ranging from Florence Nightingale to Walter Reed and William Osler had shown that medicine was a healing, helping profession. By the twentieth century in North America, foundations were beginning to pour millions of dollars into medical research, the modern hospital had taken shape, and the public had begun to expect that doctors would cure sickness.

In a sense, insulin emerged as a result of some of the institutional effects of the good image modern medicine already enjoyed in the early twentieth century. Fifteen years or so before Fred Banting went to see J.J.R. Macleod to talk about the article he had been reading, the people of Ontario had decided, through their government and public-spirited private citizens, to modernize the University of Toronto. They were particularly interested in creating a first-rate faculty of medicine in which advanced research as well as teaching would be carried out in conjunction with a great hospital. They completely rebuilt and reorganized Toronto General Hospital to be the great hospital.[1] The belief that medical research would produce great benefits for humanity had been in the air, inspiring these developments in Toronto. Few could have imagined how spectacular the reward would be.

Spectacular is the right word. Insulin did have spectacular effects, and was almost immediately hailed as a miracle "cure". Its discoverers, as we have seen, were themselves aware that medical history was being made.

The collections of press clippings kept by Banting's friends, the other researchers, and Toronto's own Insulin Committee, save an immense amount of digging as we consider how insulin and the men who found it were received.

I

Most medical people despised the press, holding attitudes not totally unfamiliar today. Reporters tended to be suckers for every quack, half-quack, over-eager scientist, or naive country doctor who thought he had a serum to cure tuberculosis, a herbal remedy for cancer, or a new surgical procedure to rejuvenate the aged. When the newspapers were not wasting space on undeserving medical stories, they were over-playing legitimate news, getting their facts wrong, and generally making a nuisance of themselves interfering in the lives and practices of busy professionals. Doctors' deep suspicion of what they read in the newspapers and even in the less-carefully edited of the medical journals, helps to explain some of the early skepticism about insulin in countries like Britain: Oh, the Americans are always curing everything; this week it's diabetes. Even in Canada and the United States it was some months before there was enough confirmation of the unlikely news from Toronto to convince wire services and the more skeptical doctors and editors that insulin was, indeed, the real thing.

The confirmation came in a typically confused way, as newspapers learned about clinical trials in their own cities and wrote them up as though a cure for diabetes had been discovered by local doctors. One of the first widespread American reports about insulin, for example, highlighted Sansum's work in California, implying that the experts at the Potter Clinic had turned Toronto's theoretical work into a practical reality.[2] A few days later, a report in the Philadelphia *Ledger*, entitled "Find Diabetes Cure Dr. Stengel Says," was reprinted across the United States. Its effect in Toronto was to produce a wave of letters addressed to Dr. A. Stengel, the University of Toronto, asking for information about the diabetes cure he had discovered. In Philadelphia the result was an angry letter of denial to the *Ledger* from Dr. Alfred Stengel of the University of Pennsylvania.[3]

The question of whether or not insulin was a "cure" for diabetes gave the newspapers all sorts of trouble. Surely if the stuff caused sick people to become totally normal again it was a cure. Even fairly knowledgeable medical people themselves had trouble with the concept, especially if they knew of the early hope that insulin really did cure diabetes. On the other hand, it was vitally important not to lead diabetics to believe that a cure was at hand, for fear of their celebrating by abandoning diet. Newspapers tended not to catch qualifications. All the Toronto newspapers published a comment made by Dr. Lewellys Barker at one of the celebrations for insulin: "Today the diabetic may if he choose, eat, drink and be merry, and tomorrow he will not die." Only one of the papers added Barker's immediate qualification, that they still had to follow dietetic rules.[4] And yet medical people themselves talked and wrote about "Victory over Diabetes", and how "No one need die today who is suffering from diabetes."[5] It wasn't a cure, but it was a conquest. Perhaps the press's trouble with doctors wasn't all the journalists' fault.

The best advertisements for insulin were the diabetics whose lives had been saved by it. The most prominent of these was Elizabeth Hughes, whose progress was chronicled across North America. She, too, disliked the reporters, whose stories brought not very welcome letters and visits from other diabetics. Writing her mother about her latest weight gain, Elizabeth added, "...please don't let on to a newspaper reporter! Haven't they been horrible though. I hate to be written up like that all over the country and I think its cheapening to the discovery. Poor Dr. Bantings even gotten to the place where doctors are beginning to kid him about advertising his discovery through me."[6]

Other patients and their relatives were delighted to talk about insulin. The Corbett family of New York City told the world of how they had bought the cemetery plot in which to bury Joe, a young school teacher, and how they would now laugh and cry when they drove by it. Frank A. Vanderlip, former president of the National City Bank, was back at his desk after insulin treatment, telling reporters that he had an appetite like a hired man and could eat even the dishes that a diabetic "dare not look in the eye." Mrs. Thomas Dixon, wife of the author of *The Clansman* (from which "Birth of a Nation" had been made), was another of the prominent Americans treated in Toronto. The *Star* reporter noted how her eyes shone as she talked of Dr. Banting and his work.[7]

The evidence of insulin's power mounted and mounted. Doctors with impeccable credentials - Allen, Joslin, Geyelin, and others - endorsed it as one of the greatest discoveries of the century, the biggest thing of the age, the beginning of a new epoch in medicine. Reporters read the clinical papers and translated them for the layman: "Tested and retested, fairy tale of science...the seal of scientific approval on the work of the Toronto doctors." As part of Toronto's planned educational campaign, detailed articles were written in popular language by medical specialists. One of the most widely circulated of these, by a leading official of the American Medical Association, concluded that the clinical data on insulin

> reads almost like the glowing accounts of the vendors of Snake Oil and Ready Relief, who used to shout their wares under the flaming torch at the village corner, but in this instance it is the report of the conservative, altruistic scientists who have nothing to sell and who

have devoted their lives and their discoveries to the service of mankind. It is true.[8]

The discoverers themselves reported only to other doctors and tried to avoid the press. They continued to be upset by what Macleod called "the uncontrollable notoriety that the whole thing has had." The high hopes raised by the publicity had put them all under tremendous pressure. Other doctors shared that view, especially during the period when their diabetic patients knew about insulin and its effects, but there was none available. (One of the first doctors to use insulin in Scotland remembered to me how his diabetic patients would hold up newspaper clippings, saying "They've got it in Canada; why can't we get it here?") The Toronto group were almost always cautious in their public statements. Interviewed in 1923 after he had given a popular lecture on alpine peaks in medicine, Macleod admitted to a reporter that perhaps insulin might be a high hill.[9]

II

There were still a few non-believers. The most vocal were the anti-vivisectionists, people morally outraged at the thought of humans inflicting pain on animals to further medical research. Insulin had come directly out of animal research, and seemed to tell heavily against the surprisingly effective anti-vivisectionist crusade (in Britain, for example, a doctor as inexperienced as Banting would probably not have qualified for the licence necessary to do animal research). But of course insulin wasn't a cure, the self-proclaimed humanitarians pointed out. Not a single diabetic had recovered. All insulin was, was a lot of unproven claims, most of which had been heard before. "Why claim this age-old American-Indian superstition of giving the organs of animals for the cure of corresponding organs in human subjects as a 'recent discovery'?" asked Dr. Walter Hadwen, one of North America's leading anti-vivisectionists.[10]

And if the mistreatment of dogs in the insulin research wasn't bad enough, look at what the drug was doing to humans. There were those poor veterans in Christie Street Hospital being literally driven up the walls by over-dosing. The anti-vivisectionists managed to get the Parkdale Branch of the Great War Veterans Association to issue a sarcastic public protest at such mistreatment:

> To us, it seems only in accordance with the prevalent grateful treatment of men whose frames have been appallingly racked on the battlefield, that they should be administered extracts that throw them into convulsions or cause them to climb the walls of the experimental chamber of torture.[11]

The veteran in question – who had had the severe hypoglycemic reaction described in the clinical paper – responded by telling reporters that he would be willing to climb the walls of the Canadian Pacific Railway building to get the benefits he received from insulin. President Falconer of the university dismissed the anti-vivisectionists, surely some of history's most misguided idealists, with the comment, "Why, these people simply don't understand what the word humanity means."[12]

III

So there had been a very big discovery in Toronto, another milestone in the march of modern medicine. Wonderful. But who had made it? Who discovered insulin? Who should get the laurels, honour, applause, tribute, immortality, thanks, etcetera, that grateful laymen, diabetics, doctors, and countrymen, were ready to bestow? This was a subject that the press would have something to say about through the way that the story was covered; but it was also being discussed in many other circles, most particularly in the innermost circle, the discovery group at Toronto.

Fred Banting and Charley Best had not even been present at the announcement of the discovery of insulin in Washington on May 3, 1922. They seemed to have disappeared into the background: Best the student assistant, Banting the untrained country doctor, both replaced by the real scientists, Collip and Macleod, and the expert clinicians, Campbell and Graham. By the end of the summer of 1922, however, the situation had reversed itself. Collip had left Toronto, and Best was working his way through medical school as director of Canada's insulin production. Macleod, too, had been away from Toronto during the crucial summer months, leaving Banting as the man on the spot whom everyone turned to during the dramatic struggle to produce good insulin and save patients' lives. To some extent Banting's revival would have happened anyway. Once the others had agreed to give him primacy in the clinical development of insulin, he was bound to have considerable prominence during the period when clinical developments outshone the experimental physiology handled by Macleod.

The change in Banting's fortunes during the summer of 1922 was nearly total. Previously he had had no connections in the scientific world outside of Toronto. Now he got to know Clowes, Allen, and Geyelin, all of whom would support him in the complex struggle for credit that lay ahead.* So

*Clowes' support is particularly evident in the letters he and Banting exchanged that summer. Banting had obviously told Clowes the whole story from his point of view. He expanded on it by arguing to the Lilly man that Macleod could not be trusted not to give the secret of the method away to a competitor. Clowes' personal feelings toward Macleod are not known, but it was certainly tactically useful to him to have Banting as a kind of personal ally, feeding Clowes with inside information about attitudes and goings-on among the leading members of Toronto's Insulin Committee.

would Joslin. After the May Association of American Physicians' meeting Joslin had written enthusiastically about the achievements of Macleod "and his co-workers." In July, however, Clowes told Joslin of the situation as he understood it, and Banting gave him offprints of his and Best's early articles. Joslin thereafter wrote and talked about the work of Banting and Best.[13]

Most important, Banting regained his self-confidence. So Dr. Banting wasn't qualified to treat diabetics, as Duncan Graham had said so many times? Well, now, let's just step across to the Athelma apartments and see if Miss Hughes is in. Or perhaps take a look at Charlotte Clarke, the amputee, or go over to Rochester and see how Jim Havens is doing. How many diabetics were J.J.R. Macleod and J.B. Collip treating? Why were Campbell and Graham so infernally slow at organizing their own clinic?

It would not have taken anything very significant to have set off Banting that fall. After all the humiliations of the past winter and spring, Banting would have welcomed a chance to show the little son-of-a-bitch, Macleod, who had or had not done the really important things. As it happened, the *casus belli* was particularly offensive. On September 6, the Toronto *Star*, in a story on the attention insulin was getting abroad, quoted a letter Professor Sir William Bayliss of University College, London, had just published in *The Times*. Bayliss, co-discoverer of secretin and author of a noted text, was one of Britain's leading physiologists and a friend of Macleod. He complained in his letter that Macleod was getting inadequate credit for the discovery, stated flatly that Macleod had devised the duct-ligation method of producing pancreatic extracts, dismissed Banting as one of the collaborators who had possibly helped in the clinical application, and concluded, "the discovery is the result of the painstaking and lengthy investigations of Prof. Macleod, which have extended over many years, and it is to him that the chief credit should be given."

Best was the first person to go to Macleod about the article. Having reason to be upset at having been neglected himself, he asked Macleod if he thought Bayliss had been fair to Banting. Macleod had had nothing to do with Bayliss' letter - Bayliss had stupidly written it without any real knowledge of the situation because he had been upset by a dispatch from Canada attributing the whole discovery to Banting[14] - and told Best he was not going to get into a newspaper controversy by doing anything to refute it. Probably trying to explain to Best that any scientist had to learn to live with misstatements in the papers, Macleod said, "Banting will have to get used to it."

To Banting, hearing the conversation at second hand, Macleod seemed to have been saying that Banting had better get used to all the credit for the discovery going to Macleod. Not many hours later Banting was in Macleod's office with Greenaway of the *Star*, asking for a correction of Bayliss' statements. According to Banting's account of the meeting, Macleod again said he had nothing to say. Banting asked Greenaway to leave the room. He told Macleod that if Macleod himself did not refute the statement there were others who would. Clark Noble came in on the pair in the middle of the discussion: years later he recalled seeing Banting, sitting in a chair opposite Macleod, with his feet on Macleod's desk, demanding an immediate denial of the report and accusing Macleod of having engineered the situation to his own advantage. Macleod finally wrote out a statement for Greenaway.[15]

It was published under the heading "Gives Dr. Banting Credit for Insulin." The article began:

> The credit for complete discovery of the Insulin extract for the treatment of diabetes was given to-day to Dr. F.G. Banting by Prof. J.J.R. Macleod.... This is an important statement. It once and for all authoritatively refutes the imputations in the London Times and some American papers that it was improbable that so young and comparatively inexperienced a laboratory man as Dr. Banting himself could have made this epoch-making discovery in the history of medicine.

Neither the *Star* reporter nor the *Star*'s readers understood how carefully Macleod had phrased his statement. It actually read:

> With regard to the letter which recently appeared in the Times it should be pointed out that Sir Wm. Bayliss is in error in stating that the idea of preparing Insulin from pancreas sometime after ligaturing the ducts originated with me. As a matter of fact, this is particularly the part of the work that originated with Dr. Banting, who in collaboration with Mr. Best, put it to experimental test in the laboratory of Prof. Macleod. As a result of the successful demonstration of the effects on diabetic animals of extracts from this source, the problem of the physiological action of Insulin was then taken up by the physiological department of the University by a group of workers, including Dr. Banting and Mr. Best and under the direction of Prof. Macleod....[16]

All Macleod was doing was giving Banting credit for the duct-ligation experiment.

The *Star*'s headline and introduction did not satisfy Macleod. Macleod's statement did not satisfy Banting or Best. There were more meetings. Banting insisted on Macleod giving him, in Macleod's words, "full credit for the discovery of Insulin as it is now used in the treatment of Diabetes." Macleod refused, and apparently made clear his belief that the *Star*'s treatment of his statement had been misleading.[17]

Best was just beginning to realize the danger of his contribution being lost sight of in the conflict between Banting and Macleod. Macleod admitted that he had not given Best due credit in his statement, and Banting issued an additional statement clarifying his "assistant's" role:

> While the idea, it is true, is mine, Mr. Best must have equal credit for the success we have attained. I never would have been able to do anything had it not been for him. We have worked side by side, sharing ideas and developing them together, and but for his unflagging devotion and enthusiasm and his patient and meticulous work we would never have made the progress we have.
>
> From the very beginning it has been a case of Banting and Best, and if our hopes are realized I desire to see Mr. Best given all the honour that would be his due.[18]

No detailed accounts of the September quarrelling have been found. Macleod's letters, with their passing references to "this fresh outbreak of Banting's"... "an extremely uncomfortable position here"... "unbelievable trouble", show clearly enough how unpleasant the situation had become. The *modus vivendi* worked out that spring had broken down entirely. All the old suspicions and misunderstandings had come back to the surface. The discovery of insulin was sitting there on the table to be fought over.[19]

Macleod found the tension almost unbearable. It disrupted his research and impinged on all sorts of matters. He was being put forward for Fellowship in the Royal Society of London that year, for example. To a suggestion that he submit his paper on insulin in fish for publication by the Royal Society, he replied:

> Banting has also criticized my placement of papers for publication, stating that his work should appear in an English journal. I have defended my policy on the ground that immediate publication was desirable. In view of all this I believe that it would only serve to fan the fires still more – and they are almost unbearably hot at present – if I were to publish my recent researches in the Transactions of the Royal Society, dearly though I should love to do so. I find that Banting has succeeded so well in sowing the seeds of distrust in me that it will be necessary for me not to take any step that could possibly be misinterpreted. If I sent this to the Royal Society he would immediately say – "I told you so, Macleod all along was endeavouring to minimize the importance of my work by its publication in ordinary journals whilst he placed his work in the most conspicuous ones he could think of", and if I should be elected to the Society after this article appeared he would claim that I sailed in under false colours.[20]

Colonel Albert Gooderham, the prominent member of the Board of Governors and patron of the Connaught Laboratories who was also chairman of the Insulin Committee, decided to intervene. Upset at the squabbling, Gooderham determined to settle the matter of credit once and for all. He asked Banting, Best, and Macleod each to prepare a typewritten statement of their understanding of the discovery of insulin. Each was asked to outline Collip's contribution (Gooderham did not bother to write Collip in far-off Alberta and ask for his views). Gooderham planned to compare the statements and then meet with the trio to harmonize them. He hoped it would be possible to clear up all misunderstandings and prepare one agreed-upon history of the work.[21]

In the third week of September 1922, Banting, Best, and Macleod sat down and wrote their accounts of the discovery of insulin. Almost sixty years later, these were invaluable sources in the writing of this book. At the time, they settled nothing. When he received them, Gooderham must have realized the impossibility of ever reconciling the conflicting claims of the three men. The same events were being described, that was clear enough, but by different personalities, with different perspectives, different emphases, and, in some cases, different memories of events.

Macleod, who wrote the longest account, was quite certain that he had always given Banting and Best appropriate support, encouragement, and advice. If he had been critical of Banting's early proposals, it was because Banting had come to him with such superficial knowledge. If he had criticized the early results and demanded better ones, it was to strengthen the credibility of the work. He at first resisted clinical testing because there was altogether too much premature clinical work in medicine. He jumped into the New Haven discussion to protect the reputation of his lab. And so on. At every step, Macleod felt he had given Banting and Best proper assistance, valuable suggestions, and adequate support. To make crystal-clear his belief that the young men should get full credit for their experiments, he had explicitly declined Banting and Best's offer to add his name to their first paper, published in the February 1922 *Journal of Laboratory and Clinical Medicine.* If anything, he had bent over backward not to claim as much as other research directors might have:

> In many, if not most, laboratories it is the custom for the "chief" to have his name on the papers when the investigation is in a subject related to that in which he is engaged and if he stands responsible for the conclusions and has participated to the extent that I did in the planning of the research. By this step I made it perfectly evident that I considered the full credit for this investigation to be Banting and Best's. This is surely what counts in questions of priority.

Macleod was concerned that Collip be given full credit for the purification of the extract: "it is unfair and unjust for Banting and Best to rob him of any of the credit by saying that they told him of the percentages of alcohol at which the active principle was soluble. Collip denies that they gave him any information that was of use in this connection and they never communicated any such to me."* Generally, Macleod stressed the large amount of research - the investigation of rabbits, of glycogen formation, of acetone excretion, of respiratory quotients - that had been done by members of a team working under his direction.

Macleod's position has not been understood or appreciated. All of the stories of an evil Macleod conspiring to steal credit are silly. The notion of an innocent physiology professor who never tried to claim any credit is also untrue. Macleod was proud of his achievement and wanted credit for it. He also wanted to make sure that other members of the team, notably Collip, were not deprived of due credit, and that the collaborative nature of this, like most scientific investigation, be properly understood. J.J.R. Macleod believed that the discovery of insulin as used in the treatment of diabetes, "has depended on the conjoint efforts of several investigators working under my direction, of which Dr. Banting was one." He saw the insulin work as a whole package, one that Banting and Best had put on the table perhaps, but that he had, with help, wrapped up, tied up, and given to the world. "Through concentrated effort, for the co-ordination of which I have been responsible, we have given to Science in little more than one year a practically completed piece of research work." That was no mean feat, Macleod realized.

Macleod welcomed the opportunity to say all of these things in writing, "If by so doing I can help to retain for the University of Toronto the reputation which it has already acquired, through the publications on Insulin, as a place where collaborative investigation among diverse groups has been successful in giving to medical science a finished piece of work within a few months' time."[22]

Fred Banting knew that he - he alone - had had the Idea that led to the discovery of insulin. He knew and would never forget that Macleod had been critical and discouraging of his work at every turn. Macleod had not believed in him or his Idea. As Banting remembered the events of 1921, Macleod had been part of his problem, not an aid to its solution. Discovering insulin was a matter of pursuing the Idea in the face of a long series of obstacles. That September 1921 interview, when Macleod told him "As far as you are concerned, I am the University of Toronto", had been particularly devastating. He had only stayed at Toronto, Banting now believed, because of Velyien Henderson. Henderson was the one person who had encouraged and stood by him when he needed help. Henderson was barely getting any credit at all. Banting tended not to remember any of Macleod's specific suggestions, or remember them as being of any value, only that Macleod had not done any of the experiments, not a single one.

In Banting's history, Collip had started work only after the important advances had been made. (Without quite realizing the implications of

*For a discussion of this question see chapter four, note 50 (page 262).

what he was saying, Banting used Macleod's reluctance to let Collip start work as further evidence of the professor's lack of helpfulness.) Macleod had treated Banting unfairly at New Haven, then Collip had broken the gentlemen's agreement by refusing to tell Banting and Best his methods, which Banting accused him of wanting to patent. Macleod had allowed the impression to be spread throughout the United States and England that he, Macleod, had originated the work. Banting was willing to credit Macleod with a "most admirable" execution of the investigation of insulin's physiological action, beginning - in Banting's mind - about February 1, 1922. Well before that date, Banting believed, he and Best had discovered insulin. In an appendix to his account Banting catalogued another half-dozen examples of Macleod showing "a lack of trust and co-operation" to him, ranging from a squabble over summer research funds to derogatory remarks Macleod had apparently made to another doctor. Banting concluded, a little hesitantly perhaps, that "All these points of difference might have been reasonably and easily explained to me had Professor Macleod wished to do so."[23]

As Macleod's student and Banting's co-worker, Best had been more or less caught between the two of them in their running quarrels. He had tended to mind his business, spend his spare time with Margaret, and take little part in the fighting. Of the three accounts of the discovery Best's was the shortest, only about a thousand words, but perhaps the most objective. It was a straightforward, sometimes almost point form, statement of who had done what. Best gave much more credit to Macleod than Banting did, confirming, for example, Macleod's claim to have suggested the use of alcohol as an extractive. He also gave more credit to Collip than Banting did, though not on the key point of methods of purification. ("In my opinion the principal work which Dr. Collip performed was to determine the highest concentration of alcohol [in] which the active principle was soluble.") There was no rehearsal of injustices in Best's account, or any of the sense of grievance that echoed through both Banting's and Macleod's documents.

Best was the only one of the three to comment directly on what, in retrospect, appears to be a vital point in the dispute. Could it be said that Macleod and Collip, seeing how promising the junior men's work was going to be, had stepped in and taken over, getting the good results and trying to get the credit, without having given Banting and Best a fair chance to do it themselves? "The work during the fall months reported in our two papers was performed entirely by Banting and myself," Best wrote. "We had the benefit of Dr. Macleod's advice, but as he states, we were given the opportunity to conclusively prove the efficiency of our extract upon diabetic animals, and as will be stated subsequently, diabetic patients, before the other members of the Physiological Staff participated in this work."

Unfortunately, neither Best's nor Banting's accounts discusses that first testing of their extract on diabetic patients. Best concluded his history by saying that he was going into less detail than he intended to; he wrote it under a momentary misapprehension that Banting and Macleod had managed to reconcile most of the details in their versions.[24]

There was no reconciliation, then or later. The meeting Gooderham had suggested was not held. It is not known how Gooderham reacted to these documents, except that he realized that they disagreed.[25] No comprehensive account of the discovery of insulin was ever prepared at Toronto. The documents were not made public (Gooderham's original copies still cannot be found), of course, and the discoverers gave no more statements to the press about credit in 1922.

IV

Relations among the principals at Toronto continued to be tense. Whenever Banting and Macleod had to settle anything together the atmosphere was either frigid or heated, nothing in between. After the summer of 1922 the only significant written communication is a long formal letter from Banting to Macleod written at the end of September, setting out the arrangements that the two of them and Velyien Henderson had reached governing Banting's resumption of research. Banting had not been able to get satisfactory space in any of the laboratory departments, he claimed, and had finally threatened to start a private laboratory outside of the university on his own initiative. Henderson stepped in and offered to give Banting space in the Pharmacology Department if Macleod would surrender a share of the Carnegie grant that had been awarded to Physiology for insulin research. Macleod went along with the proposal.[26]

While feuding with Macleod over lab space, Banting was also quarrelling constantly with Duncan Graham about the management of the Diabetes Clinic at Toronto General Hospital. He complained that Campbell and Fletcher were getting all the patients, that he was not getting paid, that his colleagues were not treating patients properly, that he did not have enough lab space. The clinic did not work well, Banting wrote in 1940: "Graham was a close personal friend of Macleod's. I could tolerate Fletcher but I could not tolerate Campbell. Graham was always absolutely fair and unselfish and I respected him because of this unselfishness and absolute honesty. But we were not friends. I could not talk to him. We did not then understand each other. I hated him, and he hated me." According to Banting, Graham at one point lost patience with him entirely, and suggested he leave Toronto and go to New York. Graham himself was having anything but an easy time directing a clinic using a spectacular new treatment that hundreds of patients were clamouring to obtain. Banting was a perpetual thorn in his side.* And in the side of his secretary, Stella Clutton, to whom Banting often complained about her boss. Reaching the end of her patience one day, she said to him, "Fred Banting, you're acting like a fifteen-year old; why don't you grow up?"[27]

Macleod and Banting continued the dispute about credit in front of their fellow scientists and by proxy. Both were in demand as speakers at gatherings of medical men. Banting's talks centred on how he got his idea and how he and Best got their first results that summer of 1921. Macleod and Collip would be mentioned, briefly, as having done valuable work in the development of the discovery. Macleod's talks always began with Banting, Best, and duct ligation, but gave much attention to glycogen and respiratory quotient experiments, the purification problem, Collip, and the rest of the team. On occasions when they both spoke, each giving his complete version, it was a long evening. People who knew what was going on did not know whether to be amused or angry.[28]

They talked and wrote privately to friends who they knew would make use of what they learned. When Macleod wrote Bayliss explaining the difficulty his letter had caused, for example, he sketched the history as he saw it, and mentioned how "greatly relieved" he was "that there are those in England who will see to it that due credit is given to all who have participated in our joint endeavours." Bayliss later published a letter in *Nature* which Macleod told Collip "puts things pretty straight."[29]† Banting had always talked freely to his friends; they in turn talked fairly freely to reporters, especially Greenaway of the *Star*, leaking details of Banting's hardships, the difficulty he had getting adequate working space, and other injustices done to him.

Banting's written statements are franker, cruder, more accusatory, and more bludgeoning than Macleod's cool, scientific prose, self-justifying as it, too, could be. Banting now hated Macleod with a passion, an attitude he

*In 1961 Best described the following confrontation between Banting and Graham: "... he came back one day to the Connaught, or the basement of the Medical Building where I was working and said 'I had a little session with the Professor of Medicine.' I said 'Tell me about it.' And he said, 'Well, he called me in and told me that I had represented certain things falsely.' And I remember saying 'What did you do?' Banting said, 'Well, he was sitting down so I went over and lifted him up by the collar and said 'Professor, are you calling me a liar?' And if he had said 'Yes' I'd have smacked him, but he said, 'No, just probably a mistake - I'm not calling you a liar.' " (FP, Dictation, Nov. 24)

†Bayliss' account is very straightforward. He mentions Banting getting the duct-ligating idea, and goes on: "Dr. Banting was then in medical practice at London, Ontario, but gave up his practice and went to Prof. Macleod's laboratory at Toronto to make the necessary experiments on animals. Here he was joined by Mr. Best, an assistant in the laboratory, by Prof. Macleod himself, and at a later date by Dr. Collip and others. The experiments were successful. In another way it was found possible to prepare active extracts.... But it was clear that these methods could only afford a small supply. Hence attempts were made to discover a means of preparation from the ordinary ox pancreas. Dr. Collip was finally successful by making use of alcohol." *Nature*, Feb. 10, 1923, p.189.

never abandoned. His most violent written expression of his feelings was in 1940, at a time in life when many of his friends thought he had mellowed. In some ways he had, but not when he was writing about the discovery of insulin and remembering those fights he had had with Macleod in 1921 and 1922:

> MacLeod...was never to be trusted. He was the most selfish man I have ever known. He sought at every possible opportunity to advance himself. If you told Macleod anything in the morning, it was in print or in a lecture under his name by evening. He was grasping, selfish, deceptive, self-seeking and empty of truth, yet he was clever as a speaker and writer. He never produced a physiologist for he took all that anyone had for his own purpose. He loved acclaim and applause. He had a selfish, over-powering ambition. He was unscrupulous and would steal an idea or credit for work from any possible source. Like all bullies, MacLeod was a coward and a skulking weakling if things did not go his way.[30]

The invective, which continues for another several hundred words, ending with "simpering coward," says more about Banting than it does about Macleod. Everyone I have talked with who knew Macleod personally – friends, students, colleagues, his former secretary – considers Banting's view of him absurd or worse. J.J.R. Macleod was a gentle, honest, dedicated scientist. He was perhaps a little shy and reserved, particularly with students and strangers, perhaps a little vain. He was by temperament a cautious scientist, not brilliant or imaginative, but sound and plodding. He liked to quote Pasteur's remark that in science chance favours the prepared mind. He was an urbane, cultivated, and dapper member of the professoriat. A common view of him as having been very authoritarian, on the German model of the "Herr Geheimrat" professor, is flatly denied by everyone who knew him, including former students, employees, and colleagues who worked with truly authoritarian German professors.

Macleod was bewildered by Banting and his ferocity. The quarrelling seems to have been deeply troubling to him. He never put his deepest feelings about Banting on paper or talked frankly to friends about Banting and the discovery period. In later years it was Mrs. Macleod who would drop the occasional remark about "that horrible Doctor Banting who made our life so miserable in Toronto." After studying Macleod's correspondence and talking to people who knew him, I believe that at bottom Macleod was contemptuous of Banting for his ignorance as a researcher and for the crudeness of his manners, dress, and language. Macleod believed, I think, that Banting and Best would not have come close to insulin without his and then Collip's help.

It is remarkable, in a way, that Macleod seldom did more than hint at this attitude in his letters and articles. He never said nearly as much as he could have about Banting's scientific ignorance, the weaknesses of Banting and Best's experiments, the problems with the first clinical tests, and, above all, the fact that when Banting had an open field in front of him to develop his extract in the fall of 1921, the best suggestion he had been able to produce was the idea of pancreatic grafts. Macleod slid over so much of this in his many accounts of the discovery that it is possible to read them as giving more credit to Banting and Best than was either necessary or Macleod believed they deserved. It is almost as though he was protecting his younger researchers from the full glare of critical scrutiny of their work. I have found only one instance of Macleod telling a fellow scientist that Banting and Best would have gone off on the wrong track in 1921 without his advice. He said as much to August Krogh, the Danish Nobel laureate, on Krogh's visit to Toronto in November 1922. The importance of that conversation will be discussed in the next chapter.

V

The devastating criticism of Banting and Best's work came from England, in the letter that Dr. Ffrangcon Roberts published in the December 16, 1922, issue of the *British Medical Journal.* Having studied Banting and Best's first two substantial papers (those in the *Journal of Laboratory and Clinical Medicine*), Roberts set out to review the steps leading up to the production of insulin in Toronto. The work began there with Banting's hypothesis that it was necessary to protect the internal secretion of the pancreas from the powerful external secretion, the proteolytic enzyme trypsin, by ligating the pancreatic ducts to cause the trypsin-producing cells to atrophy. Roberts declared that the hypothesis was simply false. "Now it is one of the best established facts in physiology," he wrote, "that the proteolytic enzyme exists in the pancreas in an inactive form - trypsinogen - which is activated normally on contact with another ferment, enterokinase, secreted by the small intestine." Roberts allowed that trypsinogen is also activated when a pancreas is cut out and begins to deteriorate, but this happens only slowly and can easily be prevented by chilling. Given these facts, there was no physiological basis at all for Banting and Best's duct-ligation experiment. They had undertaken a cumbersome, time-consuming process to forestall enzyme action which would never take place.

Keeping that in mind, as well as the possibility that the good results obtained in Toronto may have disproved "established facts" about trypsinogen (i.e., proved that active trypsin *is* found in the pancreas), Roberts examined Banting and Best's experiments carefully and critically. In passing, he pointed out some of the factual disparities between the charts and text in the first paper, as well as the apparently abnormal condition of some of the dogs. His main target, though, was the experiment Banting and Best had run on August 17 and 18 using whole gland pancreas. Using their published figures, Roberts showed that the experiment (discussed earlier on p. 76) demonstrated that extracts made from whole pancreas were more effective and more lasting than those made from degenerated pancreas. Banting and Best's own evidence showed the incorrectness of their working hypothesis. Instead of realizing this, they had drawn the "astonishing conclusion" that the whole gland extract was weaker, and believed that the experiment reinforced their hypothesis.

Their attempts to exhaust glands of the external secretion by means of secretin-stimulation were meaningless, Roberts argued, because they had no means of showing that exhaustion had actually taken place. "To establish their point Banting and Best have to show that the gland which they say is exhausted really is exhausted. This can only be done by demonstrating the absence of the three ferments by the ordinary methods. This they have neglected to do." If anything, their secretin experiment also disproved the main hypothesis, for, without realizing it, they were again showing that extracts of whole pancreas were potent. The one thing Banting and Best had not directly tried was the crucial experiment that would have verified or nullified their hypothesis: instead of *assuming* it, they should have tried to *prove* that there was no active principle in extracts made from a normal pancreas.

Roberts drew attention to more problems with the experiments, such as inadequate data on blood sugar patterns after pancreatectomy but before injection, and summarized Banting and Best's situation:

> Having therefore failed to establish their main thesis, but encouraged by a complete misreading of their results (I challenge any unbiased person to read the paper carefully and come to any other conclusion), Banting and Best then proceed to investigate further methods of preparing a hormone free from the destructive action of ferments. They tried foetal pancreas...no comparison has been made between foetal and normal adult pancreases.

Then they had completely changed their methods, adopting alcohol as an extractive and suddenly moving from foetal to normal pancreas. Perhaps they had concluded that alcohol did the job of destroying the (non-existent) proteolytic enzymes, but they never proved this by comparing alcohol with aqueous extracts. Instead, they had turned to a totally different aspect of the problem, concentrating on how to produce a non-toxic rather than a non-inactive extract. This problem had been solved, but, Roberts pointed out, "What Banting and Best have failed to realize is that in so radically changing their method they have abandoned the principle from which they started and which they never proved."

What did this set of experiments, for which Roberts had not one word of commendation, amount to? The experiments led eventually to insulin. *But,*

> The production of insulin originated in a wrongly conceived, wrongly conducted, and wrongly interpreted series of experiments. Through gross misreading of these experiments interest in the pancreatic carbohydrate function has been revived, with the result that apparently beneficial results have been obtained in certain cases of human diabetes...whatever success the remedy will have will be found to be due to the fact that the hormone has been obtained free from anaphylaxis-producing and other toxic substances. The experiments of Banting and Best show conclusively that trypsin *qua* ferment has nothing whatever to do with it.[31]

Macleod's British correspondents apologetically alerted him to the critical article. In mid-January he wrote Dale that Roberts' letter "has, I think, been overlooked by Banting and Best, and I see no object in calling it to their attention at present." The next summer, in a major lecture on insulin, Macleod himself answered Roberts:

> The criticism has been made that Banting and Best's experiments in which simple extracts of duct-ligated pancreas were used formed no essential step in the investigations which have given us insulin. I need scarcely reply to these criticisms. They were apparently made without any appreciation of the real obstacle that stood in the way of development of the subject - namely, convincing evidence that an antidiabetic hormone does actually exist in the pancreas - and to Banting and Best is due the credit of furnishing this by experiments of a different type from those of their predecessors. We owe much to the initiative, skill and patience they displayed in completing this first essential step in the investigation.[32]

This was a revealing response. Instead of dealing with Roberts' specific criticisms of Banting and Best's hypothesis, experiments, and conclusions, Macleod was saying that these were all beside the point. Banting and Best's work, Macleod stated, had provided "convincing evidence," that there was an anti-diabetic hormone in the pancreas. That ended the "first essential step" in the investigation. Then, in this lecture and most of his other accounts of the discovery, Macleod went on to describe the rest of the steps, notably Collip's isolation of the active principle and the investigation of its physiological effects by the team of workers at Toronto. In Macleod's mind, the whole importance of Banting and Best's experiments had been in convincing Macleod and the others of the team that the internal secretion was there to get. They then went and got it. If anything, Roberts' criticisms reinforced Macleod's view that Banting and Best's work was only one part of the discovery process.

The only other response to Roberts was H.H. Dale's letter in the next issue of the *British Medical Journal,* claiming that it was out of place for Roberts to belittle "the simple, honest record" of Banting and Best's experiments. Like Macleod, Dale did not address Roberts' substantial points. Insulin had, after all, been discovered. "And, if it proves to have resulted from a stumble into the right road, where it crossed the course laid down by a faulty conception, surely the case is not unique in the history of science. The world could afford to exchange a whole library of criticism for one such productive blunder..." Dale also seemed to say that if only Roberts had realized how unqualified Banting and Best were to do good work, he would have been more charitable:

> He did not know that the work he attacks was the first, unaided attempt at research by two young enthusiasts; that one spent half the war as a combatant, and the rest, after being seriously wounded, as a medical officer in England, while the other has not even yet completed his student course. He had no conception of the personal sacrifice and heroic labour in which their enterprise involved them. Working thus on their own initiative, without the invaluable help and co-operation given later by the head of the laboratory,... they may have wandered along a wrong trail for a time, though this has yet to be proved. It may be that they made an unnecessary detour, before finding themselves at the point where E.L. Scott had stopped. The important point is that Scott did stop, and that Dr. Roberts would not be writing about his work now if Banting and Best and the other Toronto workers had not gone much further.... It may be that the enthusiasm, which carried them further, was fired by an imperfect interpretation of their earlier results. If so, the mistake will be cleared up in time by others working more calmly and with more experience, and the truth will emerge.[33]

Dale's letter was much too censorious. Perhaps it reflected a distaste for scientific controversy; perhaps Dale wanted to protect the people he had met and liked in Toronto; perhaps he wrote out of concern that intense criticism of the Toronto work could upset the still uncertain patent situation. Nonetheless, Roberts' criticisms were fair comment on Banting and Best's work. They were not pointlessly destructive, but centred on a factual point of some importance in physiology about whether Banting's trypsin-antagonism hypothesis had been proven. The critical examination is also significant for those trying to assess Banting and Best's contribution to the discovery, as opposed to Macleod's and Collip's. Roberts did not realize that at the time. Nor, having made his points, did the young scientist have any taste for a public dispute with the powerful H.H. Dale. He wrote nothing more on the subject. His critical article was referred to a few times in European surveys of the insulin literature, a few times more in anti-vivisectionist propaganda, and then forgotten.

For the next thirty years no one else studied Banting and Best's experiments carefully and critically. Banting himself never seems to have read or known about Roberts' criticisms, and he did not know enough physiology to correct his own errors; Best believed, probably correctly, that Banting went to his grave secure in the knowledge that his great Idea, the duct ligation to prevent trypsin action hypothesis, had been the breakthrough leading to the discovery. The pandora's box Roberts had opened and Dale had slammed shut was only finally reopened in the late 1940s when Joseph H. Pratt of Boston undertook his detailed study of the insulin research, culminating in the paper he published in 1954.

In that paper, and in a much longer, more sharply phrased draft of it, Pratt raised all of Roberts' points (actually he had arrived at the same conclusions independently before learning about Roberts' article) and added some new criticisms of Banting and Best's work. He was not sure, for example, that they had succeeded in ligating ducts properly, for what they said were degenerated pancreases seemed to produce a surprising amount of extract. Their surgical hypothesis, that ligation causes acini to degenerate while the islets remain intact, was also technically wrong; both groups of cells degenerate, though usually at different rates. Pointing out how difficult it is to do a complete pancreatectomy, Pratt also questioned whether Banting and Best's animals, especially the dog Marjorie, were actually diabetic. Such data as the Canadians had provided on the most accurate index of diabetes in the animals, the D:N ratio, suggested that the dogs were not. Finally, Pratt argued, Banting and Best had not found as many toxic side-effects in their dogs as researchers like Zuelzer, Scott, and Kleiner had reported, because they had not looked for them. Fever was the most commonly reported side-effect; Banting and Best had not published any temperatures of their dogs (as we have seen, they only once checked for fever, found it, and did not check again). When Pratt made up an aqueous extract of pancreas, following Banting and Best's original method, it brought on a mild fever in a normal dog.[34]

Unlike Roberts, Pratt was interested in the problem of apportioning credit for the discovery. (In passing, Pratt scored heavily against a generation of textbook authors who had discussed the duct ligation, trypsin elimination process as physiological gospel.) Taking account of the work of the predecessors, and noting Banting's own admission in 1929 that his and Best's results were "not as encouraging as those obtained by Zuelzer in 1908,"[35] Pratt emphasized the multi-step nature of the research. He particularly stressed the contribution of Collip. Banting and Best had taken the work to the point Scott and Zuelzer had reached. Collip, by producing the first non-toxic extract, had gone beyond.[36]

VI

Roberts' and Pratt's criticisms of Banting and Best's conception, execution, and interpretation of their experiments were, for the most part, well taken and unanswerable. (W.R. Feasby's attempted answer to Pratt was a pathetic piece of work, a nearly incoherent combination of nit-picking, special pleading, unwarranted *ex cathedra* claims, and – particularly unfortunate from Banting and Best's point of view – appeals to the authority of H.H. Dale.[37]) Had these critics been able to go beyond the published papers to the notebooks, and discovered the errors and other problems in the research as described in chapters three and four of this book, they might have been even more severe in their judgments. The evidence indicates that Fred Banting and Charley Best were, as H.H. Dale said, "two young enthusiasts," engaged in their first attempt at research. They did wander along a wrong trail; their enthusiasm was fired by a misinterpretation of their early results; insulin did result from "a stumble into the right road, where it crossed the course laid down by a faulty conception."

On their own, Banting and Best were not experienced and knowledgeable enough to have carried their work through to a successful conclusion. They badly needed Macleod's advice. Indeed, Macleod's real failure as a scientist in the 1921 research was not, as Banting thought, in his being so critical of their results, but in his apparent failure to notice the many flaws in what were not, in fact, very competent experiments. Some of this was understandable. Roberts notwithstanding, Macleod and many other physiologists saw merit in the hypothesis that digestive enzymes in the pancreas destroyed the internal secretion.[38] They did not trouble to reconsider the hypothesis after the experiments, which seemed to have succeeded, had gone on to a far more exciting stage. The excitement at what the experiments resulted in – insulin, with all the opportunities and challenges it opened up – seems to explain why Macleod joined almost everyone else in being less than critical of Banting and Best's work. As indicated earlier, he may also have been kind to the young enthusiasts, saying less in criticism of their work than he could have.

Why he did not offer such truly useful criticism as correcting their error about Paulesco, however, is more difficult to explain. Probably it was just an oversight; perhaps, as commonly happens with even the best-informed professors, Macleod had not yet read Paulesco. Most of Macleod's attention in 1920 and most of 1921 was being given to his substantial university duties and to his own experiments on anoxemia.

Even so, without Macleod's directions in the spring of 1921 Banting and Best might never have prepared an effective extract of any kind. There *was* a problem with enzyme action in pancreatic extracts, but Banting's great idea did nothing to overcome it. It was the immediate chilling of all the pancreatic material, as suggested by Macleod, that stopped self-digestion of the fresh pancreases by the activated enzymes. Then, in October and November, Macleod appears to have stopped Banting and Best from becoming side-tracked in futile grafting experiments. Macleod had first suggested to the pair that they use Scott's method of extracting with alcohol. At exactly the same time as Collip was making very rapid progress in December, Banting and Best were failing repeatedly to produce pancreatic extracts with any potency at all. Their clinical test on Leonard Thompson was a failure. While hypothetical statements are always unprovable, there is no good reason to assume that Banting and Best possessed the experience or the skill to purify their extract on their own. In any case, as Best wrote in 1922, they had had that chance before the task was given to Collip. Banting's impatience to get something that would work clinically on diabetics was probably the greatest single factor pressuring Macleod to expand the team.

Banting and Best alone did not discover insulin. Their work was part of the discovery of insulin. It was not the whole discovery. Banting and Best began the process that led directly and without significant interruption to success at Toronto. But it was a multi-stage or multi-step process, to which Collip, Macleod, and perhaps others made vital contributions. It is particularly important to repeat that Banting's great Idea, duct ligation, played no essential part in the discovery. Except in the sense that it got Banting and Best making pancreatic extracts in Macleod's lab. So many of their extracts were potent, from whatever kind of pancreas they used, that everybody decided there had to be an internal secretion at work. In December 1921 and January 1922, the team isolated the internal secretion (in potent enough form to prove that there was an internal secretion; in a strict sense isolation did not occur until the production of pure crystalline insulin some years later).

Another way of arguing to the same conclusion is to ask at what point in the Toronto research it could be said that insulin had been discovered. If it had been discovered when Banting and Best's first extracts lowered the blood sugar of their dogs, then priority for the discovery belongs to Zuelzer, Scott, Murlin, Kleiner, Paulesco, and others, who had all done as much, or more, earlier. If it had been discovered when an extract had anti-diabetic effects on a human, although also having toxic side-effects, Zuelzer again - as Banting indicated - had done this earlier. But both of these concepts of discovery are very thin, begging many questions about what could be claimed to have been discovered.

There are really only two tenable views of the "moment" of discovery. One is that insulin had been discovered when a non-toxic preparation of it reduced the cardinal symptoms of diabetes in a human being. That happened with Collip's insulin in January 1922. And it was the distinction Toronto used in its patent hearing to distance itself from Zuelzer and the others with their unworkable extracts. According to both insulin patents and patients, the discoverers were Banting, Collip, and Best. To them we should probably add Macleod.

The other view, following Darwin's maxim of credit going to the man who convinces the world, is that insulin had been discovered when convincing evidence of its existence had been presented. There is a possible argument that this criterion leads back to Minkowski and von Mering in 1889. But the much stronger argument is that it leads to the May 3, 1922, presentation to the Association of American Physicians. At that time, the Toronto team of Banting, Best, Collip, Campbell, Fletcher, Macleod, and E.C. Noble, presented evidence of the existence of insulin which their peers accepted. On the basis of authorship of the critical paper, every one of the seven was part of the discovery team.

A not uncommon layman's view of the discovery holds that without the Toronto work the world might still be without insulin. This is impossibly unlikely. The internal secretion of the pancreas had been "discovered" theoretically back in 1889; its practical isolation and therapeutic use was only a matter of time, determination, ingenuity, technical skill, and resources. Many of Toronto's predecessors, including Zuelzer, Scott, Murlin, Kleiner, and Paulesco, did have active pancreatic extracts – that is, extracts containing insulin. None of them, however, had been able to purify their extracts sufficiently to eliminate their toxic properties and convince the medical world that the internal secretion had been obtained. If the experiments leading to success had not been begun in Toronto, they would almost certainly have been soon tried somewhere else. Perhaps Paulesco would have purified his extract; this can never be known, though my judgment is that his limited resources, primitive techniques, and theoretical misconceptions would have held him back. Possibly Murlin at Rochester, or Scott at Columbia, both of whose interest was renewed by Paulesco's publications, would have published the great paper. Perhaps someone else might have reread Kleiner's 1919 paper carefully, and thought about where it led. Perhaps Frederick Allen's search for a pancreatic extract, begun in the early months of 1922, would have been successful.* Without Toronto, insulin might have been discovered within five months, or within five years, certainly no more than that. These months, or years, of course, meant the difference between life and death for Leonard Thompson, Jim Havens, Elizabeth Hughes, and thousands of other diabetics.

Why was insulin discovered at the University of Toronto rather than somewhere else? To us, and sometimes to themselves, the discoverers were

*In 1949, Allen himself said he did not think his work would have succeeded. It appears that he was working with a pancreaticoduodenal serum, apparently on the theory, which seems to have influenced Knowlton and Starling, and Murlin, that secretin, which was necessary to activate the external secretion, was necessary to trigger the activity of the internal secretion.

using incredibly primitive apparatus in dingy, smelly rooms, and with little help from a penny-pinching university. Actually, by the standards of the time, the surprising thing was that the University of Toronto had the resources (such as the dogs and their quarters) to sustain major animal research, and that it had the money and prestige necessary to assemble the team that discovered insulin. Most North American and European universities in the early 1920s were not as fortunate as Toronto. The recent technological advance in the micro-estimation of blood sugar was another vital factor. But, most important, Toronto was the one place where a total determination to find the internal secretion of the pancreas was coupled with the technical expertise to do it. Fred Banting, aided indispensably by Best, provided the determination. Macleod, like Carlson, Allen, and many other experts, had his doubts; Banting, the novice, believed. Whatever the results of the experiments, Banting considered they were good, and urged Macleod to go on to the next stage. Banting possessed unshakeable, unscientific faith.[39]

To those who understand the university world of the early twentieth century, or readers who noticed how Zuelzer was treated at the University of Berlin in chapter one, the surprising aspect of J.J.R. Macleod's handling of Banting is that Macleod gave him so much, endured so much, and finally saw clearly enough the importance of the work that he added his and Collip's, and then Campbell's, Fletcher's, and Noble's expertise to the team. The tragedy of the interaction was that the compound of powerful personalities necessary to produce the great scientific advance was so unstable. The team was impossibly volatile. Its members were literally fighting about the discovery of insulin on virtually the day it was made.

It was partly a problem in human relations, and there is a temptation to see Macleod as a classic example of a professor too busy, or too authoritarian – or, as some who knew him believe, too shy – to handle a difficult "student" with sensitivity and finesse. Professors usually have to bend a good deal in their relations with students, but even in these latter days of student power and teaching evaluations, there are limits. Some students are simply impossible to deal with. Macleod could have done better, perhaps, but only a superman could have led the untutored, insecure, bull-headed Banting through to insulin without major troubles. It would have been like going through a whole Canadian hockey season without allowing a single fist-fight.

CHAPTER NINE

Honouring the Prophets

Most people were not interested in or equipped to understand the fairly technical distinctions and subtle arguments involved in any accurate or fair apportionment of credit for the discovery of insulin. To all but the experts, the story seemed clear. Fred Banting's idea led to the discovery of insulin. He had the idea, he should have the credit.

As well, the cultural predisposition of the ordinary North American was to honour one or two heroic individuals as inventors or discoverers. And to be particularly impressed by discoverers who turned out to be just ordinary men - ploughboy geniuses - winning their success after heroic struggles against adversity. Banting's story was perfect: the wounded veteran, the failing small-city doctor, the great idea late at night, nothing but discouragement from the establishment, only a young student helper, grinding poverty, imaginative experiments under the worst conditions - perhaps even having to steal dogs to keep going - and then brilliant, spectacular success. As many writers commented, the discovery of insulin had all the ingredients of a fairy tale or a novel: "A story of bitter struggle, discouragement and scientific greatness, the romance far surpassing the most thrilling fiction tale of the day." Except, they always added, that it was true.[1]

In 1923, Canada, the United States, and then Britain discovered the shy young discoverer of insulin, Frederick Banting. The acclaim began with a standing ovation for Banting when the Federation of American Societies for Experimental Biology met in Toronto at the end of December 1922, carried on through the cheers of six hundred fellow doctors at the New York Academy of Medicine, accolades from Canada's prime minister and the leader of the opposition at a banquet in Ottawa, a triumphant homecoming in Alliston, Ontario, the first life membership ever given by Toronto's Canadian Club (before a thousand cheering businessmen), and luncheon upon luncheon, talk after talk, as everyone wanted a chance to pay tribute to the glory of insulin and its discoverer. In the summer of 1923 the young Canadian went back to the Old World to be honoured, speaking to Britain's and Europe's most distinguished medical men, being received in a special audience by King George V. He came home to Canada and Toronto to open that year's Canadian National Exhibition. There was so much publicity and glory that Banting and his advisers had to consider carefully whether he was violating the profession's rules outlawing self-advertisement.

Banting was shy, inarticulate, and ordinary. He disliked giving interviews or speeches of any kind, and invariably gave them badly. But admiring writers had no trouble finding the diamond in the rough. Was Dr. Banting quiet and shy? Then he must be wonderfully modest and humble. A doer, not a talker, saving his talents for the lab. Silent, like Calvin Coolidge. Was he physically undistinguished, except for a slight stoop? Yes, but surely it was clear that from behind his glasses "looked forth a pair of eyes which even in their most casual glance gave the impression of penetrating beneath the surface of things and reading secrets not revealed to ordinary eyes." Did he come from a very ordinary background of Simcoe County farm stock? Just fine, for "on average soil great characters grow," and wasn't it wonderful how much he loved his mother? Was his school record mediocre? Fine, for all great discoverers - Darwin, Bacon, Pasteur, Lister, and now Banting - were actually patient, determined plodders, not brilliant dreamers. So Banting hated being interviewed, to the point of insulting reporters: call this "refreshing rudeness," and write that you can't help liking him anyway.[2]

Suppose that Banting gives a dreadful speech at an international congress of surgeons in London, a virtually inaudible performance by a completely graceless country doctor, one which looked particularly bad when compared with the fluent prose of the suave European surgeon on the same program. Use it as an excuse for a comparison of European and American upbringings:

> Where a continental student makes the art of love his besetting interest from the age of eighteen, his American companion takes just as naturally to sports. But one acquires a facility of bearing in the presence of ladies, an exaggerated superficial courtesy, accompanied often by profound inward contempt. The other acquires a more robust, vigorous demeanor towards men, an esteem for deeds and a scorning of fine phrases. He may also wear an exterior of awkwardness toward the comparatively unknown sex, but he remembers his mother and makes a good husband. It is an infinitely better breed for a new country which has still large open spaces.[3]

That session was a nice illustration of the irrelevance of sophistication. Banting had been paired with Dr. Serge Voronoff, who was presenting the latest news on his experiments in transplanting portions of monkey testicles into the scrota of old men to restore their youth and sexual prowess. Voronoff's "monkey gland" cure for old age was one of the most publicized, and satirized, of all the medical exaggerations and quackeries of the decade. All the green young Canadian had to talk about was insulin. Now how could a person have discovered insulin and not be a genius, even if only of the awkward, mother-loving, North American breed?

"Banting is greatly in the lime-light here," Macleod mentioned earlier in a letter to Collip, "and seems to bask in it."[4] Similar observations were made by Roy Greenaway, the *Star* reporter who saw so much of Banting, and by others who thought he was getting more than his share of the glory for insulin. In fact, Banting's attitude was not so much a basking in the limelight, but rather the deep ambivalence caused by his believing so completely in his own myth. On the one hand, he was a shy, unsophisticated country boy who hated speeches, banquets, formal dress, and reporters. At times he became thoroughly sick of the attention he was getting. Wanting nothing so much as to be left alone to tinker in a lab somewhere, he would hide from reporters, or just refuse to have anything to do with them. On the other hand, he believed more deeply than even his most ardent admirers in the story of his heroic labour to give birth to insulin. Being as ambitious as any normal person, and more insecure than most, Banting felt he deserved all the recognition he got, especially if it might otherwise go to Macleod. At times Banting's hatred of Macleod and his paranoia about being deprived of credit – which usually included a belief that Macleod was engineering a conspiracy against him – led Banting into active involvement with the group of friends and wellwishers who were trying to advance his interests.

Most of the time Banting was able to resolve his conflicts by adopting the sense of a "duty to the public" that many prominent figures come to live by. After a luncheon in May 1923 he noted on his desk calendar, "I sometimes wish most for a luncheon or dinner at which I did not have to speak, a conversation at which diabetes was never mentioned, a postman without letters to answer, a telephone which did not wring, a bedtime with my work all caught up. But one must develop the phylosophy of 'the greatest good to the greatest number'."[5]

So he went on playing his role as the humble genius. Sometimes he slightly misplayed it. The Canadian prime minister, Mackenzie King, noted in his diary after the Ottawa dinner for Banting and Best, "They sat on either side of me & we had a good talk together, both modest young men – the modesty perhaps overdone, eg I asked Banting where he lived and he said in an attic & Best was quick to say he could not join the Royal Society not having the money to pay for it – but both are good types." Had the very observant prime minister looked down, he would have noticed a further example of Banting's gaucherie in the blue pants he was wearing with his black tuxedo. He had forgotten to bring pants with him to Ottawa and had been unable to rent a pair, finding that they had all been reserved by people coming to the dinner for him.[6]

I

Many of Banting's friends and many other Canadians thought that a humble genius who had given the world insulin should be recognized by something more tangible than applause or life membership in Canadian Clubs. No Canadian scientist had achieved as much as Banting had; none had even come close. Banting had done it all in Canada, too, turning down all the glittering attractions of fame and fortune in the United States. Thousands of highly educated young Canadians were leaving their native land for the United States in these years. How wonderful that Banting had stayed in Canada to give the world insulin- and was still staying in Canada after becoming so famous. And possibly...just possibly...if we don't give him the honours he deserves, he'll take one of those big-paying American jobs we know he's been offered.

As early as the autumn of 1922, some of the Toronto newspapers began to wonder why the University of Toronto was not giving the discoverers of insulin the kind of attention and recognition they were starting to get from the outside world. Later in the year some Toronto doctors began to organize a campaign to obtain the Nobel Prize for Banting; it seems to have fizzled out in the realization that it might be premature, and that attempts to apply pressure to the Nobel trustees could be counter-productive.[7]

Early in 1923 a much more general movement began to have Banting, and perhaps Best, honoured. On February 27, in the House of Commons in Ottawa, a Conservative Member of Parliament from Toronto, T.L. Church, injected into a discussion of financial estimates a plea that the government of Canada give substantial financial aid to distinguished Canadian scientists like Banting and Best. "When you think of the numbers of our brightest professional men who are leaving Canada for the United States and England, I think you will agree that it is time the government of Canada did something to encourage scientific discovery and work of this nature, and announce its policy on the subject," Church said. The minister responsible for the Department of Health avoided the issue by claiming that insulin was still in the experimental stage. While he was willing to look into the subject, he was a little worried that if money were given to Dr. Banting, it would have to be given to many other students and scholars who had achieved valuable scientific results.

Two days later, in the legislature of the province of Ontario in Toronto, the Conservative leader of the opposition, Howard Ferguson, suggested that the province take up the idea of honouring Banting if the Dominion would not. An annual sum should be set aside to support Banting in his research. The premier of Ontario, E.C. Drury, promised to look into the matter.[8]

There was already a movement afoot at the university to provide for Banting's future (he still had nothing more than a temporary appointment relating to his duties at the diabetes clinic, which would soon close down). On February 26 he noted on his desk calendar that he was considering accepting a full-time research appointment in the institution if it were offered him. "Will *not* work under or with or have anything to do with Prof. Macleod," he added. On the day of Ferguson's question in the Ontario legislature, Banting noted it, and added revealingly in his calendar:

> I would like to propose to the house that they should give the University $10,000 per year for a chair of research & leave out the personal element.
>
> Macleod is jealous as h___ and raises the objection to DeFries that Collip his dear diciple in selfishness will be left out if the government take any step in such a manner as they now propose.
>
> Prof. Henderson saw the president today and layed before him the plan of a research appointment for me. The President concurred in the usual way of agreeing in full but not acting in full.
>
> They will likely dilly dally till the opertunity is lost.[9]

The prospect of Banting being honoured caused (and had possibly been created by) a flurry of activity by some of his influential friends. Inside the university Velyien Henderson seems to have been the most active. In the wider political world the most energetic of Banting's friends was Dr. George W. "Billy" Ross, the Toronto physician and former lecturer of Banting's, who is listed in the 1912 edition of *Canadian Men and Women of the Time* as "the inventor of a serum for the cure of tuberculosis, 1909." Ross's invention having not quite worked out since then, he developed an admiration amounting to hero worship for young Freddie Banting, whose anti-diabetic serum was the real thing. Ross also had good political connections and know-how, his father having been a Liberal premier of Ontario. As the prospect of national honours for Banting developed, Ross determined to do all he could to influence his fellow Liberal in Ottawa, Mackenzie King. On March 13 he told Banting of his "plan for money from Dom'n gov't." On March 14, with Banting's consent, Ross wrote to all of the leading American diabetologists and to Charles Evans Hughes, saying that Canada was considering honouring Banting and, in effect, asking for testimonials on behalf of Banting and insulin. Banting himself wrote to Antoinette Hughes, asking her to give Ross's letter to her husband for his earliest possible attention.[10]

Ross, or one of Banting's other friends[11], also approached the vice-chancellor of the university, Sir William Mulock, an eighty-year-old Liberal statesman of immense influence, particularly with his former protegé, Mackenzie King. Mulock did not know Fred Banting personally, but did know the family, apparently having years earlier lent Banting's father the money to buy what became the family farm.[12] Sir William read Banting's account of the discovery (Banting had probably given Billy Ross a copy) and passed it along to Mackenzie King. Howard Ferguson, the Ontario politician, also received a copy. No one seems to have circulated Macleod's, or Best's accounts. Mulock became another ardent admirer of Banting and his native Canadian genius. Sir William and some others had already been talking about raising a trust fund to provide for Banting, in the same way that rich Canadians often raised trust funds to provide for some of their prominent political friends, such as Mackenzie King himself. But they were now urging King to have the government do some or all of the job:

> The good to the world resulting from Dr. Banting's discovery is simply incalculable. It is recognized throughout the world as the product of a Canadian Brain, and it seems to me fitting that Canada as a whole should identify with it by making a substantial gift towards one of the greatest benefactors in all ages, to the human race....
>
> ...from your government's standpoint, it seems to me that Canadian pride, Canadian gratitude and Canadian dignity, would be best satisfied if the whole needs of the situation were met, in the name of the people of Canada, by a grant from the Dominion Government.

Mulock had been Canada's postmaster general when Guglielmo Marconi, the inventor of wireless telegraphy, had been ready to leave England for the United States because of problems with the British government. Mulock intervened to help Marconi then. Banting's account of his difficulties reminded Mulock of Marconi and his troubles. Besides, Banting had a great future in store for him. In another letter to King, Mulock urged that

> Banting in the matter of research has the research instinct. In part is I think almost a genius in matters of research work. Those who know him so describe him, a simple, unaffected, shy and thoughtful man, contemplate his solving the cancer problem.
>
> Is it not worth taking the chance of enabling him to devote himself to the work?[13]

II

Whatever happened to Charley Best? The newspapers sometimes remembered to mention him in the same breath with Banting. Banting often, but not always, remembered to give him a great deal of credit. Many of Banting's friends, and many others at the university, did not rank him as a partner of or co-discoverer with Banting. To them, Best was a student assistant, doing useful work perhaps, but clearly much junior to the man who had had the idea. The campaign by his friends to have Banting honoured almost completely ignored Best, who did not himself have powerful friends or patrons in Toronto.

The situation did not go unnoticed. "Got a rotten letter from Charlie's aunt," Banting wrote in his desk calendar on March 10. Lillian Hallam, with whom Best had lived in Toronto during the summer of 1921, but who had now moved to Saskatchewan, was tired of hearing accounts of the discovery of insulin which mentioned Banting always, Collip sometimes, and Charles hardly at all. She wanted to know the truth. "So now I ask you," she wrote Banting,

> are you responsible for the present preparation as it is being administered to patients? What was Charles' part in the discovery? If he is as responsible as yourself why is your name always put ahead in big letters while his is added "with" instead of "and"? What is at the bottom of this whole thing? Why is just enough credit given to him to keep people from asking Questions? What is it about your part that makes it so much more valuable?

Banting replied a few days later:

> Mr. Charles Best is not "my associate" in the sense that the other workers have been, but it is my sincere wish that he be known as a partner in this work. There is no one, including Charles himself, who feels worse than I do when his name is not mentioned with mine in connection with Insulin. He has worked with me from the very first, and because of his honest efforts and enthusiasm, even before there was such a thing as Insulin, he has become part and parcel with me in working out this problem....
>
> With regard to Dr. Collip, Charles and I both feel that although he did contribute splendidly to the work, the manner in which he made his contribution has lost for him any personal gratitude from us....
>
> The reason that Charlie's name has not been mentioned with mine is possibly due to the fact that it is I who has had to lecture and present papers on the Clinical part of the problem and the newspapers have at times used my name separately.[14]

Late in March, Premier Drury of Ontario came to Banting's office to see his research and discuss with him the province's course of action. Banting told Drury, and repeated in writing a few days later, that his and Best's names should be coupled together. They had worked together from the very beginning, they were close friends, and they still wanted to work together.[15]

Banting's attitude to Best in the 1921-23 period is difficult to determine, partly because all of Banting's later accounts, especially those given verbally to his friends, are coloured by Banting's later coolness towards Best. At all times Banting credited Best with having stood by him when he most needed help – a reference to the incident in the spring of 1922 when Best persuaded him to come back to the lab (in later life Best claimed there were two of these incidents). During the 1920s Banting also freely admitted that he could not have carried out the experiments without Best's help. At no time, however, including the letters just cited, did Banting credit Best with specific ideas or proposals that advanced the research. This explains, I think, why Banting often acquiesced in being singled out for special glory. To Banting, Best was the comrade who had been through the wars with him, had seen it all, and endured it all, and had come to the older man's aid when he was lying wounded. Sometimes, in Banting's mind, this made Best his equal partner; at other times, it seems to have made him his faithful batman.

Neither Banting nor Best gave any thought to Collip, whose annoyance that spring at not receiving credit for his insulin work was being countered only by his excitement over glucokinin. In his academic publications Collip went out of his way to identify himself as a co-discoverer of insulin. In June he read a paper, "The History of the Discovery of Insulin," to the Pacific Northwest Medical Association which, although scholarly and veiled, was his equivalent of the histories of the discovery that Banting, Best, and Macleod had written for Gooderham in September. Privately, Collip was listing his contributions for the benefit of his friends and posterity, and apparently urging Macleod to speak out on his behalf. Collip never revealed the inner history of the discovery period at Toronto, but summarized his relations with those people this way:

> There are some people in Toronto who felt that I had no business to do physiological work. Against this I would say that when I entered the collaborating group early in December 1921 it was with a view of putting my whole effort into the pushing forward of the research irrespective of any water-tight compartments. The result was that when I made a definite discovery my confreres instead of being pleased were quite frankly provoked that I had had the good fortune to conceive the experiment and to carry it out. My own feelings now in the matter are that the whole research with its aftermath has been a disgusting business.[16]

Collip's contribution to the insulin work was well known at the University of Alberta, where he was one of the most illustrious members of the faculty. Albertans decided to honour their province's contribution to the discovery, and in May 1923, held a banquet for Collip in Edmonton and a luncheon in Calgary. At the Calgary affair the president of the city's Canadian Club complained that Banting and Best had been getting such a large share of the glory because of the attention they had been getting in the press of the Eastern cities.[17]

III

The testimonials Billy Ross had solicited on Banting's behalf were given to both Mackenzie King and E.C. Drury. Allen, Joslin, Woodyatt, Williams, and Wilder all enthusiastically supported government recognition of Banting's achievement. Insulin was magnificent, they wrote. Banting (and his associates) had clear priority in its discovery. Banting had great potential as a researcher. Honouring him would be to honour science, encourage research, and honour Canada itself. The clinicians had been asked about only Banting. None of them chose to raise any other name in his testimonial. No other doctors or scientists were asked their views. Charles Evans Hughes in his letter told Mackenzie King something of Elizabeth's story, saying "I cannot adequately express my gratitude for Dr. Banting's work, and I trust that he will receive the recognition which is his due." This was one of the letters King read to his cabinet when they discussed honouring Banting. Another was from Billy Ross, saying that Banting would likely get the Nobel Prize and it was only fitting that he should be first honoured by his own rather than a foreign country.[18]

Early in May 1923, the Ontario government announced that the University of Toronto was establishing the Banting and Best Chair of Medical Research, a special non-teaching professorship to be held by Banting. An annual grant of $10,000 was to pay Banting's salary, support his research costs, and fund Best in his research. A special appropriation of $10,000 was passed to reimburse the discoverers for the discovery period; Banting gave Best $2,500 of the money. The only opposition to Ontario's honours had come from the anti-vivisectionists.

In Ottawa, the Canadian cabinet had approved in principle the idea of some kind of annuity to honour the discovery. (Until just a few years earlier, honours would have come as a matter of course in the form of titles from the monarch, but as a result of public outrage at some of the titles given during the war the Parliament of Canada had decided in 1919 to ask the king not to grant titles to Canadians.) When Mackenzie King wrote Sir Robert Falconer telling him of the decision, he added, "I doubt if it would be possible to go beyond Dr. Banting in this matter to recognize also the services of Mr. Best. Whilst Dr. Banting has chivalrously identified Mr. Best with the credit which has come to himself, I assume that there is no doubt that what is of greatest significance in the discovery is due primarily to Dr. Banting." King asked Falconer for his opinion.[19]

The opinion of Falconer, Sir Edmund Walker, Albert Gooderham, Sir Joseph Flavelle, and the other pillars of the University of Toronto, Mulock excepted, is still unclear. Outsiders to the scientists' struggles, discreet and businesslike, the university's governors probably hoped the principals could straighten out awkward affairs like this themselves. Such evidence as there is suggests that Gooderham, who had commissioned the 1922 accounts, decided that Banting, or Banting and Best (Collip called them B^2), deserved most of the insulin glory. The one public statement of a governor's view was given by T.A. Russell, an automobile manufacturer with a long and intimate involvement serving the university. Russell was a member of the Insulin Committee and had taken great interest in the work:

> I had always known that Dr. Collip had some part in the discovery, but, of course, Dr. Banting, as I understand it, is the man who, in a sense, was the inventor. It was his idea and we looked upon him as being primary in connection with it.... The work was too technical to know anything about the relative parts that each took in it.
>
> I understand that Dr. Banting had the idea and Mr. Best and Dr. Collip contributed materially to the working-out of the idea with the suggestions that they made in regard to it. Of course, Professor Macleod's connection with it was well known as a man who had complete knowledge of physiology and as to what had been done in this field of research, but in my analysis of the part played by each I might be all wrong.
>
> If somebody came to our plant here with an idea that appealed to us we would give him a chance to work it out. We would place engineers at his disposal and would supply him with necessary tools of steel and aid him with suggestions without which he might fall down, but we would regard him as the inventor. The important thing is that it is a wonderful discovery, bringing hope and life to thousands of people.[20]

Sir Robert Falconer, who was considerably better equipped to understand the subtleties of the situation, may not have shared this confusion of discovery/isolation with invention. But that cannot be known, for he was a master of discretion and fence-sitting. Falconer's reply to King's inquiry covered almost all possibilities: "There is of course the case of Mr. Best, and furthermore Dr. Collip of Alberta, who did valuable work on the chemical side in connection with its refinement. Whether you should recognize these gentlemen in addition to some extent is a matter for you to decide."[21]

The Banting admiration society was unrelenting. Ross pelted King with letters, arguing that the magnitude of Banting's discovery warranted more consideration than Ontario's action, telling the prime minister of Banting's rejection of the million-dollar American offer for insulin "at a time when he was on the verge of starvation for want of funds." Ross and his friends arranged to have both the Academy of Medicine, Toronto, and the Canadian Medical Association pass resolutions urging the Dominion government to honour Banting. The Academy of Medicine also declared that Banting and Best had priority in the discovery of insulin.[22]

On June 27, 1923, the Canadian House of Commons unanimously accepted a resolution moved by the prime minister, seconded by the leader of the opposition, to grant Dr. Banting, in recognition of his discovery of insulin, a lifetime annuity "sufficient to permit Dr. Banting to devote his life to medical research." The sufficiency was $7,500 a year, a very large sum in 1923. In the brief parliamentary discussion there were references to Banting's great personal sacrifices, the extent of which would probably always be unknown to the public, and to his selflessness in turning down offers to make large sums of money from his discovery. Members of Parliament believed the Canadian government had a duty to recognize great achievements by distinguished Canadians, but were also pleased that the annuity might make possible more great discoveries by Dr. Banting. None of them knew that of all members of the insulin team Banting was the least likely to make further discoveries. Of these others, only Best was mentioned in Parliament, and this only when T.L. Church asked what was going to be done for him and the other Canadian discoverers. No one bothered to answer.[23]

The reaction to the award was near-unanimous approval from outsiders, such as newspaper editors. Only the anti-vivisectionists had objected. On the inside, Banting's friends were delighted at a job well done. Sir William Mulock wrote Billy Ross, for example, that "with this endorsement of the Dominion Parliament, Banting is not likely to be robbed of the credit for his great discovery."[24] Other insiders were not so happy. From one quarter a powerful letter was sent to Mackenzie King by C.A. Stuart, chancellor of the University of Alberta:

> ...there exists in Alberta, among the medical profession and the public generally, as well as among the scientific men in the University of Alberta, a very strong impression, amounting to a firm conviction, that the work of Dr. Collip of the University of Alberta, who assisted in the research which resulted in the discovery of insulin, is being entirely and quite unfairly ignored by the Toronto people.
>
> ...The recognition by the Federal Parliament of Banting alone only will, I fear, tend to increase the feeling of injustice and dissatisfaction which I know is abroad among the public of Alberta with respect to the entire absence in the East of any recognition of Dr. Collip's share in this discovery.[25]

From Toronto itself a long telegram came to the prime minister from J.G. Fitzgerald, the director of the Connaught Laboratories, to whom a very upset Charles Best had gone on hearing of the news of the annuity. "Banting and Best worked together from the beginning on the research problem which led to the discovery of insulin," Fitzgerald wired (not completely unambiguously). "Best was not an assistant but a collaborator.... The names of Banting and Best are inseparably connected in the original scientific literature.... Banting has energetically supported Best's share in the discovery."[26]

Knowing the telegram was probably too late, Best poured out his disappointment and anger in a letter to Banting, who was en route to England:

> ...It was rather disconcerting to me, after the way my side of the story has been supported, especially by you, to have the Government acknowledge you as the discoverer, with no reference whatever to my help. However, this is an old story now.
>
> I can see plainly the way it happened. Dr. Ross could just as well of had the thing come through Banting and Best. Obviously he wrote to Allen, Joslin etc asking their opinion of your work. Their replies, which you have probably not seen [actually Banting had seen them], were fine. If he had asked their opinion of our work, they could have spoken equally well of your originality in starting the work and of our progress and the whole thing would have come out o.k. You say that Dr. Ross is a friend of mine. I can not see it. If it had not been for you he would never have connected me up in the academy thing. Perhaps, however, the idea was to keep well outside of the range in which Collip figured. That could have been done all right.[27]

Banting's cousin, Fred Hipwell, had taken over most of his private practice. Banting wrote him about receiving Best's letter and of Best being upset at not being remunerated. "I wish to h they had instead of me," Banting wrote. "I scarcely know how to answer him. It worries me." Hipwell had heard from Best, too, but like many of Banting's friends, had little sympathy for the young man. "I am afraid some one of us is going to have to put him in his place soon," Hipwell wrote Banting.[28]

In the meantime the annuitant had written Best saying how sorry he was that it had all worked out this way. It seemed too late to change anything with Ottawa, "only this I can assure you that you will be looked after in some way." As for himself, Banting concluded, "All I want in the world at present is to get down to work quietly and uninterruptedly in a lab. Any person can have any damned thing they like if I can only be left alone. I have some new remote ideas in a new field and am going to give up practice and everything pertaining to Insulin, and am sick of it all."[29]

Mackenzie King responded to the complaints by pointing out that both Collip and Best had been considered, but there was no possibility of Parliament honouring more than one man. "In associating Dr. Banting's name with the discovery of Insulin," King wrote, "the Government [was] only following the general consensus of professional and scientific opinion, of which it was necessarily obliged to take account." Then he concluded, "I am sure Mr. Best and Dr. Collip would be among the first to approve the course which the Government has taken."[30]

IV

Word of the annuity had come to Banting in England, where he was not quite the medical idol he had become in North America. He was a fish out of water in British medical and social circles, and the whole visit, he remembered, was one of the most trying ordeals of his life. A young Australian student, Howard Florey, who saw him at a meeting of the Physiological Society at Oxford remarked on him as "a most poisonous looking fellow."[31]

J.J.R. Macleod, by contrast, was on his home ground that summer, a well-established member of a scientific community most of whose members tended to believe that the discovery of insulin by Macleod and his young associates was the crowning achievement of the Scotsman's years of research in the field. Despite the damage Banting was doing to Macleod's reputation in Toronto, he had won high honours recently, having been elected to the Royal Society that winter and in the spring been awarded the University of Edinburgh's prestigious Cameron Prize, given for distinction for therapeutics.

(Macleod's response to news of the Cameron Prize can be read as containing an element of dissembling: "The work on insulin, as you know, has been the outcome of a joint effort by several of us and I feel a little embarrassed at accepting this prize on that account. However, I suppose the award was made after full consideration of these facts and with full knowledge of them and on that account I will feel that I am justified in accepting it."[32] Macleod was ambitious enough and sure enough of his contribution to the discovery, that he was not going to reject a personal honour for it, no matter how often he wrote about team work.)

Macleod was particularly at home at an international meeting of his fellow physiologists, the Eleventh International Physiology Congress, held in Edinburgh in the last week of July. Invited to give the keynote address on insulin at one of the general sessions, Macleod began by thanking the congress on behalf of "my collaborators, as well as myself." His lecture was an exhaustive account of earlier work on pancreatic extracts, the discovery, insulin's therapeutic and physiological action, the problem of assay, and the preparation and clinical characteristics of the hormone. It was a major *tour de force* of this intricate and exciting new development in endocrinology. Banting and Best's work was given three of Macleod's fifty-eight paragraphs, and it is difficult to read his lecture without a sense that this was about the right balance. Banting, who was in the audience, jotted in his desk calendar that Macleod had been "very fair, but not at all unselfish." Later in the day at a sectional meeting on insulin both Banting and Macleod gave papers on their recent research. Now Banting was less charitable. "Macleod showed lantern slides of 'his' work which was mostly negative results but voluminous. He has a diarrhoe of words and experiments & constipation of ideas and results."[33]

The sessions that day were particularly interesting to a delegation of professors from the Caroline Institute in Stockholm, Sweden. Among its other duties, the institute was responsible for awarding the Nobel Prize in physiology or medicine. Rumours from Edinburgh that the discovery of insulin was being considered for the 1923 Nobel Prize were correct. The chairman, secretary, and other members of the institute's Nobel Committee were there as part of the investigation to determine whether the discovery was worthy of the prize.[34]

Early in 1922 the Caroline Institute's Nobel Committee had sent out its annual requests for nominations of individuals worthy of receiving a prize for the discovery in physiology or medicine which, in that year, had, in the words of Alfred Nobel's will, "conferred the greatest benefit on mankind." Nominations came in during December 1922 and January 1923. After the usual flock of self-nominations by inventors of cancer cures, publicity seekers, and the simply naive (such as the Canadian police magistrate, Emily Murphy, who felt her exposé of opium use on Canada's west coast merited the 1923 prize) were tossed out, the committee found that a total of fifty-seven individuals had been nominated for the award, many of them by several distinguished scientists. The prize could be awarded to more than one but no more than three of these men.

Frederick Banting had been nominated for the discovery of insulin by G.W. Crile, a distinguished professor of surgery in Cleveland, and also by Francis G. Benedict, a leading researcher in problems of metabolism. Mentioning that "probably no one thing in medicine has stirred the physicians in the United States as much as the development of this pancreatic extract," Benedict made a point of expressing his belief that none of Banting's co-workers had contributed anything like an equal share in the researches.

J.J.R. Macleod was also nominated. Professor G.N. Stewart, a Canadian-born friend and former colleague of Macleod's at Western Reserve University, also a formidable figure in American physiology, based his nomination of Macleod on the discovery of insulin by Macleod "and the young collaborators whose work he has directed" as the culmination of his years of investigation into carbohydrate metabolism.

There was also a nomination of Banting and Macleod jointly. It came from August Krogh, the Danish Nobel laureate who had visited Toronto that November, met the people involved, and was now working on insulin himself. Krogh nominated the Torontonians for the discovery of insulin *and* their exploration of its clinical and physiological characteristics. From his own most recent work on the hormone, Krogh could verify that insulin was a discovery of vast theoretical scope and great practical importance, exactly the kind of discovery Nobel had hoped to honour. The one difficult question, he wrote, was how to apportion credit for the work in Toronto:

The publications so far regarding insulin are the results of a collaboration among several authors, but I really think that the prize should go to Macleod and Banting, and the other authors be passed by.

According to the information I personally obtained in Toronto, and as is also contained in the publications, though not so distinctly, the situation is that credit for the idea for the work that led to the discovery unquestionably goes to Dr. Banting. He is a young and apparently very talented man. But he would surely never have been able to carry out the experiments on his own, which from the beginning and at all stages were directed by Professor Macleod. The other authors should be considered as Macleod's and Banting's collaborators, but there is reason for specially mentioning the chemist J.B. Collip. He has made a very important contribution in the method of producing insulin in a major practical way, beginning with the adult animal pancreas. But I do not think that is sufficient ground for the award of a prize.

Macleod's special contribution in the experimental work has been only partly published at this time (Jan. 31, 1923). It is about locating insulin in the pancreases of several species of fish (and thereby proving the character of the hormone), and then exploring insulin's action on the total system and the respiratory quotient, and ongoing exploration of its action on carbohydrate metabolism – these explorations all show clearly the action of the hormone, though as yet there is no explanation of its action.

In April the committee reduced the horde of nominees to a short list of nine, counting Macleod and Banting as one, whose work would receive special appraisal or investigation. Perhaps because there were two nominees for insulin, it was decided to have two appraisals of the discovery, one by John Sjöquist emphasizing its physiological importance, another by J.C. Jacobaeus on its practical application. In addition, the committee's secretary, Goran Liljestrand, wrote a special report on the insulin sessions of the congress at Edinburgh.

Nobel appraisals are detailed, expert studies of the work of the nominees. Sjöquist and Jacobaeus read very widely in the publications on insulin. While the former attended the Edinburgh congress, the latter went to Copenhagen to see the results of clinical tests there and to meet other European specialists who were using insulin. Both investigators were particularly concerned to find out whether enough was known about insulin, both experimentally and clinically, to justify the very major claims being made about it. Past experience had taught Nobel nominators and examiners that it was almost impossible to assess the results of medical discoveries so quickly. One year was almost always too soon to tell. Often it took ten or twenty years or more for the true importance of a fundamental discovery to be realized.

In their several-thousand-word reports, the examiners described the work with considerable thoroughness. Sjöquist discussed the work of several predecessors, especially Zuelzer, but seemed particularly impressed by the investigation of the respiratory quotient, glycogen formation, and other experiments, including some of Banting's recent work on insulin in the blood (which Banting had discussed in Edinburgh). Sjöquist seemed to see these follow-up studies as vital proof that the discovery was soundly buttressed in physiological investigation. Jacobaeus reported on the most recent conclusions by clinicians, and cited several European and American experts on insulin's value. Both examiners concluded that the discovery of insulin was of fundamental importance, worthy of a Nobel Prize. The Liljestrand report on Edinburgh was very factual, not arriving at any conclusions or recommendations.

Who should get the prize? Sjöquist accepted the suggestion of dividing it between Banting and Macleod:

> Banting had the distinction of having had the idea and the initiative. If you look at the publications and comments by observers, Macleod has been the leader of the scientific work, which has been done in his laboratory, and it is beyond doubt that without his major contribution this discovery would not have had the importance it now has. I should also say that it was not a coincidence that Banting went with his idea to Macleod, who had earlier made many very important studies in carbohydrate metabolism.

Jacobaeus was more puzzled by the difficulty of assessing Macleod's role, but reached the same conclusion:

> Dr. Banting, who undeniably first had the idea and did the exploratory work, has the first claim to the prize. On the other hand it is difficult to judge Macleod's contribution. It is not clear from an examination of the literature. Macleod, who is the head of the physiological institution in Toronto, has worked before with blood sugar studies. Banting came with his idea to Macleod and completed his work on insulin under Macleod's guidance. It has been said to me that it is very possible that the discovery would not have been made or at least not made as quickly, were it not for Macleod's guidance. It is even said that Banting was about to make an experiment which would not have led them to the goal, until he was corrected by Macleod.
>
> The question is, therefore, whether Banting alone should be awarded a prize, or if it should be given to Banting and Macleod. I conclude that Banting and Macleod should share the Nobel Prize.

Others on the short list had been deemed worthy of a Nobel Prize. Fortunately the committee had two prizes to dispose of, having postponed any decision the year before. On September 22, 1923, the Nobel Committee decided to recommend the award of the 1922 prize to A.V. Hill and Otto Meyerhof for their work on muscular action, and the 1923 prize to Banting and Macleod.[35]

The recommendations had to go to the Nobel Assembly, which at that time consisted of all faculty members of the Caroline Institute, for final approval. There was no problem with the Hill-Meyerhof recommendation. But at its October 11 meeting, the Assembly decided that there were difficulties with the Banting-Macleod recommendation. Having been challenged, it was sent back to the committee for reconsideration.

Professor Alfred Pettersson had objected most strenuously to a Banting-Macleod award. He formally explained his objections in a letter to the committee. "It is quite clear to me that a fundamental requirement in awarding a person a Nobel Prize is knowledge of what part the person has actually taken in the work being honoured," Pettersson wrote. He quoted the references to Macleod in the two appraisals, and went on,

> During the time I have participated in the awarding of the Nobel Prize, the justification for the award has never been based on hear-say evidence from unknown persons, on statements like "it is beyond doubt", on things that are thought of as "very possible". In my opinion, it is very necessary that the Assembly adhere only to verifiable facts. Otherwise the Assembly risks the development of unpleasant discoveries at a later date. I also point out a certain contradiction in Professor Jacobaeus's final judgment about Macleod's part in the work relating to insulin production. Banting is said to have been ready to make an experiment that would not have led to the goal, and to have been corrected by Macleod. But before that, Jacobaeus writes that Banting came with his idea to Macleod and worked through to insulin under Macleod's direction. If the work was totally under Macleod's direction, then Banting could hardly be made responsible, at least not alone, if they, in the beginning, started out on the wrong road.

The committee met again, reconsidered, and reaffirmed its recommendation. In a formal letter to the Assembly it named the provider of "hear-say" evidence as August Krogh, and emphasized that he had originally made the joint recommendation based on his visit to Toronto. Pettersson was wrong to interpret the difficulties in apportioning credit mentioned by the examiners as an indication of any hesitancy or doubt on the committee's part, its members wrote. They described Banting's coming to Toronto with his idea, and quoted the explicit statement in Banting and

Best's first paper that the work was done under Macleod's direction. The short published statement about the work (the abstract of the New Haven paper), they added, carried Macleod's name as an author. The committee went on:

> At the international physiological Congress in Edinburgh in July this year, where Banting was present, it was Macleod who in his formal lecture at the congress's opening summarized the situation regarding insulin. He started his lecture with these words, "Speaking for my collaborators as well as myself..."
>
> Krogh, who personally visited Toronto and there for a time followed the work, discusses the prize-award very thoroughly and concludes that Macleod's part in the work merits the prize.
>
> From studying the relevant literature, Sjöquist, as written in his investigation, has reached the firm opinion that the idea was Banting's alone, to be sure, but that it was Macleod's guiding hand that helped Banting's idea reach such a happy culmination in the beautiful result which we now see. It is beyond doubt, according to Sjöquist, that the award should go to Banting and Macleod together.
>
> The undersigned were at the physiological congress in Edinburgh and the Committee's chairman and secretary had the opportunity of attending Macleod's formal lecture and also two short papers of Banting's and some discussion about the discovery of insulin. The information we received there confirms what has just been said.
>
> ...it is not possible to make a more thorough investigation of this discovery and the relative parts of Banting and Macleod, nor is it necessary.

The Nobel Assembly had a special meeting to discuss the recommendation on October 18. No record of that discussion survives. But the force of Pettersson's objection to Macleod seems to have been blunted by the fact that he was advancing it largely to further a somewhat quixotic crusade to have the prize awarded to R. Pfeiffer, a German bacteriologist who thought he had discovered the cause of influenza. Pfeiffer's "discovery" had been made back in the 1890s, was of current interest because of the flu epidemics of 1918–20, and was later shown to be erroneous. Nobody but Pettersson seems to have supported him for the prize. On October 25 the nineteen assembled professors of the Caroline Institute voted by secret ballot to award the 1923 prize to Banting and Macleod.

V

Fred Banting drove down to Toronto from Alliston early in the morning of Friday, October 26, after spending a day with his parents. He got to the city about nine o'clock and went straight to his office. Hearing his telephone ringing, he went inside, tucking his morning paper under his arm. An excited friend was on the phone: "Congratulations...where have you been...trying to get you...have you seen the newspapers?"

"Calm down and tell me what you're talking about."

"You damned fool, didn't you know you and Macleod got the Nobel Prize?"

"Go to hell."

Banting hung up and opened his paper. There it was - the Nobel Prize, and Macleod! Macleod! Macleod!

> I rushed out and drove as fast as possible to the laboratory. I was going to tell Macleod what I thought of him. When I arrived at the building Fitzgerald was on the steps. He came to meet me and knowing I was furious he took me by the arm. I told him that I would not accept the Prize; that I was going to cable Stockholm that not only would I not accept but that they and the old foggy Krogh could go to hell. I defied Fitzgerald to name one idea in the whole research from beginning to end that had originated in Macleod's brain - or to name one experiment that he had done with his own hands. Fitzgerald had no chance to talk...

Nobody had ever seen Banting quite so angry. "He was furious," an eighty-year-old lady who was there that day told me, her voice rising to imitate him. "Oh, he was furious," she repeated, clenching her fists. "He could have torn the whole building down...Oh, he was helling and damning..." When Fitzgerald was finally able to interrupt Banting's tirade he told him that Colonel Gooderham was waiting for him in his university office. Banting went in to see Gooderham. "The weight of his presence cooled me down. He was one man whose calm and strong personality always reminded me of my father."

Gooderham congratulated Banting, told him to get to Stockholm on the first boat so he could get the prize in person, and offered to pay all his expenses. Banting replied that he was going to turn down the Prize. Gooderham, according to Banting,

> was one of the few men who knew the whole story and he said words to the effect that he understood my feelings and that he agreed with me but that there were other considerations that must be taken into account. I must think first of my country - what would the people of Canada think if the first Canadian to receive this honour were to turn it down? Then there was science to consider - what would the world think of scientists who would because of differences of opinion disagree about a Prize. I had not thought of this aspect of the situation. He did not ask me to decide immediately but asked me not to

do anything rash & "better wait 24 hours."

Banting did not need the twenty-four hours. On the spot he decided to share the cash and the credit too.[36]

Best was in Boston that day to address the Harvard medical students. He had not heard of the prize. After his talk, Elliott Joslin got up and read a telegram just received from Banting: "At any meeting or dinner please read following stop I ascribe to Best equal share in the discovery stop hurt that he is not so acknowledged by Nobel trustees stop will share with him."[37]

J.J.R. Macleod heard about the prize on his way back to Canada from Britain. When he landed in Montreal on November 2, he was met by a *Star* reporter anxious for his side of the Nobel story. Macleod had heard about Banting's decision to share with Best. He was going to take a few days to think about how he would dispose of his share, he told the reporter. "You may be sure, however, that my decision will be in no way influenced by the action of others."

If Macleod was being quoted correctly, the reporter had caught him in one of his iciest moods. "It was very handsome of Dr. Banting to divide that amount of money. It is very handsome indeed. A fine thing to do. But Dr. Banting is a very wealthy man now." The reporter also asked Macleod if he had seen a statement Banting had just made in London, Ontario, that he had been given a time limit of six weeks to make the experiments for his discovery. "I have no doubt that every statement Dr. Banting makes is accurate in every particular," Macleod answered.[38]

Either that day or within the next two or three, Macleod telegraphed Collip from Montreal asking the chemist to share his half of the prize money. Collip accepted. On November 7 Macleod gave a statement to the press:

> It would be invidious and quite unnecessary to try to dissect or divide up the work on insulin among the various men who were engaged in it.
>
> The University of Toronto has been given a great deal of credit for this discovery and it would like to emphasize that it is team work that did it. We found that we were engaged on a work that appeared to have in it great benefit to mankind and our aim was to hurry it along as fast as we could to completion. Other work was dropped while this was proceeded with. It was on this basis of understanding that Dr. Collip, who was on leave of absence from Alberta university, came into the work with us.
>
> Dr. Collip made a very important contribution to the work and his share was equal to that of the others.

When the reporters pointed out to Macleod that he had said nothing about his own share in the work, he laughed and said he was only "the impresario - the managing director."[39]

The *Star's* journalists seem to have smelled another story underneath the Nobel Prize story, but were unable to ferret it out, perhaps because the reporter now assigned to it, a future Nobel laureate himself named Ernest Hemingway, was both overworked and unhappy in Toronto. Earlier, a seventeen-year-old University of Toronto student, Charles Stacey, who was a cub reporter for the *Varsity*, almost got the grand story when Best told him that Banting was thinking of making his anger public and would give him, Stacey, the statement. Stacey was dazzled at the prospect of a world scoop, but when told that Banting had changed his mind accepted the decision and probed no further.

Banting refused to comment on Collip's part in the work. Privately there may have been yet another angry confrontation between Banting and Macleod. Macleod had said that Collip's share was "equal." A day later the last public word from the new Nobel laureates was a correction by Macleod:

> The statement that Dr. Collip was entitled to an equal share of credit for his part in the work was not quite properly phrased. It might be more accurate to say that he is entitled to a fair share of the credit. I would be glad to correct any misapprehension. If I used the word "equal" I should not have done so.[40]

Privately, J.J.R. Macleod was unrepentant. A few months later he wrote a friend in Scotland that by dividing his share with Collip, "I think I have succeeded in getting people here to realize that his contribution to the work as a whole was not incommensurate with that of Banting. It is of course sad that it should require such drastic methods to persuade people of this fact but it could not be helped, it was the only thing to do under the circumstances."[41]

Both Banting and Macleod received many letters of congratulation. One of Macleod's first was from August Krogh, who was "greatly pleased that they have not in Stockholm taken a formal point of view but recognized explicitly your great share in the great discovery... we feel proud indeed in counting you among our personal friends."[42] Rawle Geyelin, on the other hand, wrote Banting saying how disgusted he was with the award to Macleod and the nerve Macleod had shown in accepting it. Geyelin had drafted a letter to the press attacking the Nobel trustees as either ignorant or woefully misinformed. Did Banting think he should publish it?

Banting had finally calmed down. On November 10 he wrote Geyelin that although he agreed with his views, the letter should not be sent. "While I feel that the whole thing has been a great injustice to Best, and whereas I cannot understand Professor Macleod in this matter, I would beg of you not to publish this letter because the University of Toronto and Science in general would be discredited for their rangling. At the present time the outburst of indignation is subsiding, and any additional controversy would do only harm, since nothing can actually be done about the award."[43]

Banting's acceptance of the situation was realistic, for he knew that the Nobel award was immutable. Stockholm never explained, changed, or apologized. Consequently the Nobel Committee heard nothing about the outrage their award had caused among Banting and his friends. There is no comment in the Nobel records about the division of the cash (which turned out to be about $24,000). This was the prize winners' business. The 1923 prize had gone to Banting and Macleod. There is no foundation for stories that Banting had somehow ordered Sweden to put his name before Macleod's.

The Nobel Committee did receive furious letters of protest from other quarters. From Berlin, Georg Zuelzer, knowing he would never have success or fame, made a pathetic plea for some recognition of his priority. From Bucharest, Nicolas Paulesco, outraged at what he believed was Toronto's theft of his work, demanded justice from the Nobel Committee. The protests were ignored.[44]

VI

The University of Toronto recognized its Nobel laureates with a special convocation on November 26, at which Banting and Macleod were each awarded the honorary degree of Doctor of Science. The ceremony was followed by a glittering banquet for four hundred people in the Great Hall of Hart House. The after-dinner speeches were glowing tributes to insulin, Banting, Macleod, sometimes Best, and to the university whose facilities had made it all possible. There were pleas to the public and to governments for more money to support research, and to Canadians and the university to be worthy of their inheritance. It was an evening of self-congratulation and harmony, punctuated only by the class yells of the members of Meds 1T7 and the music of the jazz orchestra hired for the occasion.

Some of the insiders must have smiled to themselves as Banting gave his most generous thanks to Best and Velyien Henderson, while Macleod spoke as the representative of all the co-workers in the army that had conquered diabetes. Lewellys Barker, who almost exactly two years earlier had been the first person outside of Toronto to hear of the work, summarized what most reasonable men felt when he said, "There is in insulin glory enough for all."[45]

CHAPTER TEN

A Continuing Epilogue

J.J.R. Macleod left the University of Toronto in 1928 to go home as Regius Professor at the University of Aberdeen. People who knew him give conflicting accounts of his reasons for returning to Scotland. In many ways it was a natural move, but there are also stories about how unpleasant Macleod's life was made in Toronto as a result of Banting's unrelenting hatred. In one version, Macleod is said to have felt he had to leave or take legal action against Banting. When the university held a farewell dinner for Macleod, Banting not only refused to attend, but is said to have requested that there be an empty place set for him at the table. A friend of the Macleods who went to the train with them the day they left Toronto noticed that the professor was shuffling his feet. Asked why he was doing that, J.J.R. Macleod said, "I'm wiping away the dirt of this city."[1]

Macleod was received in Aberdeen as a great physiologist, the man who had, with his associates, discovered insulin. While it was known that there had been "troubles" in Toronto over insulin, there was no first-hand local knowledge of the details, and the professor himself politely turned aside inquiries from curious students and colleagues. He built a beautiful home, its walls lined with the Canadian paintings the Macleods had collected. But he never talked about the Toronto days. Most of Macleod's Aberdeen days were spent in nagging pain from a severe arthritic condition. He died in 1935 at age fifty-nine.

Macleod's replacement as Professor of Physiology at Toronto was twenty-nine-year old Charles Best. Best had sped through the medical course at Toronto, graduating near the top of his class in 1925. He had married Margaret Mahon in 1924 and the beautiful couple sailed to England, where H.H. Dale was waiting for the young Canadian to do postgraduate work with him at the National Institute for Medical Research. When Dale had first visited Toronto in September 1922, he had been favourably impressed by Best's work on insulin production and his potential as a scientist. Dale's advice to Best had been to get out of the limelight and adulation in Toronto and get the thorough training he needed to supplement the insulin adventure. When Toronto was canvassing for a successor to Macleod, Dale and others, including Macleod, recommended Best as one of the most promising physiologists anywhere.

Fred Banting was able to get back to his lab in 1923, but he never really got out of the limelight. He was barely at work that fall when the Ontario Minister of Health created an enormous newspaper fuss by announcing that Banting was on the verge of discovering something greater than insulin. Banting never got the advanced training or the wise advice that might have guided him into fruitful scientific work after insulin. He should have spent the rest of his life practising medicine. Instead, he was a well-to-do research professor (his annuity and professorial salary totalling $13,500) with no teaching obligations and all the research funding he needed. As students clustered around him, the Banting and Best Chair grew into the Banting and Best Department of Medical Research, a unit separate from the rest of the university. It was Banting's own little kingdom, a happy land peopled by colourful, often hard-drinking, students and cronies of Banting's, some of whom were adept scientific tinkerers. All the unpleasant administrative work was handled by the faithful Miss Sadie Gairns, who had become Banting's first research assistant in 1922. In 1930 the university named a new medical building the Banting Institute. While the Banting and Best Department was housed on one floor, the "Institute" had no real existence at all, a situation causing no end of confusion both inside and outside the university.

Most of Banting's research after insulin was directed at finding a cure for cancer, mainly by way of experiments trying to produce resistance to Rous's sarcoma (a virally induced tumour). Banting tried very hard to duplicate what he thought was the insulin experience: *viz,* the having of a great idea, thinking up the ingenious approach that would solve everything. In his many talks on medical research he always emphasized the ideas, not the training, that researchers brought to their work. Except for some interesting and now controversial work on silicosis, Banting's ideas did not pan out. Throughout his life, of course, the press and the public continued to hope that Banting would repeat the triumph of insulin. H.H. Dale, later Sir Henry Dale, told the story of a reporter who called to confirm a story about a great discovery by Banting in Canada: "Is it possible that Dr. Banting has found a cure for metabolism?" the reporter asked Dale.

Fred's love for Edith Roach suffered several more tempestuous episodes. According to those who liked Fred but not Edith, she was all too willing to marry him now that he was the discoverer of insulin. Those who knew Edith well remember that she demurred all the more strongly, saying that everyone would think she was only marrying him for his fame. In 1924 Canada's most eligible bachelor was suddenly swept off his feet and married by a doctor's daughter from Elora, Ontario, the attractive and very sociable Marion Robertson. Edith married many years later. The Banting marriage was a disaster, ending in 1932 in a sensational, some said scandalous, divorce. This aspect of his life, his attempts to cope with his notoriety, his growing interest in painting and the arts, his travels, and his professional adventures, are dealt with in my forthcoming biography of Banting. Fred Banting was more interesting and more successful as a man than as a scientist. He was the kind of person who fulfills Joan Didion's definition of a literary character – "ambiguous and driven and revealing of his time and place."[2]

Banting matured and mellowed in the 1930s. He grew out of many of the hatreds of the insulin days, becoming close friends with Bert Collip and coming to revere Duncan Graham as a father-substitute. His hatred for Macleod did not diminish. To it was added a dislike of Best. It grew out of the friction naturally resulting from the two of them having to share power and influence in the hothouse worlds of the University of Toronto and Canadian medical research. Banting disliked Best's ambition; Best could not respect Banting as a scientist.

When a Conservative government of Canada briefly resumed accepting titles for Canadians in 1934, Banting was honoured with a knighthood, becoming Sir Frederick Banting, K.B.E. It was the most incongruous of all his honours, and he knew it. "Next person who calls me 'Sir' will get his ass kicked," he once said at a meeting of medical men in England. Fred Banting was nothing if not one of the boys.

When war resumed in 1939, Banting had just married again, to a technician in his department, Henrietta Ball. He tried to enlist as an ordinary medical officer, but was pressured to serve as co-ordinating chairman of Canada's wartime medical research effort, and appears to have done this job well. While in London in the winter of 1939-40, consulting with his British counterparts, he filled in weekend hours writing the long account of the discovery of insulin referred to so often in these pages. It contains some excellent passages, but Banting's history is rambling, unpolished, and was never checked for accuracy – more a documentary source than a history in its own right.

Banting returned to Canada in the spring of 1940. On February 20, 1941, he took off from Gander, Newfoundland, aboard a Hudson bomber, en route to England for a second time. The plane crashed in Newfoundland. Banting died in the wreck. There was much speculation about the reason for his embarking on such a hazardous mission, and on the cause of the crash. Was Banting carrying vital military intelligence? Was his plane sabotaged by Nazi agents? An accurate account of Banting's last mission will be found in the biography.

Best and Collip were the two members of the insulin team who went on to long careers as productive researchers. Best, his associates, and his students at Toronto, continued Macleod's work on the properties of insulin, did basic studies of the dietary factor choline, and developed the important anti-coagulant, heparin. After Banting died, Best inherited his mantle as the most prominent living discoverer of insulin. He also took charge of the Banting and Best Department, which was eventually integrated into the faculty of medicine. In 1953 the university erected the Best Institute next door to the Banting Institute, doubling the confusion caused by naming buildings institutes. In the later years of his life Best was showered with honours by grateful diabetic associations, medical bodies, and universities. The Nobel Prize, which he felt he ought to have shared in 1923, eluded him. Bouts of serious depression appear to have accentuated an obsessive concern for credit and glory. Charles Best retired in 1967. He died in 1978 of an illness brought on by hearing the news of the death of his oldest son.

J.B. Collip recovered from the dead end of glucokinin to do intensive pioneering work on the isolation of the parathyroid hormone, in his spare time adding a medical degree to his list of letters. In 1927 he almost went to the Mayo Clinic in the United States, but changed his mind – turning down what is said to have been a "staggering" salary[4] – and the next year accepted the chair of biochemistry at McGill University. Collip's McGill years were a whirlwind of endocrinological research as he and his students were in the forefront of the isolation and study of the ovarian and gonadotrophic and adenocorticotrophic hormones. Just as Banting always hoped he would repeat his great Idea, Collip always hoped he could isolate another magic hormone, and there was a long succession of products produced in co-operation with drug companies – parathormone, then Emmenin, Premarin, and brands of ACTH. None of them came close to insulin, of course, but in trying so hard and so variously (with a fairly high degree of failure and error, as well, as in the glucokinin affair), and by concentrating his restless energy in one field, Collip made himself by far the dominant figure in the history of endocrinology in Canada. Endocrinological research generally had received an enormous stimulus everywhere by the discovery of insulin. The next spectacular therapeutic advance in the 1920s after insulin, the use of liver extracts for pernicious anemia, came out of the heightened interest in organ therapy caused by insulin and was also insulin-dependent in the sense that one of the discoverers, George Minot of Harvard, was a severe diabetic.

Fred Banting spent his last night in Montreal with Collip before going on to Gander. Collip was deeply shaken by the news of his death. In an obituary tribute to Banting a few months later he made his last statement about his own role in the discovery of insulin, writing that his contribution to the team was "only that which any well-trained biochemist could be expected to contribute, and was indeed very trivial by comparison with Banting's contribution." Banting in those last years had told friends that he and Best wouldn't have achieved a damned thing without Collip.[5]

In 1947 Collip became dean of medicine at the University of Western Ontario, by then a considerably better medical school than in the days of Banting's demonstratorship a quarter-century earlier. Collip served as dean for fourteen years and then continued as head of the Department of Medical Research at Western. He died in 1965 at age seventy-two, just after finishing another of his marathon drives across the North American continent. In later years Collip was very reluctant to talk or write about the discovery of insulin, saying that the truth was to be found in the scientific publications and might emerge after they were all dead.

Others were not so reticent about claiming credit for the discovery. Georg Zuelzer, for example, often referred to himself in public as the discoverer of insulin. A refugee from Nazism, he emigrated to the United States in 1934 and spent the last years of his life practising medicine there. According to Best, Zuelzer once came to Toronto and insisted on his priority to those who would listen to him. Years earlier in Germany, Minkowski had publicly disposed of his claims by saying, after listening to Zuelzer, "I too share with Doctor Zuelzer the regret that I did *not* discover insulin."[6] Zuelzer died in a home for the aged in New York in 1952.

Late in 1923 E.L. Scott published a brief claim to priority in the discovery of a method of extracting the active principle of the pancreas ("the discovery of the curative power of 'insulin' has been open from January, 1912, to any one who cared to repeat and extend my work"). At the same time, however, he wrote Banting congratulating him and Macleod as the "logical recipients" of the Nobel Prize, and stating that they had always given him, Scott, "all the credit that was coming to me." Scott had made his claim, he wrote, because "I was after another man who I have very good reason to believe is perfectly willing to accept credit due not only to me but others including yourself."[7]

That man, Murlin, had applied for a patent on his anti-diabetic pancreatic substance in July 1923. When it was finally granted in 1925, the patent protected as "discovery" of no use to anyone. Murlin, who never did understand how Toronto had beaten him to insulin, continued to work furiously to discover something better than insulin, and did have at least the satisfaction of discovering the second (but far less significant) islet-cell hormone, glucagon. In the late 1950s Murlin and Kramer joined the discussion Pratt had sparked about the insulin research by publishing an account of their work, claiming a place in the history books.[8] To the end of his life, I was told by acquaintances, Murlin believed that the people in Toronto had somehow stolen his work. E.L. Scott's widow, and the Romanian friends of Paulesco, also claimed that credit had not been given to those who had really done the work. Paulesco himself gravitated to proto-Nazi, anti-Semitic politics before his death in 1931. Israel Kleiner, who in 1919 was closer to success than any of them, made no claims at all.

The outsider whose life's work and reputation was most affected by the discovery was Frederick M. Allen. Before insulin, Allen was *the man* to reckon with in diabetes, his work and methods dominating his field to an extent seldom equalled in therapeutics. After insulin, Allen was just another diabetologist, the proprietor of a high-cost institute many of whose patients no longer needed its services. Allen was also a man with few friends, many enemies, and no university or foundation support.

He turned much of his attention to hypertension, becoming one of the first to prescribe low-salt diets for high blood pressure. But he also kept on with diabetes research, and in 1927 announced the discovery of a new treatment for diabetes. It was a preparation made from mulberry or blueberry leaves, a refinement of an old Austrian folk remedy which had been brought to Morristown by a visiting Austrian scientist, one Dr. Wagner. It had the advantage over insulin of being able to be given orally. Allen named his preparation "Myrtillin," applied for a patent on it, an d entered into an agreement with E.R. Squibb & Son to develop it.

Myrtillin went the way of glucokinin and other vegetable-derived hypoglycemic agents discovered in the 1920s. The action of the chocolate-coated anti-diabetic pills was too slow, erratic, and toxic to be suitable for humans. The Myrtillin failure left Allen deeply in debt to Squibb's, who in 1930 brought suit to recover the money they had loaned him.[9] They were among a host of Dr. Allen's creditors that depression year. He waged a heroic struggle to keep his half-empty Physiatric Institute going, only to be finally evicted in 1936.

Frederick Allen drifted from one little-known hospital to another, carrying on his animal experiments in basements, outbuildings, once in rented space in a public stable. A fanatical researcher to the end, he had gone on from hypertension to studies of the use of refrigeration in surgery and then to cancer research, never having adequate funding, becoming increasingly isolated, bitter and paranoic about his treatment at the hands of the medical Establishment. Allen's life, like Captain Ahab's, was a tragedy of American individualism. His pathetic unpublished autobiography ends as he is meditating on how fortunate another outsider, Frederick Banting, was, to have his work accepted and his later research so well supported. When he had first ment Banting in the summer of 1922 and heard his story, Allen had offered him a job at the Physiatric Institute. Frederick Allen was pursuing his research in the basement of the Pondville State Hospital in Massachusetts when he died in 1964 in his eighty-eighth year. The last word as his autobiography breaks off is "independent."[10]

I

The Insulin Committee of the University of Toronto continued to administer the basic Banting, Best, and Collip patent, as well as important later patents (particularly that assigned to the university by Albert Fisher and D.A. Scott for protamine-zinc insulin) until they all expired in the 1960s. One of the important unresolved issues during the discovery period had been whether the committee would stand by its original decision to collect royalties from licensed manufacturers. In Britain and Europe, where even the decision to patent was thought highly questionable, there was intense resistance to the idea of paying royalties. Among the group at Toronto both Banting and Macleod, particularly the latter, opposed the royalties; at a proposed rate of 5 per cent of the retail selling price of insulin the University of Toronto's share was a good return for any profit-minded inventor. The businessmen on the Insulin Committee seem to have argued that some royalty was obviously necessary to pay the committee's costs, especially if it had to go into court to defend its patent rights, and that any surplus could properly and ethically be devoted to research in the university. While the committee finally stopped trying to collect royalties outside of North America, it always received substantial royalties from insulin sales in the United States and Canada.

Between 1923 and 1967 the University of Toronto's royalties from insulin totalled $8 million.[11] The Insulin Committee's costs were not very high – it never had to go to court to protect the patents – so a considerable surplus was available to support research. According to an agreement worked out in 1923, half of this surplus went to research directed by Banting, Best, and Collip (Collip's share being paid to the university that employed him), the other half to the University of Toronto's general research funds.[12] The sums involved are trivial by today's standards, but were very substantial in the early years. Royalties averaging over $180,000 annually in the 1930s went a long way in those depression years – and the figures do not include other significant grants for research projects that came from the Lilly Company, which maintained close ties with Toronto into the 1960s. The one technically improper use of insulin profits at the university was the spending of several thousand carefully laundered insulin dollars in 1941 to help Oskar Minkowski's widow flee from Germany to Argentina. This was one of Charles Best's efforts to help European scientists and their families in the 1940s.

Eli Lilly and Company sold more than a million dollars' worth of insulin in its first year of marketing, and never looked back. Insulin did more than any other single product to transform the company into a giant in the American pharmaceutical industry. The relationship with Toronto which, on balance, had worked well in the interests of everyone involved, gave the company a dominance in American insulin production which it easily maintained into the twenty-first century and the age of insulin manufacture by genetic engineering. The introduction of insulin, using extensive clinical testing and physician education in an age before government regulation, had been a great credit to both the company and the university. As well, insulin had established the scientific credentials of the company and given it a reputation as a pioneer in collaborative work with university researchers, many of whom, including the American discoverers of the treatment for pernicious anemia and Canadians experimenting with vincristine derivatives to treat leukemia, came to Lilly with their ideas.

Canada's insulin was supplied by the Connaught Laboratories, wholly owned by the University of Toronto, and was the staple of the company's development into a major pharmaceutical house in its own right, the largest Canadian-owned drug company. In 1972 the University sold Connaught Laboratories, putting the proceeds into a special fund to support research.[13] Connaught's insulin division was eventually sold to Novo-Nordisk, the Danish-based insulin company that had evolved out from August Krogh's visit to Toronto in 1922 and was now competing around the world with Eli Lilly.

Talk of honouring the insulin discovery by raising a private endowment to fund research became a reality in Toronto in 1925 when the university's prominent governors, led by Sir William Mulock, launched the Banting Research Foundation. In a whirlwind fund-raising campaign, characterized by the raising of extravagant expectations about the possibility of curing cancer and other dread diseases, the foundation's capital goal of $500,000 was easily met.[14]

There was considerable hope in the early years that the Banting-to-insulin progress could be repeated. Several practising physicians brought their bright ideas to Toronto, where, thanks to the Banting Research Foundation, and Banting himself, they received more generous support than Professor Macleod had ever given Fred Banting. But Fred's classmate, Beaumont Cornell, did not solve the riddle of pernicious anemia and the liver extract proposed by Dr. MacDonald of St. Catharines to reduce high blood pressure also failed to work, dashing everyone's early high hopes. That matter also created some embarrassment and nearly ended up in the courts when a professor from the University of Western Ontario accused MacDonald of stealing his ideas. Medical history, too, has a tendency to repeat itself as farce.

Through the years many other more qualified researchers received valuable aid financial aid from the Banting Research Foundation. It lives on as a small, but historic and proudly independent granting agency. Organized support for scientific research in Canada, from both public and private sources (mostly the big American foundations) was still so primitive before World War II that the money generated by insulin was probably the largest pool of Canadian capital supporting medical research.

Insulin had given such prominence to the University of Toronto, particularly in the field of diabetes work, and created so much support for more research in Toronto, that it was later wondered why the city did not become a world centre of diabetes research and treatment. There were several reasons. Macleod's departure crippled the university's insulin research capacity. The one trained clinician on hand when insulin was discovered, Walter Campbell, ironically nicknamed "Dynamite," was slow-talking, slow-moving, fundamentally unenterprising. The clinician who took over the limelight in Toronto during the discovery period, Banting, was not a good diabetologist, and in any case decided his future lay in developing great ideas to cure other diseases. Generally, discoverers and clinicians alike shared the view that insulin had licked diabetes. When insulin became widely available, the special diabetes clinic at Toronto General Hospital closed down. You gave insulin to the world and went on to some other great thing. It happened that the next product given to the world by Toronto medical researchers was pablum, invented at the Hospital for Sick Children. More than 85 years after the discovery of insulin Canadian medical researchers had not won a second Nobel Prize.

By contrast, Elliott Joslin had devoted his life to the treatment of diabetes. He also realized that the disease was far from conquered by insulin. He considered insulin the end of one era in diabetes management, not the end of diabetes. With boundless energy, a deep sense of mission, and considerable public relations skill, Joslin continued to expand his facilities and his staff in Boston, becoming the 'master clinician' of diabetes,[15] leaving at his death in 1962 a major establishment, the Joslin Clinic of ongoing national and international significance. It was the kind of legacy Frederick M. Allen had hoped to create through his Physiatric Institute and that the Toronto discovers of insulin did not think it was necessary for them to create. In the 1950s Charles Best did encourage diabetes research, and in the late 1970s the University of Toronto decided to concentrate its expertise, build on its traditions, increase its research effort, and honour Banting and Best, by creating the Banting and Best Diabetes Centre. Toronto once again became a major player on the diabetes field.

II

Leonard Thompson, the first person brought back from the edge of the grave by insulin, died on Easter Monday, April 20, 1935, in Toronto General Hospital. Thompson was twenty-seven years old. He had lived a more or less normal life, holding down a steady job as an assistant in a drug and chemical factory, taking eighty-five units of insulin daily. He was not a very well-controlled diabetic, and was in particular difficulty during the tenth anniversary of the discovery, apparently from excessive celebration. Once in 1932 he was brought into Toronto General in a coma and only barely survived.

The story that Thompson's death was caused by a motorcycle accident is incorrect. In his final illness a bout of influenza led to pneumonia complicated by severe acidosis. He died in a coma in an oxygen tent. Leonard Thompson's pancreas was small and partly atrophied, with few islet cells. The irreverent young staff at the hospital suggested that it should be mounted over the front door of the Banting Institute. When Fred Banting met the medical student who had done the autopsy, Burns Plewes, he asked, "Did that poor boy remain on a high-fat low carbohydrate diet all these years?"

"Yes," Plewes answered.

"Did he have any fun?"

"Yes, he had some fun. He used to get drunk nearly every weekend."

"Well, I'm glad he had some fun."

Leonard Thompson's pancreas was preserved and is displayed as item 3030 in the anatomical museum at the Banting Institute.

Jim Havens, the first American to receive insulin, became an artist. He specialized in woodcuts, his work was widely exhibited, and he was eventually elected to membership in the National Academy of Design. He married in 1927, had two children, and worked closely with Elliott Joslin on experiments with insulin. Havens had more than his share of illness, but controlled his diet and insulin well and lived a fairly normal life. He was beginning to experience some of the complications of his condition when he died of cancer in 1960 at age fifty-nine.

Elizabeth Hughes graduated from Barnard College in 1929 and the next year married William T. Gossett, a talented young lawyer. The couple moved to Michigan where Gossett rose through the legal department of the Ford Motor Company, becoming vice-president and general counsel and serving a term as president of the American Bar Association. Elizabeth was prominent in civic affairs and voluntary work in the Detroit area, while raising three children, all born by caesarian section. She was overweight for several years in the late 1920s, took up smoking, and would have the occasional cocktail, but in fact controlled her diet rigidly, eventually dropping back to a bit below normal weight and giving up smoking. She was athletic and an indefatigable world traveller, but never went anywhere, of course, without her insulin.

Elizabeth remembered the years of starvation before insulin as a "nightmare" from which she had awakened in Toronto to lead a normal life. She put the nightmare years behind her, and made her life as normal as possible, telling no one of her diabetes and insulin dependence. Even William Gossett did not learn her secret until a week after they had become engaged.

After first studying Elizabeth Hughes' medical records in the Banting Papers, and then finding him through biographies of Charles Evans Hughes and *Who's Who,* I wrote to W.T. Gossett asking, in effect, when his wife had died and of the later course of her diabetes. The reply came from Elizabeth Hughes Gossett herself, alive and in good health fifty-eight years and some 43,000 injections after first receiving insulin in Toronto.

She was distressed that I had been able to locate her and discover her secret. She agreed to see me only after I promised to give her a pseudonym and disguise her identity in this book. In November 1980, we spent a Saturday together at the Gossett home in Birmingham, Michigan. Elizabeth was a slim, attractive, husky-voiced lady, somewhat wizened, and grey-haired, of course, but with none of the debilities of the legs or eyes that often plague diabetics in their old age. She was perfectly alert mentally, rather more intellectually supple and wide-ranging than many people half her age. She was just back from a six weeks' tour of China. In the 1970s, concerned to perpetuate her father's greatest work, she had been the guiding spirit founding the Supreme Court Historical Society.

At the end of our day together Elizabeth loaned me the letters she had written to her mother from Toronto in 1922. As agreed, I disguised her identity in the early drafts of this book, inventing "Katharine Lonsdale," the diabetic daughter of a prominent American political figure. On April 25, 1981, Elizabeth Hughes Gossett died suddenly of a heart attack, the condition perhaps brought on by sixty years of diabetes. She had said she would have no objection to my writing freely about her after her death.

Several months after the publication of the first edition of this book I learned that one of the "living skeletons" brought to Toronto in July 1922 (see p. 144) was still alive. Teddy Ryder, the five-year-old son of a New Jersey engineer, received his first insulin from Banting in Toronto on July 10, 1922. He weighed 26 pounds. In 1983 Ted Ryder lived quietly in retirement in Hartford, Connecticut, still taking his insulin, suffering no major problems from his diabetes. His mother, Mildred Ryder, alert and healthy at 92, had vivid memories of their trip to Toronto in the summer of 1922, including someone's comment at the station in New York: "I feel so sorry for Mildred; you know she'll never bring that child back alive."

In 1990 Ted Ryder, age 73, returned to Toronto for the launch of a display about the discovery of insulin and the naming of the J.J.R. Macleod Auditorium on the site of the old medical building. Ted and his girlfriend (after the death of his over-protective mother he was having a wonderful late-life romance) came to dinner at our home. In July 1992 Ted became the first diabetic to live seventy years on insulin, having taken some 60,000 injections. He died of old age and complications in April, 1993, giving the residue of his estate to the University of Toronto to be used for medical research. While Ted was the last of the original Toronto patients (he and Elizabeth Hughes outlived all of the discoverers of insulin), remarkable, inspiring stories of the longevity of insulin-dependent diabetes have continued to multiply. The early years of the current century witnessed the honouring of the first 75- and then 80-year veterans of insulin use.

III

But those stories could be misleading. The discovery of insulin did not lead to more than a few medical miracles like the cases of Elizabeth Hughes and Ted Ryder. The prosaic reality was that diabetics' cup was still half-empty. For all the diabetics to whom insulin became the staff of life itself, there were others who could not afford it or were too proud to take means tests to get it as charity. There were diabetics in the 1920s whose doctors had yet to learn about insulin or were too conservative to use it. There were, and in parts of the world still are, diabetics who never knew they were diabetic, having access to no doctor at all.

Some of those who were given insulin used it recklessly, assuming they could eat and drink all they wanted, so long as they covered themselves by upping the dose. The type of doctor who came to specialize in pills and bills probably contributed to this attitude, finding that prescribing insulin was the fastest way of processing his diabetic patients (many of whom, with type 2 diabetes, would not need insulin and might be better off without it). Insulin's influence on medical practice in that respect could be harmful. It was one of the first of the truly powerful and effective weapons in the ordinary physician's "arsenal" against disease. The filling-up of that arsenal, with the sulfonamides in the 1930s, then penicillin and the other antibiotics in the 1940s and 1950s, then drug upon drug, made many physicians too confident of their powers, and many laymen too certain that their doctors had a quick fix for every sickness. The trouble with being able to work miracles, virtually raising people from the dead, is that it tends to replace one kind of religion with another, one set of priests with another.

As a research achievement, the discovery of insulin was almost too perfect. The appearance of insulin out of nowhere, it seemed, as the "cure" for diabetes may have fostered, as it certainly did in Toronto, a belief that next week or next year – or, as a friend wrote to Banting, "every few minutes"[16] – the doctors would come up with another cure. Surely it was just a matter of a little more time and a little more money before Banting or somebody else unlocked the secrets of heart disease or cancer. It was only gradually realized that the discovery of insulin was not a model of how medical research would develop. On the other hand, if you looked at the discovery not as the overnight achievement of unsung genius, but as the culmination of a world-wide, thirty-year search involving hundreds of researchers spending millions of dollars and sacrificing thousands of animals, perhaps it was more typical. The problem was that hardly anyone looked at the discovery of insulin in this way.

It was gradually realized that insulin had not solved the problem of diabetes. Diabetics who got the insulin they needed and then balanced their diets and their insulin as carefully as possible could not regain physiological normality. Artificially supplied insulin, the need for which could only be estimated by crude, inaccurate urine and blood tests, could not truly compensate for the missing pancreatic function. In the years after insulin, as Chris Feudtner shows poignantly in his 2003 book *Bittersweet: Diabetes, Insulin, and the Transformation of Illness,* diabetes took a heavy toll in impaired vision, kidney disease, hardening of the arteries with a variety of circulatory problems, and other so-called degenerative complications. As Joslin once put it, the era of coma as the central problem of diabetes had given way to the era of complications. The "miracle" of insulin had been to multiply the life expectancy of an early-onset or type 1 diabetic twenty-five fold. The statistical reality was that this total life expectancy remained considerably less than that of people whose pancreas functions normally.

There were also stunning, distressing ironies. Because insulin enabled diabetics to live and propagate, and because the disease has a strong hereditary component, the effect of the discovery of insulin was to cause a steady increase in the number of type 1 diabetics. Far more seriously, a rising calorie intake in the twentieth century, generating in some countries what was characterized as an epidemic of obesity, along with increased longevity almost everywhere, began to create in practically every country in the world massive numbers of new late-onset or type 2 diabetics. At the beginning of the twenty-first century diabetes, in its several manifestations, was more prevalent, posing more medical problems, than it had been before the discovery of insulin, and all the projections were that the situation would get much, much worse.

The clichés are worth repeating. The discovery of insulin was only the beginning. Or it was the end of the beginning. Diabetes is a far more complicated disease than anyone realized in the early 1920s. The appearance of insulin raised as many questions as it answered. Every part of every answer raised more questions.

The high incidence of complications, even among insulin-users, was substantially responsible for the founding of support organizations, such as the American Diabetes Association, the British Diabetic Association and the Canadian Diabetes Association in the 1940s, and the gradual resurgence of research. In the 1960s, impatient American parents of diabetic children organized the Juvenile Diabetes Foundation with an evangelical determination to replace insulin therapy with a cure for diabetes.

In some ways progress seemed inevitable. The structure of the insulin molecule came to be understood, generating pioneering knowledge and several Nobel Prizes. Whereas the Toronto researchers had worked in almost total ignorance of how insulin makes possible the body's combustion of nutrients, discovery of the cellular mechanisms of insulin and glucose transmission made it possible to differentiate forms of diabetes. The discovery of other metabolic hormones, ranging from glucagon through leptin, began to fill in the picture of hunger and the responses it triggers. On a practical level, by the 1970s vast improvements in glucose monitoring made possible much more precise insulin dosage and diabetic control – and in the 1990s the Diabetes Complications and Control Trial firmly established the value of tight control for insulin-users. The rise of diabetes education from the 1960s through the 1990s, a key way of empowering insulin-users and attempting to minimize the impact of type 2 diabetes, was vitally important in reducing the ravages of diabetes. Therapeutic breakthroughs such as laser surgery for eye problems, kidney dialysis, and the assault on heart disease provided vital treatment for complications.

From time to time there were fears that supplies of animal pancreas would not be adequate to meet growing demand for insulin. While the Japanese did briefly make insulin from whales during World War II, there were never serious shortages of beef or pork pancreas. All concern about the insulin supply faded in the 1980s when first Eli Lilly, then Novo-Nordisk, began selling genetically-engineered human insulin, the first and still one of the greatest triumphs of recombinant DNA technology. Soon it was possible to create "designer" insulins and insulin analogues, and other substances to facilitate insulin's action, the aim being to make the achievement of diabetes control easier.

The need to inject insulin was a problem that could also be tackled by designing better injection systems, ranging from painless insulin "pens" to high-tech insulin pumps or "artificial pancreases." Eighty-five years after diabetic patients began calling for an alternative to injection as an insulin-delivery system, pulmonary delivery of inhaled insulin finally became technically safe and feasible. In the meantime surgically centred teams of researchers (led by a group based at the University of Alberta) had finally been able to achieve Banting's dream of transplanting islet cells into the bodies of human diabetics, enabling some of them to go without insulin for significant periods.

The search for something better than insulin had started with Collip's glucokinin and Allen's myrtillin in the 1920s, and it continued through the development of several generations of oral hypoglycemics. While most new drugs proved more effective than the ones they replaced – some were even thought to stimulate partial regeneration of beta cells in the pancreas – none was able to replace the need for insulin injections in type 1 diabetes. For millions of type 1 diabetics around the world, and increasing millions of type 2 diabetics, injections of the hormone are just as vital today as they were for Leonard Thompson, Jim Havens, Elizabeth

Hughes, Elsie Needham, Ted Ryder, and their fellow sufferers so many years ago.

Health care workers hope some day to turn insulin therapy and some forms of diabetes into matters of only historical interest. Can a vaccine be developed to protect against the viral triggers of type 1 diabetes? Can determined public health campaigns – a new war against obesity, paralleling the twentieth century's assault on tobacco use – induce in populations the good dietary habits that will reduce the incidence and increase the control of type 2 diabetes? Can human weakness, human hungers, be overcome?

Thousands of researchers, spending billions of dollars every year, are attacking all of the questions relating to diabetes and insulin, while many more thousands of clinicians and educators serve in the trenches of diabetes care. The difference between the world-wide effort of the twenty-first century and the events in Toronto in 1921-23 is like that between the exploration of space and the flight at Kitty Hawk in 1903. Nameless astronauts now fly space shuttles; the Wright brothers won the immortality. The immortality of the discoverers of insulin was particularly deserved, in the sense that for diabetics it was much more than a matter of just getting a new hormone off the ground. All the later questions, even the current ones, are secondary to the one answered in 1921-22 in Toronto. With insulin, the stone was rolled away, and diabetes became a matter of the quality of life, not the speed of death.

Fred Banting, J.J.R. Macleod, Bert Collip, and Charley Best knew they were making medical history. Their struggle for credit was fired by each man's desire to have his place in history, to have the only kind of immortality open to us. This is surely not an ignoble aspiration. But perhaps the group in Toronto misjudged both their situation and posterity's viewpoint. They did not realize that those who understood history would eventually come to honour all of them. Above all, we would honour their achievement.

Notes

Complete citations for the manuscript collections and publications referred to in these notes are provided in the list of sources. Some of the citations in these notes to cases, files, etc. in manuscript collections may be no longer accurate if, as has been the case with the Insulin Committee records, they have been moved or reorganized. The following abbreviations are used in these notes:

BP: Banting Papers
CP: Collip Papers
FP: Feasby Papers
IC: Insulin Committee records
MP: Macleod Papers
MRC: Medical Research Council records
UWO: University of Western Ontario

Introduction: What Happened At Toronto? (Pp. 11-19)

1 Roberts 1922.
2 Dale 1922.
3 Pratt 1954; Pratt manuscript.
4 Feasby 1958.
5 Murray 1971; also Murray 1969.
6 International Diabetes Federation 1971.
7 Bart 1976; also Pavel 1976.
8 Décourt 1976.
9 Correspondence regarding the document accompanies the Pratt manuscript at the University of Toronto, and is also contained in the CP, UWO and in the Feasby papers.

Chapter One: A Long Prelude (Pp. 20-44)

1 Quoted in Wrenshall, *et al.* 1962, p. 36.
2 Osler 1915, p. 438; Allen 1919, p. 9.
3 Allen 1919, pp. 10, 27.
4 Ibid., pp. 24-25.
5 Ibid., pp. 31, 37.
6 Ibid., pp. 42-45.
7 Quoted in Allen 1919, p. 29; the speaker was Sir William Gull.
8 Opie 1910, p. 317.
9 Houssay 1952.
10 This account of early work in endocrinology relies heavily on Young 1970; also Biedl 1913; Sharpey-Schafer 1916; and, for the state of endocrinology in that year, Hoskins 1921.
11 J.J.R. Macleod estimated 400 in an interview, BP, 47, p. 89. For summaries of the use of extracts, see Allen 1913, p. 813f; Macleod 1926, pp. 55-56; Kleiner 1959; Cheymol 1972; Opie 1910, p. 102; Dewitt 1906, p. 193.
12 Cheymol 1972 states that the foetal extract was tried. But see Allen 1913, p. 815: "Apparently no one has ever tried the interesting possibility of feeding the glands of new-born or foetal animals, in which the islets have

a relatively high development and little external secretion is present." Rennie and Fraser 1907.

13 Zuelzer 1923. For the experiments, see Zuelzer 1907, 1908; Zuelzer, Dohrn, Marxer, 1908. On Zuelzer generally, see the useful short study by Mellinghoff 1971.

14 Zuelzer 1908, p. 316 (translation).

15 Forschbach 1909 (translations).

16 Mellinghoff.

17 E.L. Scott Papers, case 2, Carlson to Scott, Oct. 7, 17, 1911; A.H. Scott, 1972; Scott 1912; Magner 1977. The view advanced by Scott's widow, that Carlson tampered improperly with Scott's conclusions, is not warranted. The extreme claims in her book (A.H. Scott 1972) are undercut by the published papers and by the unpublished letters in the E.L. Scott papers.

18 Scott 1913; Scott papers, Carlson to Scott Oct. 17, 1911.

19 Macleod 1926, p. 59.

20 Murlin and Kramer 1913, 1916, 1956.

21 Macleod 1913, pp. 90-92.

22 Allen 1913, pp. 813, 815, 816.

23 Ibid., p. vi.

24 Allen and Sherrill 1922A, pp. 394, 415, 410.

25 Allen 1919, pp. 410-11.

26 Ibid., pp. 184-85, 376, 248.

27 Ibid., p. 202.

28 Allen and Sherrill 1922A, p. 419.

29 Allen 1919, p. 411.

30 Ibid., p. 579.

31 The only published account of Allen's career is Henderson 1970. Henderson is also the custodian of Allen's papers, which include a revealing manuscript autobiography. The Physiatric Institute is also described in the Toronto *Star*, July 20, 1923.

32 Woodyatt 1921; Allen 1923.

33 Joslin 1917, p. 471; also Joslin 1919, p. 18.

34 See the very misleading chart on the first page of the early editions of Joslin's diabetic manual.

35 Kienast 1938.

36 Joslin 1946; Wilder 1946, Campbell 1946.

37 Allen 1913, p. 813.

38 Myers and Bailey 1916.

39 The paper is Kleiner 1919; the quote is from Kleiner 1959; see also Kleiner and Meltzer 1915. Attention is drawn to Kleiner's work in Goldner 1972A, B, C. The problem of animal facilities is Dr. Goldner's conclusion, letter to the author.

40 Paulesco 1921B. Most of the relevant Paulesco material is reprinted in Pavel 1976, the best introduction to the controversy.

41 Allen 1919, p. 595; MP, Allen to Macleod, Feb. 8, 1922.

42 BP, Elizabeth Hughes file, Elizabeth Hughes letters to her mother.

Chapter Two: Banting's Idea (Pp. 45-58)

1 Toronto *Telegram*, Jan. 18, 1923.

2 Flexner 1910; for the background of reform in Toronto, see Bliss 1978.

3 Banting 1940, p. 9.
4 Ibid., p. 12.
5 Very little was ever written down about Banting's romance with Edith. Except for a brief account in Hipwell 1970, stories of their tangled affair have been passed on orally. The tip of the iceberg may surface in the very unreliable biography by Harris, 1946. My judgments are based partly on these sources, partly on interviews, particularly with Edith's cousin, Spencer Clark. For Starr advising Banting to go to London, see Stevenson 1946.
6 He was actually a Bachelor of Medicine (M.B.); Toronto did not award an M.D. as a first degree in medicine until a few years later. Banting received his M.D. in 1922.
7 Banting 1940, p. 13; BP, 26, Account book.
8 Banting 1940, p. 14; see also note 5 above.
9 Banting 1940, p. 21; Barr 1977.
10 BP, 5, Lecture notes, therapeutics; Banting 1940, p. 7.
11 Banting 1940, p. 16. Banting apparently subscribed to the journal. Both Stevenson and Harris write that Banting had borrowed the copy of the journal from Western's library. They were misled by the fact that years later, during a visit to London, Banting autographed and underlined the library's copy of that issue, to show the article's importance. Canadian Diabetic Association, Scrapbooks, undated clipping from the London *Free Press*, mentioning the visit and the underlining.
12 Banting 1940, p. 17.
13 There is a bare possibility that Banting's whole account of the inspiration by the Barron article is a confused reordering of events. In a letter to Banting of Dec. 14, 1920 (BP, 1), C.L. Starr suggests that if Banting had not seen "an article in the November issue of *Surgery, Gynecology and Obstetrics* on this subject," he might look it up. By the date of that letter Banting had spoken to and written Starr about his idea. Starr had also spoken to Macleod after Macleod had talked to Banting. Is it not odd that Starr would not have known how important the Barron article was in inspiring Banting's proposal?

A second odd piece of evidence is Macleod's statement in his 1922 account, describing their first meeting, that Banting "formed this idea while reading in a textbook of surgery." The reliability of Macleod's statement is considerably reduced, however, by his having prefaced it with the clause, "as stated in the first paper published on the work." In fact, in the first published paper Banting and Best cite the Barron article as the source of Banting's inspiration. Nevertheless, Macleod might have been buttressing an accurate memory of what Banting said to him with an inaccurate memory of the citation in the first paper.

It would be unlikely but not totally surprising if new evidence came to light showing that Banting did not read the Barron article until after he had become interested, as a result of his lecture preparations or other reading, in searching for the internal secretion.
14 Banting 1929. In his 1940 memoir, using quotation marks, Banting wrote that he wrote: "Ligate pancreatic ducts of dog - wait eight to ten weeks for degeneration. Remove remnant and extract." Stevenson 1946 changes the wording slightly: "Tie off pancreas ducts of dogs. Wait six or eight weeks. Remove and extract" (p. 67). Stevenson gives no source. Stevenson and Banting 1929 are the two wordings given in *Colombo's Canadian*

Quotations, where Stevenson's version is given first and called by Colombo "the fourteen most important words in medical research in Canada" (p. 33).

15 Academy of Medicine Notebook.

16 This account of the next day is based on Banting 1922 and 1940. It follows Banting's version of when he spoke with Tew, rather than Tew's memory (that Banting talked to him on the evening of Oct. 30), which is relied upon in Stevenson and Harris.

17 Banting 1940, p. 23. Banting said he spoke to Starr, W.E. Gallie, D.E. Robertson, and L.B. Robertson. It may be that he spoke to some of these after he had seen Macleod.

18 Macleod 1922/78. Comments on the validity of this hypothesis are reserved until pp. 203-8.

19 Banting 1940, p. 23 (in the 1922 version: "To my disappointment he did not seem at all interested"); Macleod 1922/78.

20 Macleod 1922/78. In 1940, p. 24, Banting added, "He told me that he had worked for fourteen years on carbohydrate metabolism and had given it up and was commencing on anoxemia." This is plausible, but lacks supporting evidence.

21 For the later statements, see Banting 1929, 1940.

22 Academy Notebook, June 9, 14, 1921.

23 Allen 1913, pp. 834-36.

24 Banting 1922. In 1940, p. 23, Banting described his repeating his ideas this way: "A hot iron gives off steam when cold water is thrown upon it. It was the first time I had ever seen the famous professor and I was not overpowered with either the man or his knowledge of research. I told him that I did not care who had worked nor how long and that I wanted only ten dogs and an assistant for two months."

25 Macleod 1922/78: "I would place every facility at his disposal and show him how the investigation should be planned and conducted." Banting 1922: "He consented."

26 Banting 1922.

27 Macleod 1922/78, appendix, Banting to Macleod, Nov. 21, 1920.

28 Banting 1940, p. 25. Starr had trained Banting. Is it noteworthy that he apparently did not call Banting an excellent or skilful surgeon?

29 BP, 1, Starr to Banting, Dec. 14, 1920. Compare Macleod 1922/78: "I told Dr. Starr that although it was taking considerable chances I thought the research was well worth proceeding with."

30 Banting 1940, p. 24. Also Banting 1929: "The next four months were spent in reviewing the available literature." In his 1922 rough draft he wrote, "The idea grew and as time went on I devised more experiments. These were always written in my black book which I kept for writing down ideas and the place of reference." This passage was not in the 1922 final draft, which does not mention any reading or experiments that winter.

31 Macleod 1922/78, appendix, Banting to Macleod, March 8, 1921.

32 BP, 1, Starr to Banting, May 2, 1921.

33 Stevenson 1946, p. 71. Stevenson cites no source and dates this conversation before November 7. A consideration of all the circumstances suggests the conversation might have been plausible and held later; no other evidence exists, however.

34 Academy Notebook, Banting to C.S. Sherrington, March 8, 1921:

My plan is to cut the spinal cord in the lower thoracic or lumbar region in new-born kittens and dogs. Following this I wish to make a study of the reflexes of the hind-limb at various periods of growth of the animal. I am anxious to see how these reflexes may compare with those of the normal animal.

It appears to me that such observations would indicate whether such a movement as walking was developed as a reflex or as a "voluntary" (cortical) action.

Miller had advised Banting to write. Banting kept Sherrington's March 21 reply that the question had "some promise" and offering advice about technique.

35 Macleod 1922/78, appendix, Macleod to Banting, March 11, 1921. Compare Banting 1940, p. 24: "But he said that no one could look after the dogs and no one was interested, so I had better wait...."

36 My interviews, the Hipwell memoir, and Harris's account all support the view that the engagement was broken off *before* Banting insisted on coming to Toronto, and therefore not because of his determination to work on the internal secretion. But that determination, when it became clear, probably did not improve relations, for it was a sign of Fred's continued unwillingness to settle down.

37 Stevenson 1946, p. 73; see also Dale 1946, pp. 37-38.

38 Banting 1940, p. 18.

39 Ibid. But the date is open to question. In his 1922 statement Banting does not mention a trip to Toronto on April 26, and implies (using in his rough draft the same image of turning the key and taking a lone suitcase) that he made his trip on May 14. It is natural, however, that Banting would have taken a few days in Toronto as a break from teaching and to firm up the details of the work with Macleod. The commencement date of April 14, 1921, given in his Nobel Prize lecture (Banting 1925) is clearly wrong.

40 BP, 1, Starr to Banting, May 3, 1921.

41 Macleod 1922/78. As Macleod remembered it, they talked about using extracts. The Academy Notebook, however, suggests that they probably also discussed using transplants of degenerated pancreas.

42 Macleod 1922/78, note 3; Collip 1916, 1920, 1921. There is some unclarity about which meeting Collip attended. It might have been the June meeting when Macleod gave Banting the suggestions recorded in Banting's notebook on June 9 and 14.

43 The wording of Macleod's statement (1922/78, note 5) may be further evidence that an assistant had not been talked about earlier: "I gave Dr. Banting all the assistance I could in planning the details of the experiments, in searching the literature and, finding that he was entirely unfamiliar with the chemical methods necessary for the proposed investigation, I offered him the assistance of one of my research fellows, C.H. Best, who had been trained in this work."

44 See E.C. Noble's unpublished account, dated October 1971, in the Noble Papers for the statement that Macleod introduced Banting to Best and Noble together. Best 1922 simply refers to meeting Banting early in May (Best also says that in the lecture Macleod mentioned that Banting's idea might lead to the development of "an efficient pancreatic extract"). Later (Best 1972), Best thought that he and Banting met in the autumn of 1920

and at that time discussed the work Banting wanted to do in Macleod's laboratory. There is no evidence of this meeting in any of the 1922 accounts, including Best's, although it would be natural for either Banting or Best to mention it. No other account, including those of Best, refers to an earlier meeting.

45 Banting 1940, p. 25.

46 Noble account, 1971. Macleod 1922/78 apparently contradicts this, referring to Best only. But Best 1922, supported by both Noble and Banting, makes clear that he and Noble were to divide the time.

47 In most of his later accounts of the discovery (the exception is one interview, Stalvey 1971), Best did not remember the coin-tossing incident.

Banting 1940, p. 25, mentions the desire to avoid a break in the summer holidays.

The sources contain a bewildering set of contradictory statements about how many weeks the assistants were to work. Some accounts say four weeks each, others say three. Banting gave both versions. In one of Best's accounts he refers to four weeks and two weeks. And so on.

The confusion stems from the fact that only six weeks remained between mid-May and the usual beginning of holidays, but Banting proposed to work for eight weeks or more. The final arrangement is not clear. There were many possible ad hoc arrangements that the trio could have worked out. Perhaps Best and Noble were each to give Banting three weeks' time in a four-week period. Perhaps Banting thought he would not need help during or after a certain period. The fact that Best knew he would be going off for the last two weeks in June for militia training may have further complicated the arrangements.

Best raised another issue in later accounts of the beginning of the work by stating that he worked without pay and that he considered himself to have volunteered to work with Banting. See, for example, the statements in Best, *Selected Papers*, p. 5; also Best 1972; also Wrenshall, *et al.* 1962, p. 61.

While it is true that Best later volunteered to do Noble's stint, and probably did it without thought of payment (though Macleod afterwards arranged that he and Banting were paid for their time), all of the 1922 sources clearly point to the conclusion that Best was assigned to assist Banting as part of fellowship duties for which he was being paid. He and Noble received an extra $85 cash for their work in June and knew they would be fellows in 1921-22 at a salary of $800 each (MP, Macleod to Falconer, May 31, 1921; Macleod to F.A. Mouré, June 3, 1921). There is no evidence that Banting, although heavily in debt and short of ready cash, asked for, expected, or was offered payment, according to the original arrangement. He was the volunteer outsider with the idea. Best was his paid assistant.

48 Banting 1922, rough draft.

49 This uncertainty, plus the impossibility of dating Banting's discussions with Macleod, makes it impossible to state precisely when the Toronto work that led to the discovery of insulin formally began. Banting left London on Saturday, May 14. He began work in Toronto on either Monday, May 16, or Tuesday, May 17. May 17 was the date of the first experiment.

50 Banting 1940, p. 25.

Chapter Three: The Summer of 1921 (Pp. 59-83)

1 Unless otherwise noted, all details concerning the animal experiments are taken from Banting and Best's original notebooks. These consist of Banting's original notebook at the Academy of Medicine in Toronto, and the several joint notebooks used by Banting and Best which are in the Banting Papers at the University of Toronto. There are also some loose notes accompanying the original charts of the experiments in the Banting Papers. Day-by-day summaries of the experiments which I compiled as background to these chapters have also been deposited in the Banting Papers.

2 See Hédon 1909 for a description of the operation. In several of his accounts Best mentions that the first job was to look up this and other literature.

3 Academy Notebook; Best 1942, 1974; FP, 1921 file, notes of a conversation with C.H. Best, Jan. 20, 1956.

4 It is not clear whether Macleod was responsible for this suggestion. Best 1922 states that Macleod had suggested a D:N ratio higher than they ever obtained, but whether it was the very high 3.65 figure (which was common to animals made glycosuric by a very different chemical procedure, but well above the 3.00 rough average for depancreatized dogs) is unclear. See also Best 1942: "Banting had understood from Professor Macleod" that 3.65 was necessary for complete diabetes. There could easily have been a misunderstanding.

5 The sources do not disclose whether they planned to use autogenous or heterogenous grafts, or both.

Dr. F.C. MacIntosh, professor emeritus of physiology at McGill University, who read an early draft of this manuscript, suggests a plausible, slightly more speculative, variation of my interpretation of the research plan:

> As I see it there weren't to be two approaches, but a single one in which each stage, if successful, led logically to the next. This was to be the basic sequence: First, to demonstrate, by planting a working Hédon remnant into a diabetic animal, that one dog's internal secretion could prevent another dog from getting diabetes. That accomplished, to show that a ground-up pancreatic remnant could do, for a brief period, what the intact remnant had done over a longer period. Finally, to replace the emulsion (which I guess would have been pretty toxic if given by vein) by one or other kind of extract: the difference between an aqueous emulsion and an aqueous extract being essentially that the latter had been cleaned up by filtering or centrifuging it to remove the larger particles. This stagewise approach would reflect Macleod's caution. It would ensure that the most critical observations would be made on dogs that had been kept in good shape for some time after subtotal pancreatectomy, rather than on dogs that might be deteriorating from a combination of causes - surgery & infection & diabetes. And it might delay the final push to get the hormone cleaned up and bottled until the surgery had become routine, Macleod had returned from Scotland, and Collip was available to help with the biochemistry. As a sop to Banting, there would be quite a lot of varied surgery. And probably Macleod agreed that if the transplants failed repeatedly, the experimenters could try emulsions or extracts.

6 Margaret Mahon Best ms, "The Discovery of Insulin," privately held; FP, ms biography of Best.

7 That second 386 was a puzzling dog. It had not healed quickly after its first operation on May 31, occasionally showing a high blood sugar. After the second operation, apparently completing the pancreatectomy, its blood sugar stayed in the mildly diabetic range of .20 to .30, its D:N, after the first surge, became insignificant, and on the day it died, June 27, its blood sugar was a normal .116. The autopsy showed open wounds, but no infection and no trace of pancreas. It is most likely that both the operation and the autopsy missed portions of pancreas - notoriously easy to miss in both procedures - which had sustained partial pancreatic function in the dog.

8 Banting 1940, pp. 27-28.

9 Best 1942, p. 389, refers to "a great deal of discussion of the results of the glucose and nitrogen estimations." Also FP, Best to Feasby, May 6, 1957, in which Best writes (p. 11), "I persuaded Fred with difficulty that Macleod was wrong about D/N ratios. This threatened to wreck our partnership because Fred was a stubborn man. He eventually thanked me repeatedly and warmly."

10 Banting 1940, p. 26. Some of the phrasing used in the contemporary accounts and some oral accounts leads to a possible interpretation that Noble simply did not appear at the beginning of July, that Best filled in for him, and only then decided to stay on.

11 Banting 1940, p. 45; Cody Papers, G.W. Ross ms, "History of the Discovery of Insulin," undated but probably 1941.

12 Catgut is not mentioned specifically in the notebooks, but is referred to in Best 1942. The notebooks do show that silk was used for the second series of ligations. In the Cameron Lecture (Banting 1929), Banting gave a more detailed and slightly divergent account: "We chloroformed a couple of the dogs which had their pancreatic ducts ligated.... Careful examination showed that the ligature was still present in a bulbous sac in the course of the duct. It was therefore necessary to operate on all the duct-tied dogs a second time and to exert particular care as to the tension put on the ligature. If the ligature was applied too tightly gangrene developed immediately underlying the ligature and a serous exudate laid down on the surface over the ligature resulted in the recanalisation of the duct. If applied too loosely the duct was not blocked. We, therefore, in some cases applied two or three ligatures at different tensions."

13 Best's letters to Margaret Mahon were not available for this study. But the references to the work in the lab in these letters have been excerpted and are quoted in Margaret Mahon Best, "The Discovery of Insulin," and also in two summaries compiled by C.H. Best and W.R. Feasby in the Feasby Papers, 1921 file. There are minor dating errors in the summaries.

14 BP, 1, Best to Macleod, Aug. 9, 1921; for the preparation of the extract see the notebooks, also Banting and Best 1922A, 1923, Banting 1923A.

15 Banting 1940, pp. 25-26; Banting and Best 1922A; Banting 1929.

16 BP, 1, Best to Macleod, Aug. 9, 1921.

17 Ibid.; also chart in BP. The notebooks are riddled with errors concerning this dog. Its first-stage operation was on July 22. There is no record of a second stage, and on July 30 Best noted that dog 406 was dead. And yet 406, severely diabetic, receives extract on Aug. 1. It appears that the second stage was done on 406 but not recorded, and that the dog recorded

as dead on July 30 was actually 407, which had been done on July 26 and never again appears in the records.

18 Best 1922.

19 BP, 1, Banting to Macleod, Aug. 9, 1921.

20 Notebooks. Banting and Best 1922A incorrectly lists the doses as five cc.

21 See note 13 above.

22 BP, 1, Banting to Macleod, Aug. 9, 1921; BP, 22, index card list of "Ideas," dated Aug. 8. Banting also listed "Alcohol extraction" on the card, but did not put it in the list for Macleod.

23 BP, 1, Banting to Macleod, Aug. 9, 1921.

24 Details from Ross manuscript, Hipwell 1970, interview with Mrs. Fannie Lawrence.

25 Hipwell 1970.

26 Note 13 above; also notebooks. The rabbits are not referred to in the notebooks.

27 Banting 1940 and various Best accounts, particularly those in FP, National Film Board files; also Hipwell 1970.

28 Banting 1940, pp. 39-44; also Banting 1929. An account of an encounter with an anti-vivisectionist regarding this dog in Banting 1940 does not seem a plausible occurrence at that time; it probably happened a year or two later.

29 Banting and Best 1922A.

30 Another factor contributing to the error stemmed from their having made three preparations of whole gland extract. The neutral extract, given on August 17, had the marked effect. Acid extract, given early the next morning, had a slight though still noticeable effect; alkali extract had none. By making the three separate preparations they had precluded giving serial injections, although this had been their common method. For discussion of the hypothesis, see pp. 203-8.

31 Notebooks; Banting and Best 1922A. In the first secretin experiment they also stimulated the vagus nerve to try to obtain more pancreatic juice after secretin stimulation.

32 Banting 1929, 1940.

33 Banting 1940, p. 42.

34 Banting 1940, p. 18. It is not clear where Banting got a car. His 1940 account seems to show that he did not have a car when he came to Toronto and that the story of his selling his old Ford that summer to raise money is inaccurate. He actually sold his beat-up old Ford in the autumn of 1920. Of course he might have bought another car by August 1921. More likely, he had borrowed one, perhaps the university's "Pancreas car," referred to later.

35 Interview with Dr. Ian Anderson.

36 BP, 1, Macleod to Banting, Aug. 23, 1921.

37 The notebooks show some evidence of this. In their first paper Banting and Best promised a more detailed description of the histological sections in a subsequent communication: "Suffice it here to note that the pancreatic tissue removed after seven to ten weeks' degeneration shows an abundance of healthy islets, and a complete replacement of the acini with fibrous tissue" (Banting and Best 1922A). No records of these sections can be found and nothing more is said in any subsequent communication. None of the tissue had been subjected to more than seven weeks' degeneration.

38 Macleod's letter must have seemed so out-of-date that Banting, a year later, apparently told people he had never received the letter. See Macleod 1922/78, Note 4.

39 Compare Banting 1940, p. 37: "I have always advocated that no person should be allowed to operate on a dog, who has not received training as a surgeon."

40 Perhaps significantly, they never again identify their extract as "Isletin." The term is not used anywhere in their notes, or in any other source after August 9.

41 The mess with cat extract had caused them to resolve to centrifuge all future extracts (this is the only change introduced from their original methods of preparation, although some of the doses of extract were being made slightly acid), but they made their first injection of this new extract before it had finally settled. The dog had a marked reaction. It is not clear whether centrifuging was regularly used.

42 Banting 1940; Macleod's return is noted on Banting's 1922 desk calendar, BP, 26.

43 Stevenson 1946, p. 89; the notebooks indicate bowel problems with dog 92; Noble papers, Robert Noble interview with E.C. Noble, April 12, 1977; interviews.

44 Personal communication.

45 Falconer Papers, 71, Henderson to Falconer, Sept. 21, 1921; Banting 1940, p. 29.

46 Macleod 1922/78; Best 1922.

47 Banting 1922. Also Macleod 1922/78: "I found that Banting and Best were dissatisfied with the facilities at their disposal, and early in October they formally demanded of me that I improve them."

48 In Banting's 1922 account he implies that Macleod relented and made specific commitments before the meeting ended. In his 1940 account the change of heart is not evident until the next morning. Macleod 1922/78 is not helpful on this point, but mentions Banting's threats to go to the Mayo Clinic or the Rockefeller Institute.

49 FP, National Film Board file, Best dictation, transcript March 20, 1956.

Chapter Four: "A Mysterious Something" (Pp. 84-103)

1 Banting 1940, p. 31; Banting 1922.

2 Falconer Papers, 71, Henderson to Falconer, Sept. 21, 1921; Macleod 1922/78, appendix, Macleod to Falconer, Sept. 30, 1921. In his memoirs and afterwards Banting warmly thanked Henderson for providing this position, thereby, Banting claimed, making it possible for him to stay in Toronto. There is no direct discussion of this job arrangement in the sources, but the documents give the impression that at some point Henderson consulted Macleod about the idea of appointing Banting in Pharmacology. Macleod would have supported this enthusiastically as an excellent way to keep Banting in Toronto without having to put up a fight with the administration, which he might not win, for a special position and extra money for his own department. There is no evidence that Macleod ever turned Banting down for a job in Physiology.

3 Banting 1940, pp. 35, 49-50.

4 Macleod 1922/78. Banting 1940, p. 46: "I asked Macleod if Collip could join us and work on the biochemistry aspect but Macleod said there was 'plenty of time for that later on if necessary'." Banting 1922: "I was very anxious that the work advance more rapidly. I asked Professor Macleod three or four times if Dr. Collip could do portions of the work, but he advised against it."

5 BP, 26, desk calendar 1922. In his 1922 account Banting says Connaught gave them three calves for exhausted gland experiments. There are no records of these.

6 Notebooks. In Banting and Best 1922A the ligation period is incorrectly stated as six weeks.

7 Notebook notations on October 5 and 9, read in conjunction with the note on the sources of the extracts in Banting and Best 1922A, lead to no other interpretation than this. That this complete confusion had occurred cannot be realized from the publication alone; in it the relative strengths of the different extracts are stated, but not the relative degrees of degeneration of poor Towser's pancreas. It is just possible that an error in the notebooks is corrected in the publication.

8 BP, 22.

9 BP, 22. This is the only index card in Best's handwriting in the group. There is always the possibility, of course, that Banting did the reading and dictated the note to Best.

10 Paulesco 1923A.

11 This is the second part of the error in Banting and Best 1922A, and it does not follow from either Best's index card or the Paulesco article. There is a bare possibility that a missing second index card, plausible given the ending of the back of the first card, might have contained further Best notes on Paulesco leading to the error.

12 BP, 22, index cards. The article was McGuigan 1918.

13 BP, 1, Banting to Starr, Oct. 19, 1921.

14 BP, 26, desk calendar 1922.

15 Several people to whom Banting told his version of the discovery recalled how emphatically he mentioned Macleod having made them repeat their work. The most explicit reference is in Macleod's 1926 account, which seems to refer to exactly this stage in the history of the experiments: "In view of the large amount of work which had previously been done in this field, it was considered advisable to make certain of the anti-diabetic effects of the extracts, as judged by the behaviour of blood sugar and urinary sugar, before proceeding to investigate their influence on other symptoms of diabetes, such as glycogen formation, ketosis, and changes in respiratory metabolism" (p. 65). The various unpublished accounts in the E.C. Noble Papers also stress Macleod having made them repeat their summer's experiments.

16 Scrapbooks of Mrs. C.H. Best, Scrapbook No. 1.

17 Banting and Best 1922A.

18 BP, 39, typed diary entry, Nov. 14, 1921.

19 Banting 1922.

20 Macleod 1922/78.

21 Macleod 1922/78 is the only source giving this origin of the idea of the longevity experiment. But no other account contradicts this or gives any other version of its origin. I assume that "H.B." Taylor in Macleod's

account is a typographical error for N.B. Taylor, a member of the Department of Physiology at that time.

22 Banting's accounts of his reasoning correlate unusually smoothly. See Banting 1922, 1925, 1929, and Banting and Best 1922B, 1922C. Undated index card notes on Laguesse and Carlson & Drennan are in Banting Papers, 22. In most accounts Banting mentions that only later did they find an article by Ibrahim showing that trypsin/ogen/ is not present in the foetus during the first third of pregnancy. It would not be surprising if his memory was incorrect on this last point.

23 Banting and Best 1922B, 1922C.

24 Banting and Best 1922B.

25 See Banting and Best 1922B for the first precise statement of quantities: "we placed 50 grams of tissue in 250 cc. of saline, macerated and filtered; 15 cc of this solution were then diluted to 250 cc. with saline. A 15 cc dose of this solution reduced the percentage of blood sugar in a 10 kilogram dog from .40% to .15% in three hours." The context indicates this was done on November 20-22. The notebooks do not record these quantities, nor any injection showing these results.

26 Banting and Best 1922B.

27 Banting and Best 1922C.

28 The chart in Banting and Best 1922C contains all the data compiled on dog 27, with only one error. The experiment was being very poorly conducted, and it would be difficult to show that the pancreatectomy, despite autopsy findings of no pancreas, had actually made the dog diabetic. On Dec. 2, for example, a blood sugar reading 69 hours after the last injection of extract was .15.

29 Macleod 1922/78.

30 My summary chart of the results is now in the Banting Papers. I am grateful to Professor O. Sirek for suggesting this approach to the early results.

31 Macleod 1922/78.

32 MP, folder marked 342, Joslin to Macleod, Nov. 19, 1921; reply Nov. 21.

33 BP, 2, Clowes to Banting, Sept. 25, 1934.

34 Notebooks and Banting and Best 1922C. The latter seems incorrect in dating the dog's second reaction as December 4, rather than December 3.

35 The use of this name for dog 33 seems to have arisen only some years later. There are no contemporary references.

36 Best 1922.

37 For preparation methods see Banting and Best 1922B, 1922C. For the use of warm air see Macleod 1922/78. The description of the injection to dog 33 in Banting and Best 1922C, p. 6, contains quite incorrect figures. The first injection of 11 cc. on dog 33 did not cause a fall from .30 to .15 in one hour; the actual drop was from .35 to .30. The impressive drop in the first 45 minutes, from .26 to .14, had actually taken place the day before on dog 23.

38 Banting 1940, p. 46; also Banting 1922, in which he mentions having persuaded Velyien Henderson to intercede with Macleod, and Macleod having told Henderson, in effect, "that the scientific world would think he was silly if he gave up his work on Anoxaemia for the investigation of the extract, since its success had not yet been proved." Banting places this discussion in January, but it must have been earlier.

39 Collip Papers, Medicine, transcript of E.E. Shouldice statement, October 4, 1958. The statement contains some factual errors and I am not sure of its reliability.

40 Falconer Papers, 71, V.J. Harding to Falconer, Sept. 27, 1921; on Collip generally, see Barr and Rossiter 1973; for his love of research, see Collip Papers, Medicine, Ms of Addison Lecture, July 1948.

41 FP, 1922 file, Best, "The Discovery of Insulin," Osler Oration - July 12, 1957. See also FP, "Hot" file, Best to Feasby, May 6, 1957; and FP, Best to Sir Henry Dale, Feb. 22, 1954. In his 1940 account, p. 48, Banting alludes to Best being suspicious of Collip at an early stage in the collaboration.

42 I have located five documents by Collip describing his contributions to the insulin work. There are two published articles, Collip 1923K, 1924; an undated, signed statement, "The Contribution Made by J.B. Collip to the Development of Insulin While he Was in Toronto 1921-22" in his papers in the medical library at the University of Western Ontario; and two undated second pages of letters written from the University of Alberta, both apparently written in 1923. One of these is in a collection of Collip papers privately held; the other was found in what were then unsorted files of Collip papers in a storeroom in the Biochemistry Department at the University of Western Ontario. Collip also considered that Macleod 1926 contained a true account of the discovery.

In his signed statement Collip dates the rabbit experiments as beginning on December 12.

43 See Macleod 1922/78 for his claim to have made the suggestion, which Collip then acted upon. This is not contradicted in the literature except by Best 1922, "there is considerable doubt as to whom belongs the credit of the idea." He adds, however, "Dr. Collip was first to put the idea into effect."

44 Banting and Best 1922 C gives an incorrect number to the dog, an incorrect volume of extract, and mis-states the extent of the decline in blood pressure. It is also almost certainly incorrect in stating that whole beef extract was used, for the notebooks seem to show that the first preparation of this extract was not made until Dec. 12. The type of anesthetic used in the experiment is not stated.

45 Collip's accounts all omit the Dec. 9 experiment by Banting and Best on the anesthetized dog and convey the impression that the work on the liver began with Collip's experiment on the normal dog. Possible explanations of this omission are that Banting and Best's experiment was so inconclusive as to be virtually meaningless, and/or that the train of thought about the liver, set out in the text, did not occur until Collip repeated the experiment. The notebooks show, however, that Banting and Best did remove specimens of dog 23's liver for glycogen estimates, and in their paper 1922B, though not 1922C, they present the liver hypothesis as following from this experiment. There is no record, on the other hand, of the glycogen estimates actually having been done after this experiment.

My ordering of the evidence seems consistent with all of the accounts, and particularly Macleod's 1922/78 summary: "It was agreed that the time had now arrived when it would be advisable to ascertain the glycogen content of the liver and other viscera in depancreated dogs treated with the extract. Dr. Collip undertook to do this and to observe the excretion of acetone bodies in these animals prior to death." Collip later found out that potent batches of extract did lower the blood sugar of anesthetized dogs.

46 Banting 1940, 1922; Best 1922. The dog they refer to must be the Airedale described in Collip 1923K. There may have been another dog experimented on at the same time.

47 Macleod 1922/78. Best 1922: "I had been desirous of performing this experiment at an earlier date, but had been unable to obtain the apparatus."

48 BP, 26, desk calendar.

49 The notebook lists Dec. 16; in 1922C and on the chart the date is Dec. 15.

50 A month or so later it became a matter of dispute, as we will see, as to who had first discovered the insolubility of insulin or an insulin-compound in 95 per cent alcohol. In their second published paper, written in February, Banting and Best mention that this extract of Dec. 15 or 16 was also washed in 95 per cent alcohol. This washing is not recorded in the notebooks. Nor is there any record in the notebooks or the published papers of a procedure, described by Banting in both his 1922 and 1940 accounts, in which - in Collip's presence in the 1940 but not the 1922 account - 95 per cent alcohol was added to an 80 or 70 per cent alcoholic solution of the active principle and the solution became opalescent, a very fine precipitate developing.

Nor is there any record in the notebooks or the publications of an experiment first described in Best and Scott 1923D, p. 712, and attributed to Banting and Best, in which, after evaporation and washing with toluene, "they treated an aliquot portion of the dried residue with 95 per cent alcohol. The mixture was filtered and the filtrate evaporated. The residue was dissolved in saline solution. Administration of this solution produced no effect upon the blood or urinary sugar of a depancreatized dog. A saline solution of the material which did not dissolve in 95 per cent alcohol, however, definitely lowered the blood sugar and diminished the sugar excretion of the same animal."

The notebooks do show that on December 21 Banting and Best tried an extract "of the crystal first deposited by the alcohol ext. drying in warm air current." This memory, of a precipitate formed at a point where the solution was nearly pure alcohol, may have formed the basis of Banting and Best's claim to have discovered the insolubility. Except that the precipitate proved inactive in the test animals.

The other possibility is that the experiment described by Best and Scott was done early in January. But the complete absence of notes in Banting's otherwise apparently quite complete sequence, and of any references in the 1922 publications, is puzzling.

51 Gilchrist, Banting, and Best 1923.

52 BP, 22. In his 1922 account Banting also mentions this test, dating it December 24 and indicating that the mode of testing was for glycosuria.

53 Collip 1923K; also Collip Papers, Medicine, "Contribution of J.B. Collip to the Discovery of Insulin..."

54 See Collip 1923K, p. 8, for details of this experiment. The ketone readings are in Banting, Best, Collip, and Macleod, 1922B.

55 Collip 1923K. See also Macleod 1926, pp. 101-102: "the percentage of glycogen found in the liver after the animal had been given very large quantities of sugar along with insulin was so large that it was difficult to determine with accuracy; it was apparently over 20 percent." This wording, combined with a more general neglect of this result in Macleod's accounts, suggests that he may have doubted Collip's result with this first glycogen estimation.

56 Collip 1924.

57 In his 1922 account Banting states that Collip told Macleod the results of this experiment, implying that he and Best only learned later, presumably on the train. This became another item in Banting's indictment of Collip and Macleod. It might well have happened, perfectly innocently, if Collip ran across Macleod before or during Christmas week. In any case Collip, who considered Macleod the director of the work, would naturally tend to report to him first.

Chapter Five: Triumph (Pp. 104-28)

1 Banting, Best, and Macleod 1922.

2 Banting 1940, p. 48.

3 Macleod 1922/78. Joslin 1956 makes the same point in a more kindly way: "The possibility of mistakes in the work was fully exploited by those who discussed the paper in a skeptical but on the whole a sympathetic way." Another eye-witness wrote later in 1923: "Whether it was the excessive modesty of the speaker, or whether due to a somewhat apathetic attitude on the part of the audience, Dr. Banting's address made little impression on the members present." Funk and Harrow 1923. In 1972 R. Carrasco-Formiguera recalled the meeting this way: "I remember my amazement and disgust at some whom I had previously held as giants, but who now tried to underrate such a remarkable achievement. I also recall my pleasure at others, including those who themselves had been close to success, who acknowledged the achievement of the young Canadians, as well as the originality and importance of their findings."

4 Joslin 1956.

5 See E.L. Scott to Macleod, Feb. 7, 1922, printed in A.H. Scott 1972, pp. 136-38, in which, referring to their discussion at New Haven, Scott mentions the "marked reactions" Macleod had described the Toronto preparations as causing.

6 FP, Best dictation, transcribed March 20, 1956.

7 See the argument in Pratt 1954.

8 See Roberts 1922, Pratt 1954, and pp. 203-8 of this account.

9 FP, 1924 file, Feasby interview with Joslin, Nov. 1957.

10 MP, Allen to Macleod, Feb. 8, 1922; Scott to Macleod, Feb. 7, 1922, in A.H. Scott 1972, pp. 136-8.

11 Macleod 1922/78; Clowes 1948.

12 Banting 1922. By 1940 Banting had convinced himself that he could have handled the questions, and describes the meeting this way: "When I sat down there was considerable discussion and many questions asked. I noted each and since the discussion was from the audience and not from the platform, I had forgotten all about myself and was prepared to talk freely in reply to questions and discussions. To my surprise, however, I was not called upon by the chairman as was the invariable rule. Macleod himself responded to all questions and expostulated theories and referred to 'our work' and 'I believe' and 'I think'."

13 See ch. 4, note 57.
14 Banting 1940, p. 49.
15 MP, Macleod to A.R. Cushing, Jan. 7, 1922; Macleod 1922/78.
16 In Banting and Best 1922C they state it got 6 cc. of whole gland extract daily from December 8. The notebooks record, however, two injections of foetal calf extract, totalling 16cc., for December 8, 15 cc. of foetal calf extract on December 9, 10 cc. whole gland extract on December 15 and 16. There are no records of other injections and no charts. On December 11 the notebooks record that the dog's picture was taken. The pictures printed in Banting and Best 1922C are captioned as being nine weeks after total pancreatectomy.
17 Banting and Best 1922C.
18 Best 1922 states that all work after Christmas was carried out under Macleod's direction. Both Banting and Macleod in their 1922 accounts refer to the more formal division made in late January.
19 Best 1922: "In the winter of 1922 I spent most of my time in superintending the collection and initial concentration of material which was then handed over to Dr. Collip for completion."
20 The only explicit statement that this was Banting's expectation is in Best 1956.
21 See Collip 1923K, 1924; Banting, Best, Collip, Macleod, and Noble 1922A, 1922C.
22 Gaebler 1965.
23 Noble Papers, Noble accounts, March 12, 1977, October 1971; also Robert Noble interview with E.C. Noble, 1977, and author interview with Robert Noble 1980.
24 Mann and Magath 1921. Collip's accounts of these observations imply that he discovered the reaction and its antidote independently. If so, he was surprisingly out of touch with the most current literature.

In a Feb. 22, 1954 letter to Sir Henry Dale, (FP) intended for posthumous publication, Best states that "Banting and I had noted hypoglycaemia and had recorded this and the beneficial effects of sugar." There is no note or record of this.
25 Macleod 1922/78.
26 Tory Papers, File 504-9, Collip to Tory, Jan. 8, 1922.
27 Banting 1922.
28 Interview with R.B. Kerr, 1980, who is writing a biography of Graham.
29 Banting 1940, p. 55.
30 Macleod 1922/78.
31 In his Feb. 22, 1954 letter to Sir Henry Dale (FP), Best describes the background as follows: "Banting began to get very restless about Collip's activities and his relationship to us...and spoke to me one day about preparing some potent insulin to give to the first human case of diabetes which was to be treated on the wards of the Toronto General Hospital. Banting's statement to me was: 'The insulin which Collip is making may be somewhat freer from impurities than that which you have made and which we have given to depancreatized dogs. We know, however, that the material which you have made from whole beef pancreas, is really potent and that it gives no obvious reaction in the dogs, i.e. no local reaction.' He said: 'I think it would be much more appropriate, Charley, in view of our work together, if this first case should receive insulin made by your

hands and tested by us on dogs and on ourselves.' I had no hesitation in agreeing...."

32 Interview, July 10, 1980.

33 Allan 1972.

34 Campbell 1946, 1962; Banting, Best, Collip, Campbell, and Fletcher 1922.

35 This account of the preparation of the extract relies on Banting and Best's published description in Banting, Best, Collip, and Macleod 1922A. There are variations in the description Best gave in his Feb. 22, 1954 letter to Dale. In later accounts of the first clinical test (e.g., Best 1956), Best stressed that he would have much preferred to use foetal pancreas, that he had said this to Banting at the time, but was persuaded that they had to use a commercially available source. Had extracts of foetal pancreas been used, Best seemed to be saying, they would have been more successful. Such an extract was not used. Whether or not it would have been more successful is not known. Best sometimes cited clinical tests done many years later which showed the potency of foetal pancreas (Salter, Sirek, Abbott, and Leibel 1961), but a close examination of the method of extraction used in that study shows it was not identical to that published or noted by Banting and Best in 1921-22.

36 Campbell interviewed by Dr. Robert L. Noble, c. 1967; in Macleod and Campbell 1925, Campbell describes it (p. 68) as "a murky, light-brown liquid containing much sediment, which dissolved to a considerable extent on being warmed."

37 Macleod and Campbell 1925, p. 68.

38 Banting 1940, pp. 52-3.

39 Banting, Best, Collip, Campbell, and Fletcher 1922.

40 The summary of Leonard Thompson's medical record that I was able to obtain has a note on January 12 (or January 18) referring to an "area of induration with soft centre over left buttock." Banting 1922 refers to sterile abscesses. Best 1922 refers to "severe local reactions"; see also Banting, Best, Collip, and Macleod 1922A in which the reference to abscesses is probably to both Thompson and the dog Marjorie.

41 Collip 1923K refers to "a few patients"; Best 1922 says "two patients"; Banting 1923A, B, and 1925 all say "three cases."

42 Banting 1923A, B. Campbell interviewed by Robert L. Noble, c. 1967; Campbell states that three patients were given single injections.

43 Banting 1925. He also suggested in this lecture and later accounts that the effect of the test on Thompson was to cause Macleod to focus his whole attention and laboratory resources on the extract. In fact the informal decision to expand the team came before, and the formal agreement to divide the work came after this abortive and unfortunate clinical test.

44 Banting 1929. Campbell's view, expressed in 1946, was that "The earliest results would not have convinced anyone familiar with the variations to be expected in diabetics under treatment," but there was "some encouragement to continue." Campbell 1946, p. 100.

45 BP, 26, desk calendar, March 13, 1923; Thompson medical record.

46 See Collip 1923K; Banting, Best, Collip, and Macleod 1922A; CP, Medical, "The Contributions of J.B. Collip to the Discovery of Insulin..."

47 See *Canadian Annual Review, 1922;* Bliss 1978.

48 *Star*, Jan. 14, 1922; Macleod 1922/78; for Greenaway's role in covering insulin, see his 1966 memoirs. Although his first reference is to his article

in March, I assume Greenaway was the reporter involved in this first scoop. He might not have been.

49 Some of the early inquiries are in the Insulin Committee alphabetical and geographical files. To be fair to the press, there were other inquiries from people who had heard of the work through doctors who had been at New Haven. For Banting's concern about the article, see Macleod 1922/78.

50 Banting 1940, p. 51, says that Starr phoned Andrew Hunter, who corroborated Banting's statements about Macleod.

51 Macleod 1922/78; Banting 1940, p. 51.

52 CP, privately held, undated handwritten letter listing Collip's contributions and apparently dating the isolation of the extract on January 17. However, Alison Li, in *J.B. Collip & the Development of Medical Research in Canada* (2003), cites convincing evidence for the 19th.

53 CP, Special Collections, Collip to Dr. C.F. Martin, Nov. 23, 1949.

54 CP, Medicine, copy of remarks made by Farquharson at N.R.C. dinner in Ottawa, Nov. 14, 1957.

55 Banting 1940, p. 54.

56 FP, Best to Dale, Feb. 22, 1954.

57 Noble Papers, October 1971 account. Banting's 1922 reference was as follows:

> Shortly before January 25, 1922, Dr. Collip, who had been working in the laboratory of Dr. Harding, in the Pathology Building, announced that he had developed a process by which he could obtain an extract which contained no protein and no lipase. On being asked his methods of preparation he refused to tell them. This was a breach of a gentlemen's agreement amongst Dr. Collip, Mr. Best, and myself, as we had agreed amongst ourselves to tell all results to each other. Dr. Collip discussed this new preparation with Professor Macleod and secured the consent of Professor Macleod to keep the process a secret. I believe that Dr. Collip at this time endeavoured to patent this process, and was only prevented from doing so by Professors Macleod, Hunter, and Henderson.

A completely garbled reference to these events is also contained in a letter written to Dale on Feb. 2, 1932, by A.B. Macallum; it is in the Best file of the H.H. Dale papers at the Royal Society, London.

For years in Toronto the story of the Banting-Collip fight was the piece of insulin gossip passed on in the hushest, or most inebriated tones. An enterprising reporter for *Time* magazine printed a garbled version of it in Banting's obituary on March 17, 1941.

58 The only copy of this agreement I have located is in the Banting scrapbook, BP, 48, p. 61.

59 That Fitzgerald had these fears is mentioned in the Macallum letter, note 57 above.

60 Correspondence between Toronto's president, Sir Robert Falconer, and Alberta's president, H.M. Tory (Falconer Papers, case 74), suggests that in late January Collip had been considering staying permanently in Toronto and was mulling over whether or not to accept a position at what Falconer called "a salary we can give him." Whether his future lay in Toronto or not may have been on his mind the night of the fight. Perhaps he reasoned that if he shared his process the Toronto group would take it over and he would wind up back in Edmonton without either an exclusive process or a share in the glory. Such a line of reasoning would have been fairly prescient.

61 In re-creating these events my first instinct was to assume that Banting's confrontation with Macleod about stealing his results was part of the same blow-up that started the fight, and that the Connaught agreement was the settlement of the two confrontations. Such an interpretation is not impossible, but careful study of the wording and tone of the documents, especially Macleod's description of his meeting with Banting, suggests they were actually separate incidents a few days apart.

62 Banting, Best, Collip, Campbell, and Fletcher 1922.

63 Banting and Best 1922C.

64 Banting 1940, p. 36.

65 Ibid.

66 Banting and Best 1922C. They went on to say that they were repeating the experiment and would report further results. They never did.

67 See, for example, Banting and Best 1922C and Macleod 1926. In Banting and Best 1922B the nodule became only two mm. in diameter. The qualification is in Banting, Best, Collip, Campbell, Fletcher, Macleod, and Noble 1922, though not Banting, Best, Collip, Campbell, and Fletcher 1922. In Banting's Nobel lecture (1925) he mentions the autopsy as failing to find any islet tissue. In his Cameron lecture (1929) the autopsy is not mentioned. In his 1940 account the autopsy reveals "a microscopic group of cells so small that it was agreed that they could not be responsible for the survival of the dog."

68 Gilchrist, Best, and Banting 1923 dates the test as February 4. The more likely date, given in Banting, Best, Collip, Hepburn and Macleod 1922, was Feb. 17. Whether this establishes Gilchrist as the second person to be injected with Toronto's insulin, as Gilchrist thought he was, cannot be determined.

69 Gilchrist, Best, and Banting 1923; Banting, Best, Collip, Hepburn and Macleod 1922.

70 See Banting, Best, Collip, Macleod, and Noble, 1922A; also Banting, Best, Collip, Campbell, Fletcher, Macleod, and Noble 1922.

71 Banting and Best 1922C; Macleod 1922/78.

72 Greenaway 1966; MP, Macleod to W.B. Cannon, April 29, 1922.

73 Banting, Best, Collip, Campbell, and Fletcher 1922.

74 In his 1967 interview with Robert L. Noble, Campbell stated that he and Fletcher wrote most of the paper. Banting 1922 indicates that Collip did some of the writing. For Banting's absences, see Macleod 1922/78.

75 Banting 1940, p. 54.

76 BP, 18, Notes re Cancer Research file.

77 BP, 26.

78 Banting 1940, p. 56.

79 The clinical tests are described in Paulesco 1923B, the "aglycemia" in Paulesco 1923A. It is remarkable that there has been no discussion of these experiments, except for a passing reference in Murray 1971, in the literature generated about Paulesco's work. The zero blood sugar observation was consistent with Paulesco's hypothesis that the internal secretion of the pancreas acted as a kind of catalyst or cement on the nutrients ingested by the body, enabling them to combine to form what Paulesco called a "plasmine" in the blood. See Paulesco 1920, pp. 301-305. In his model a zero blood-sugar reading meant that the extract was totally effective. In Paulesco 1924, by which time thousands of physicians had seen hypoglycemic reactions, he is still denying that hypoglycemia causes any abnormalities.

80 Nobel archive, Miscellaneous correspondence, 1923, Paulesco to Nobel committee 21 Dec. 1923.

81 I.C., Great Britain (General) File, Macleod to Sir Edward Shafer, Nov. 3, 1922; Sharpey-Shafer 1916, p. 128; de Meyer 1909B.

82 Macleod 1922/78.

83 Joslin 1922. The discussion was printed as a supplement to Banting, Best, Collip, Campbell, Fletcher, Macleod, and Noble 1922.

84 Campbell 1962.

85 Macleod 1922/78; also G.W. Ross account. Macleod later regretted not having insisted on Banting and Best coming to the meeting. As members of the group felt at the time, the expense excuse was somewhat thin, for both Banting and Best were being reasonably well paid. My guess is that Banting had decided to stay home in a fit of role-playing pique – the poor, humble outsider – and had persuaded Best to do the same

Chapter Six: "Unspeakably Wonderful' (Pp. 129-53)

1 For Collip's responsibilities, see Macleod 1922/78. A careful reading of Best's various accounts of his work in 1922 indicates that he did not take over direction of insulin production until after Collip left Toronto.

2 The exact date of the production failure is impossible to determine. Banting 1922 states that the supply failed on February 19. This is unlikely, inasmuch as there is no reference to any shortage in the Banting, Best, Collip, Campbell, and Fletcher paper, which has results to February 1922. In a letter written on April 29 (MP, to W.B. Cannon), Macleod states that the production failure developed after publication of that paper, which means after March 22. On the other hand, the Banting, Best, Collip, Campbell, Fletcher, Macleod, and Noble paper, delivered on May 3, states that for two months it had been impossible to secure potent extracts that could be used in the clinic. These confusing statements probably reflect a complex reality, in which the production breakdown was gradual, with the extract supply being inadequate at some times, adequate at others, the small-scale methods working when the large-scale failed (as mentioned in Macleod and Campbell 1925, p. 69), and so on, but with periods when nothing worked at all. April was certainly the cruellest month.

3 They may have exchanged letters, the letters Collip is said to have burned after hearing of Banting's death in 1941.

4 See Banting 1922, both rough and final drafts. The most extraordinary version of the fight story, apparently circulating as early as 1923, was that Collip lost the knack because one night in the lab an unknown assailant knocked him out and stole his confidential notes. See FP, file C2, C.I. Reed to Best, April 15, 1955.

5 Eli Lilly Archives, XRDe, Clowes memorandum; Sprague 1967.

6 Collip 1923K.

7 Campbell 1946; Banting, Campbell and Fletcher 1923; Campbell 1922. In his 1946 account, nowhere else, Campbell stated that this patient was the

first case they had seen of recovery, albeit temporary, from coma; because death from acidosis in humans is not strictly comparable to the condition in dogs, Campbell recalled, the incident was new and important and the final proof positive of the extract's value for humans.

8 Banting 1940, pp. 56-57.

9 IC, Clowes to Macleod, March 30, 1922.

10 IC, Macleod to Clowes, April 3, 1922.

11 CP, private, Banting, Best, Collip, Macleod, and Fitzgerald to Falconer, April 12, 1922.

12 IC, "K" file, Kendall to Macleod, April 10, 1922; no copy of Macleod's letter to Kendall on April 3 has been located.

13 Falconer Papers, box 81, C.H. Riches to Governors, University of Toronto, Feb. 12, 1923.

14 Ibid., insulin file; CP, private, Banting, Best, Collip, Macleod, and Fitzgerald to Falconer, April 12, 1922.

15 The best description of the teamwork leading to the regained process is in Best and Scott 1923D. A notebook in the Banting Papers contains records of the experiments Banting did with Best in April, mostly using glycerine as an extractive. These do not seem to have worked.

16 Best 1922; Hare, p. 94.

17 Macleod 1922/78; CP, private, undated hand-written letter listing Collip's claims; in later years Best several times remembered that he had begun using acetone first.

18 Interview with Moloney, June 20, 1980. For the wind tunnel and the old fan, see also Lilly archives, Gene McCormick interview with Best, Jan. 17, 1969. The best description of the process of manufacture is in Best and Scott 1923D. The formula is also to be found in the Lilly archives, XRDgb, Lilly research notes, volume 1. A handwritten copy of the formula also exists in the privately held Collip Papers. It is incorrectly dated, probably by Collip himself, December 22, 1921. He probably added this date many years later and mistakenly gave the day of his great success with the glycogen experiment.

19 Exactly how the Christie Street situation developed is unclear. An April 3, 1922, Memorandum to the minister, by the Director of Medical Services, Dr. W.C. Arnold, recommending the establishment of the clinic, is in the PAC, RG 32, C2, vol. 13, Banting personnel file. Arnold recommended establishment of the clinic as an excellent and justifiable step by the ministry. But there is some evidence that he had known Banting earlier and wanted to help him out.

20 No copy of this agreement has been found, but it is referred to in the Insulin Committee Minutes, Aug. 17, 1922.

21 Gilchrist, Best, and Banting 1923; Banting 1929.

22 Havens family papers, James Havens, Sr., to E.C. Gale, March 1, 1921.

23 Williams 1922.

24 All of the documents from the time suggest that Havens was the first. In a letter to Best on March 13, 1939 (BI, Best Papers, Historical file), however, Williams casually mentioned that he began giving extract to one Lyman Bushman, a veteran, on May 14, 1922. The Havens family correspondence shows this cannot have been true, and was a slip of Williams' pen or memory. Circumstantial evidence suggests that Bushman was first given insulin in June or July.

25 Havens family papers, James Havens, Sr., to George Snowball, May 24,

1922. Also Woodbury 1962: this account of the Havens case written for a mass magazine heavily emphasizes dialogue and drama and contains several clear factual errors.

26 Havens family papers, S.B. Cornell to James Havens, Sr., June 15, 1922.

27 BP, Havens File, Williams to Banting, June 5, 1922.

28 IC, Universities file, John Howland to Macleod, May 13, 1922; also MP, R. Carrasco-Formiguera to Macleod, May 19, 1922.

29 IC, Woodyatt file, Woodyatt to Macleod, May 10, 1922; reply May 15; Clowes to Macleod, May 11; reply May 15.

30 IC, Universities file, Macleod, Fitzgerald, Banting, and Best to Falconer, May 25, 1922. There seems to be no significance in Collip not having signed this letter. He may have been out of town when it was sent.

31 This account of Lilly and Clowes draws upon Eli Lilly archives, XRDe, Clowes' research reports; XCAe, J.K. Lilly, "A Plan for promoting the affairs of Eli Lilly, 1920-1923"; Dr. G.H.A. Clowes, Jr.'s biographical study of his father; and a centennial publication about Eli Lilly and Company, Kahn 1976.

32 IC, Universities file, Macleod *et al.* to Falconer, May 25, 1922. Early drafts of the Lilly agreement are in the Banting Papers and the CP privately held. A copy of the final agreement, not differing in any important detail from the drafts, is in the Eli Lilly archives. For Clowes' early proposals, see IC, Clowes to Macleod, March 30, 1922.

33 Banting, Best, Collip, and Macleod 1922A.

34 Eli Lilly archives, XRDgb, Research notes, vol. 1.

35 IC, Potter file, W.D. Sansum to Macleod, June 15, 1922.

36 IC, Committee of Clinicians file, Allen to Macleod, June 24, 1922.

37 IC, Potter file, Macleod to Sansum, June 21, 1922; Woodyatt file, Macleod to Woodyatt, June 21. Macleod may have had second thoughts about this policy, for a week later, replying to Allen's request, he did not send details, but simply promised a reprint of the paper giving the method when it was published. This became his standard reply to similar requests in the next several months. Of course it was also possible to learn the method from Collip. It is said, for example, that Woodyatt actually learned how to make insulin from a conversation with Collip.

38 Banting 1940, p. 33.

39 BP, 1, F.A. Hartman to Banting, May 2, 1921; see Williams 1947 and BP, Havens file, J.S. Havens to Banting, June 17, 1922, for George Eastman's interest in Banting coming to Rochester. John Harvey Kellogg had also offered Banting a job in his sanitarium in Battle Creek.

40 Falconer Papers, box 76, Falconer-Blackwell correspondence; Banting 1940, p. 61a.

41 Banting 1940, pp. 61a-63.

42 Falconer Papers, 76, Falconer to Blackwell, June 19, 1922; BP, Falconer to Banting, June 29, 1922; MP, folder 342, Graham to Macleod, July 14, 1922.

43 Best family papers, M.M. Best scrapbook, Banting to Best, July 15, 1922; Toronto *Star*, Feb. 24, 1923.

44 BP, Elizabeth Hughes file, A. Hughes to Banting, July 3, 16, 1922; MP, A.S. Ferguson to Macleod, July 31, Aug. 9, 1922.

45 Havens papers, Banting to James Havens, Sr., July 10, 18, 1922; Gilchrist, Best, and Banting 1923; Banting, Campbell, and Fletcher 1922, p. 550;

also Lilly archives, XRDc, John R. Williams to G.H.A. Clowes, Jan. 27, 1958: "One day Fred Banting took me up to the Christie Street military hospital where there were 8 soldiers each suffering horribly with large abscesses in hips-buttocks. I was having same trouble with Jim Havens and 3 other cases I had here."

46 IC, Committee on Clinicians file, Macleod to Williams, June 30, 1922; BP, Whitehall file.

47 Murlin's block came when he and Kramer found that the anti-glycosuric results obtained from administering an alkaline extract of pancreas could be achieved by administering a simple alkaline solution alone. That finding led Murlin to distrust his much more important results showing an increase in the respiratory quotient of some of his dogs that received extract. See Murlin and Kramer 1916, 1956.

48 BP, Havens file, J.S. Havens to Banting, July 14, 1922; Williams to Banting, July 11. Williams did not tell the Toronto people he was trying Murlin's extract, but mentioned it in a Feb. 25, 1939, letter to Best (historical files, Best Papers, BI). The most detailed story of Murlin's work with extracts is in Murlin, Clough, Gibbs, and Stokes 1923. See also the various articles by Murlin and his associates in the *Proceedings of the Society for Experimental Biology and Medicine,* XX, 1922-23.

49 BP, Clarke file, notation by Palmer; Banting 1940, p. 76e.

50 Lilly archives, XRDc, Williams to Clowes, Jan. 27, 1958; interview with Moloney.

51 IC, Clowes to Macleod, March 14, 1923.

52 IC, University of Toronto Miscellaneous file, Eadie to Macleod, July 7, 1922; MP, Macleod to Defries, July 14, 1922; Banting 1922.

53 BP, 1, Clowes to Banting, July 18, 1922.

54 Best family papers, M.M. Best scrapbook, Banting to Best, July 21, 1922.

55 Lilly archives, J.K. Lilly to Eli Lilly, July 26, 1922.

56 BP, Clarke file.

57 Banting 1940, p. 72e; also BP, 1, Banting to Falconer, Aug. 5, 1922; Falconer Papers, box 76.

58 BP, 1, Clowes to Banting, Aug. 8, 11, 1922.

59 Lilly archives, XBLk, J.K. Lilly to Clowes, Aug. 4, 8, 1922.

60 Ibid., Aug. 8.

61 Ibid., Aug. 8, 11.

62 Ibid., Aug. 4, though this may be an error by J.K. Lilly; the historical memory at Lilly is that testing was delayed in Indianapolis until Joslin gave his first injections.

63 Lilly archives, Joslin address at the dedication of the Lilly Research Laboratories, Indianapolis, October 1934.

64 Joslin, Gray, and Root 1922; FP, 1924 file, Feasby transcript of dictation by Joslin, Nov. 22, 1957.

65 Kienast, 1938. In her recollection the author dated this event in March 1922, a very unlikely time. I believe she was wrong, and that these days in August are the most likely dates of the scenes she remembered so vividly. Possibly they took place in May when Allen went to the A.P.A. meeting.

66 BP, 1, Allen to Banting Aug. 16, 1922; Elizabeth Hughes file.

67 BP, Elizabeth Hughes file; EH to her mother, Aug. 22, 1922.

68 BP, Elizabeth Hughes file.

69 EH to her mother, Aug. 22, Oct. 1, 1922.

70 Banting 1940, p. 65a; with telling, Banting's one suit became shabbier and shabbier, covered with dog hairs and dung.
71 Ibid., p. 59.
72 EH to her mother, Aug. 22, 1922.
73 Ibid., Oct. 6.

Chapter Seven: Resurrection (Pp.154-88)

1 Banting 1940, pp. 77-8.
2 EH to her mother, Sept. 24, 1922.
3 Ibid., Sept. 29, Oct. 1.
4 Ibid., Oct. 17.
5 Ibid., Oct. 25.
6 Ibid., Oct. 14.
7 IC, Minutes, Sept. 1, 22, 1922; MP, folder 342, Duncan Graham to Macleod, Aug. (misdated July) 14, 1922.
8 MRC 1092/19, Clowes to H.H. Dale, Feb. 12, 1923; Lilly archives, XRDe, General Letter to Salesmen, #13, Feb. 20, 1923.
9 Stevenson 1946, p. 140; Gilchrist, Best, and Banting 1923.
10 Banting 1924.
11 Best Papers, BI, historical file, Williams to Best, March 13, 1939.
12 This and all other material on the clinical testing is drawn from the papers published in the November 1922 issue of the *Journal of Metabolic Research,* which was not actually published, however, until May 1923.
13 See the clinical papers and MP, Macleod to E.H. Starling, Nov. 7, 1922.
14 EH to her mother, Nov. 19, 1922.
15 Dr. Randall Sprague to author, Oct. 27, 1980.
16 Allen and Sherrill 1922B, pp. 811-16.
17 BP, 48, p. 118, undated newspaper clipping.
18 Joslin, Gray, and Root 1922.
19 Hospital for Sick Children, *Annual Report 1923,* p. 21. Testing was slow to get under way at the Hospital for Sick Children, where Banting had been a resident, apparently because Dr. Allan Brown, the hospital's head of medicine, did not like Banting. They were finally reconciled by Dr. D.E. Robertson. See Banting 1940, p. 60.
20 Macleod and Campbell 1925, p. 77.
21 Allen and Sherrill 1922B, pp. 811, 831.
22 Williams 1922, p. 734.
23 IC, Woodyatt file, Woodyatt to Macleod, Oct. 4, 1922.
24 EH to her mother, Nov. 26, 1922.
25 Lilly archives, Joslin address at the opening of the Lilly Research Laboratories, 1934.
26 MP, R. Carrasco-Formiguera to Macleod, May 19, 1922; IC, Spain file, Carrasco-Formiguera to Macleod, Oct. 5, 1922; Carrasco-Formiguera 1922, 1972. In his 1972 article he misdates the injection as October 4.
27 IC, Great Britain (General) file; MRC, 1092/5, Meakins to Fletcher, Jan. 30, 1922.
28 Interview with Sir Harold Himsworth, Sept. 30, 1980.
29 MRC 1092/23, Fletcher memorandum, July 7, 1922.
30 Ibid., Dale to Fletcher, Sept. 26, 1932.
31 Ibid., H.H. Dale and H.W. Dudley, "Report to the Medical Research

Council of Our Visit to Canada and the United States...", Oct. 30, 1922; W. Fletcher memorandum for members of the MRC, "Insulin Treatment of Diabetes", Nov. 8, 1922.

32 Cammidge 1922A. He compounded his misjudgement five months later by declaring that the insulin treatment, for the average patient, was "a will-o'-the-wisp, reliance upon which usually ends in disappointment." Cammidge 1922B.

33 MRC 1092/17, Fletcher to the Minister of Health, April 7, 1923.

34 MRC 1092/1A, O. Leyton to Fletcher, Dec. 5, 1922. A question was asked in the House of Commons as to why the council was "throwing obstacles in the way" of insulin production: House of Commons, *Debates*, Dec. 13, 1922.

35 For Paula's illness see Inge, *Diary*, pp. 72, 85, and *Personal Religion*, pp. 87f. There is no written record of the Dale-Inge exchange, but it was told to me by half a dozen people, including Dale's daughter. The one reference to Inge's inquiries in the MRC records is 1092/1A, G. Adami to Fletcher, Dec. 8, 1922.

36 MRC 1923. In addition to the supply problem, a deliberate decision had been made to concentrate on a few very severe cases rather than spread insulin thinly among many patients.

37 See MRC 1092/10, Fletcher to Sydney Holland (Viscount Knutsford), Jan. 8, March 2, 1923.

38 MRC 1092/17, Fletcher to Newman, March 28, 1923.

39 See note 35 above.

40 For Krogh's visit see IC, unsorted, Krogh to Macleod, Oct. 23, Dec. 16, 1923; Best family papers, Macleod to Krogh, Oct. 27, Nov. 7; BP, 7, handwritten note by FGB, Nov. 24, 1922, in Notes on Diabetes file; also Paulson 1975.

41 Nelken 1972; Lusk 1928, p. 650.

42 IC, unsorted, Minkowski to Macleod, Jan. 12, 1923.

43 Best Papers, BI, historical file, sent to Best by Dr. Goldner.

44 BP, 9, Best to Clowes, Aug. 28, 1923; see also IC, France file, and IC, Lorne Hutchison report to the Insulin Committee, Jan. 14, 1925.

45 Gley 1922.

46 BP, 1, Paulesco to Banting, 5 fév. 1923.

47 Macleod 1922A; Cammidge 1922A. It was another two decades, however, before it could be shown conclusively that insulin was produced in the beta-cells of the islets of Langerhans.

48 IC, Woodyatt file, Macleod to Woodyatt, Sept. 7, 1922.

49 Macleod 1922A; Dale 1959 for the estimate of Connaught's yield.

50 Connaught Laboratories Archives, HI, Best to J.C. Fitzgerald, Nov. 11, 1922; IC, Macleod to E.H. Mason, Macleod to Charles Hunter, Sept. 23, 1922; BP, 47, clipping May 17, 1923.

51 Connaught archives, Best to Fitzgerald, Nov. 11, 1922.

52 Dale 1959, p. 6: "Dudley, by systematic trial, soon found that the real key to success was to conduct all filtrations at reactions sufficiently far, in either direction, from the isoelectric point, to prevent the loss of insulin by absorption, especially on clogged filters."

53 IC, unsorted, Clowes to R.D. Defries, March 13, 1923.

54 Walden's description of his process is enclosed in MRC 1092/23, Clowes to Dale, Jan. 17, 1923. See also IC, unsorted, Clowes to Defries, March 13, 1922; Lilly archives, K.W. Wantland, "Notes on Insulin Extraction 1922", Sept. 2, 1966.

G.H.A. Clowes' research report for 1921 (Lilly archives, XRDe) mentions that Walden had been working that year on the isoelectric points of materials suspended in various preparations marketed by Lilly. Techniques involving adjustments of H-ion concentration had only recently been introduced in the early 1920s and were not yet applied routinely, which may explain why, even if he knew the method, Walden took several months to make it work with insulin. Collip's accounts indicate that he, too, had earlier fiddled with adjustments of the isoelectric point, but to no avail.

55 MRC 1092/23, Clowes to Dale, March 16, 1923; also IC, unsorted, Clowes to Macleod, March 14, 1923.

56 The cost for those who had to pay was even higher during the experimental period. Connaught charged Banting $1.00 per cc. (old unit) for his insulin; most patients used two to five cc. daily. At the Potter Clinic in Santa Barbara the cost of insulin was $10.00 per patient-day in August, reduced to $3.00 by February (IC, Sansum to Macleod, Feb. 16, 1923).

57 IC Minutes, Feb. 14, March 5, 1923. IC, Clowes to Macleod, March 7, summarizes the group's understanding of what it wanted to do with the public statements:

> Physicians are to be informed in the statement that on account of the risks associated with the use of Insulin, everything possible should be done to safeguard the patient and consequently those desiring to use it should read literature referred to, and should visit one of the group of clinics now using Insulin in different parts of the United States in order to familiarize themselves with the best means of adjusting the dosage, thus safeguarding their patients against risk of overdose and also learning how to treat acidosis and coma. The statement in question to be so worded that the whole responsibility for the use of Insulin will be put on the shoulders of the doctor, it to be made clear that only those having a knowledge of metabolism and possessed of adequate clinical and laboratory facilities could hope to handle Insulin successfully. The statement to be so worded as to frighten off utter incompetents and yet not to positively restrict the use of Insulin to those who were prepared to visit one of the clinics. This course, as Sir Robert Faulkner [*sic*] pointed out, would put the responsibility entirely up to the individual doctor and could never be construed as reflecting on the status of any member of the medical profession.

58 MRC 1092/17, Insulin Committee (Great Britain), Minutes, May 16, 1923.

59 MRC 1092/25, Dale to Dr. R. Obrian, Jan. 31, 1924; 1092/17, Fletcher to Sir George Newman, Jan. 23, 1924; 1092/17, I.C. (U.K.) Minutes, Jan. 29, 1924. Fletcher wrote (1092/2 to Elliott, Jan. 30) that the Insulin Committee had decided that "They cannot force practitioners to educate themselves, and they cannot dictate, even to panel practitioners, about their methods of work. You will realise that we are very far yet from a State Medical Service! Waste and disaster attending Insulin use are probably no greater than those attending the use of other means in other directions, e.g., the use of pituitrin at child-birth. The view taken is that the only cure for the present evil is the spread of knowledge by the papers or text-books which the experts are writing or may write, and of course more generally by improvements in medical education."

60 MRC 1092/10A, Clowes to Riches, July 20, 1923.

61 IC, Clowes to Macleod, April 14, March 14, 1923.

62 BP, 1, Banting to F.M. Allen, Aug. 31, 1922. "Adrenaline" was actually the original scientific term; when Parke Davis managed to trade-mark "Adrenalin" for the United States, American scientists had to fall back on the Greek-derived "epinephrine".

63 IC, Clowes to Macleod, Sept. 5, 1922; Clowes to Best, Sept. 12; Minutes, Sept. 22.

64 MRC 1092/23, Dale to Fletcher, Sept. 26, 1922; also 1092/23, Dale and Dudley's report to the MRC, Oct. 30, 1922.

65 IC, Digestive Ferments file; see also Armour, Harrower, and Hoaxes files.

66 See Mackenzie Wallis 1922, Crofton 1922.

67 MP, Murlin to Macleod, Sept. 25, 1922; E.L. Scott to Murlin, Sept. 8, Murlin to E.L. Scott, Sept. 25, in A.H. Scott 1972, pp. 143-4; Scott Papers, case 3, Murlin to Scott, Sept. 5, Oct. 10, 1922; IC, Minutes, Sept. 28, 30, 1922; IC, "W" file, Murlin to C.H. Riches, Oct. 7, 1922; Sutter and Murlin 1922; Murlin 1922; Clough, Stokes, Gibbs, Stone, and Murlin 1923.

68 IC, Patents United States file, Collip and Best application; Banting, Campbell, and Fletcher 1923.

69 The patent applications, amendments, etc., and most of the relevant correspondence is in the IC, Patents United States file. It includes C.E. Hughes to Hon. Thomas E. Robertson, Nov. 24, 1922. Also BP, 1, Banting to Hughes, Nov. 21, 1922, reply Nov. 25; IC, "W" file, Macleod to Woodyatt, Nov. 29, 1922.

For Williams' "espionage" see IC, Committee of Clinicians file, Williams to Macleod, Nov. 27, 1922; Lilly archives, XRD2f, Williams to Clowes, Nov. 15, 1922. James Havens Sr. was also in Washington at the time of the hearing and might have been involved.

70 Collip and Best's first application on May 22, 1922, had been for a patent on the process only. On advice from the Lilly people, Toronto decided to file a separate application in Best's name, dated June 19 (Collip was in Alberta at the time), on the anti-diabetic pancreatic product. In December the Collip-Best application was amended to include the product as well as the process. IC, Patents United States file.

71 IC, Minutes, Dec. 11, 1922; BP, 1, Banting to Falconer, Jan. 27, 1923; also BP, patent files, esp. George Schley to Eli Lilly & Co., Aug. 22, 1922.

72 IC, Minutes, Sept. 30, 1922.

73 IC, Minutes, Dec. 30, 1922; Macleod to Clowes, Jan. 2, 1923.

74 IC, Clowes to Macleod, Jan. 8, April 14, March 14, 1923. There is no doubt that the Lillys laid down the law on Iletin. Clowes tried to argue to Toronto that they had misinterpreted his personal views for a company commitment.

75 IC, unsorted, J.K. Lilly to Clowes, Jan. 3, 1923.

76 IC, unsorted, Clowes to Defries, March 13, 1923; also Clowes to Macleod, March 14.

77 IC, March 16, 1923; also Minutes, March 16.

78 IC, unsorted, Riches to Macleod, April 3, 1923.

79 IC, Minutes, April 2, 1923; IC, unsorted, Macleod to Dr. John Howland, April 2, 1923.

80 Earlier, it seems, Shaffer had consented to Lilly patenting the method. See MRC 1092/23, Macleod to Dale, Jan. 17, 1923. Shaffer was a purist who wanted to have nothing to do with commercialism or patenting. Defries'

reminiscences of that visit, related to me by Dr. Neil MacKinnon, were that Shaffer had at first flatly refused to have anything to do with Toronto's and Lilly's fights. The Torontonians might as well go home, but of course he would show them around St. Louis first and they would dine together. At the end of that meal Shaffer said to Defries: "Get rid of the lawyer." When they were alone, Shaffer said that the day together had convinced him that Toronto wasn't in it for the money, but really was interested in the good of humanity. "How can I help you?" he said. The work in Shaffer's lab is described in Somogyi 1951.

81 IC, Clowes to Macleod, April 8, 1923. While Lilly always believed that Walden's method had priority over Shaffer's, this would have been impossible to prove. The discoveries actually were independent and simultaneous, and Shaffer might have had a claim to priority by virtue of a public announcement of his discovery in December 1922. See Doisy, Somogyi, and Shaffer 1923.

82 IC, Minutes, April 15, 1923; also correspondence in the AMA file, especially Macleod to W.A. Puckner, April 20, in which he writes that the Iletin matter is "a concession to Eli Lilly & Company in consideration of their having agreed to turn over to the University of Toronto all patents applied for by them covering improvements in their original method of manufacture."

83 MP, P.A. Shaffer to Macleod, Oct. 10, 1923. IC, Lilly, Clowes to Macleod, Nov. 2, 1936, in which he suggests that the Lilly concession had saved Toronto from losing credit.

84 Collip 1923G.

85 Best and Scott 1923A. Macleod and Noble had also tried adding insulin to yeast and sugar, with no result: Macleod 1926, p. 139.

86 Collip 1923B; MP, Collip to Macleod, July 18, 1922.

87 IC, Collip file, Collip to Macleod, Jan. 22, 1923; Collip to Defries, Jan. 4; MP, folder 342, Collip telegram to Macleod, Feb. 9, 1923: "So much of my time is going into insulin production that urgent research problems are being sacrificed. A sufficient supply of insulin to care for our few cases here must now be available from Lilly. You are supplying Vancouver and others, why not Edmonton, and thus give me a fair chance at the research side?"

88 Collip 1923D, p. 520. Collip was influenced by findings of Winter and Smith in Britain to the effect that insulin's function was to create gamma-glucose, which they thought might be the essential form of sugar in all animals and plants.

89 The correspondence about yeast is in MRC 1092/13, especially two letters of Hopkins to Fletcher on Feb. 20, 1923. See also Winter and Smith, 1923C, D.

90 Best and Scott 1923A.

91 MRC 1092/19, Clowes to Dale, April 3, 1923. Lilly's thoroughness also included an experiment on a human pancreas, recovered about four hours after death. It contained no active insulin. Lilly archives, XRDqb, Laboratory notes, vol. 1, p. 57.

92 IC, clippings file, Minneapolis *Forum*, Feb. 9, 1923. The Fleischmann yeast company observed the race with interest and enthusiasm, supplying Toronto with free yeast for its experiments.

93 Collip 1923C.

94 Collip 1923E.
95 Collip 1923D, pp. 526f.
96 IC, clippings file; BP, 47, p. 74.
97 MP, folder 342, Collip to Macleod, April 21, 1923.
98 Collip 1923F, H.
99 Best and Scott 1923B.
100 MP, Macleod to Scott, April 23, 1923; Macleod 1926, pp. 292-3; Macleod and Campbell 1925, pp. 17-18; Macleod 1924B, p. 65; MP, Macleod to Dale, May 15, 1923.
101 Robert Noble interview with Harold Ettinger, c. 1967.
102 See MRC 1092/13.
103 Best and Scott 1923C; Best, Smith, and Scott 1924; Best 1924.
104 Academy of Medicine, Toronto, E.C. Noble Papers, Macleod to Noble, undated but clearly summer of 1923.
105 McCormick and Noble 1924; McCormick 1924; IC, University of Toronto Miscellaneous file, A.G. Huntsman to J.G. Fitzgerald, July 14, 1923, reply July 18.
106 IC, clippings file, *Royal Gazette* (Bermuda), May 18, 1923; Lilly archives, XRDgb, Laboratory notes, vol. 2, p. 24; MRC 1092/11, Dale to Fletcher, Jan. 1, 1924; IC, Great Britain file, Dale to Macleod, Feb. 15, 1924.
107 MRC 1092/23, Macleod to Fletcher, Sept. 28, 1923; Fletcher to Macleod, Sept. 27.
108 MRC 1092/23, Dale to Fletcher, Oct. 11, 1923.
109 Jonathan Meakins Papers, Dale to Meakins, Dec. 6, 1923; pun by D. Hannay.
110 MRC 1092/19, Dale to Fletcher, Oct. 19, 1923.

Chapter Eight: Who Discovered Insulin? (Pp. 189-211)

1 See Bliss 1978.
2 *New York Times*, Oct. 8, 1922.
3 *Ledger*, Oct. 13, 15, 1922; IC, state and alphabetical files; see also New York *American*, Oct. 22, 1922, for F.M. Allen's attempts to correct these misconceptions.
4 *Star*, and other Toronto newspapers, Nov. 27, 29, 1923. The *Star* added the qualification.
5 IC, clippings file, syndicated article by "Harley Street Doctor", May 1923; Toronto *Globe*, May 17, 1923; see also the confusion stemming from an interview with Geyelin, Toronto *Star*, Feb. 10, 1923.
6 EH to her mother, Oct. 21, Nov. 21, 1922; for coverage of her see BP, 47, p. 23.
7 BP, 47, p. 73, Corbett clippings, April 1923; *Ibid.*, Vanderlip, April 11; *Star*, May 21, 1923.
8 *Star*, June 20, 1923; *Star*, and other papers, May 19, 1923; see also Slosson 1923, which was commissioned by the Insulin Committee.
9 MRC 1092/23, Macleod to Dale, Feb. 6, 1923; interview with Professor Robert Garry, Oct. 10, 1980; BP, 47, p. 75, clipping March 14, 1923.
10 BP, 47, p. 80, undated clipping, Hadwen letter to the Toronto *Globe*.
11 Mackenzie King Papers, J2, vol 13, H1000, undated clipping.
12 BP, 47, clippings c. Feb. 1923; *Ibid.*, p. 63, postcards and clippings on

vivisection; p. 30, Falconer statement, June 1, 1923.
13 Joslin 1922; Banting 1940, p. 60, seems to condense events between May and July.
14 FP, "Hot" file, Bayliss to Macleod, Sept. 29, 1922.
15 Banting 1922; Noble papers, ms. notes on the discovery of insulin.
16 *Star*, Sept. 8, 1922.
17 CP, private, Macleod to Collip, Sept. 18, 1922; Macleod 1922; also MP, folder 342, Macleod to A.B. Macallum, Sept. 14, 1922: "Now he claims that I should in the same way give him full credit for all the work which has been done subsequent to this [duct-ligation] experiment. This I will of course not do since he has participated very little in the work, and not at all during the past six months."
18 Toronto *Globe*, Sept. 9, 1922; Macleod 1922.
19 CP, private, Macleod to Collip, Sept. 18, 1922; FP, "Hot" file, Macleod to Sir Wm. Bayliss, Sept. 13, 1922; MP, folder 342, Macleod to A.B. Macallum, Sept. 14, 1922.
20 MP, folder 342, Macleod to A.B. Macallum, Sept. 14, 1922.
21 BP, 1, Gooderham to Banting, Sept. 16, 1922; MP, Gooderham to Macleod, Sept. 16.
22 Macleod 1922/78.
23 Banting 1922.
24 Best 1922.
25 See the note Macleod added to his 1922/78 account.
26 BP, 1, Banting to Macleod, Sept. 27, 1922; Banting to J.G. Fitzgerald, Oct. 5, 1922.
27 Banting 1940, pp. 61-64; interview with Stella Clutton, 1980; see also BP, 37, Banting notes on his position, early Oct. 1922; and case 1 for his coldly formal correspondence with Graham.
28 Wilder 1963.
29 FP, "Hot" file, Macleod to Bayliss, Sept. 8, 13, 1922; Bayliss 1923; MP, folder 342, Macleod to Collip, Feb. 28, 1923.
30 Banting 1940, pp. 33-34.
31 Roberts 1922.
32 Macleod 1923.
33 Dale 1922.
34 Pratt 1954ms, p. 31a.
35 Banting 1929.
36 Pratt 1954 and 1954ms. Pratt once remarked that while Banting may have been the quarterback of the team, it was Collip who scored the winning touchdown.
37 Feasby 1958.
38 See Pratt 1954ms. It was Macleod's erroneous understanding on this point, however, that seems to have inspired Dale's oft-repeated quip that insulin could only have been discovered in a lab whose director was slightly stupid – a revealing comment in view of Dale's having publicly scolded Roberts for scornfully belittling Banting and Best.
39 The unpublished version of Joseph H. Pratt's reappraisal contains a rather different, also meritorious, summary of the steps and those responsible for each, in the discovery of insulin as a therapeutic agent:

> There were eight distinct and essential steps or stages leading to the production of an insulin that was sufficiently pure to be used in the

treatment of diabetes and which could be produced in adequate quantity to meet the urgent demand. The first of the steps was the use of alcohol in extracting the hormone from the minced pancreas. This was announced by Zuelzer in Berlin in 1907. It was rediscovered by the Toronto investigators in 1921, who did not know of Zuelzer's work. The second was a method of determining the amount of sugar quickly and accurately in a small quantity of blood. This was devised by Lewis and Benedict in New York in 1913. The third was the discovery of Kleiner and Meltzer in 1915 that an aqueous extract of the normal pancreas injected into the veins of a normal dog reduced hyperglycaemia as well as glycosuria. The fourth step was the discovery that the active principle was insoluble in 95 per cent alcohol. This was made by Banting, Best and Collip in December 1921. The fifth was the preparation by Collip of the first relatively non-toxic insulin to be used in the treatment of diabetes (January 1922) with success. The sixth step was the physiological assay based on Collip's observation (February 1922 [actually December]) that insulin in adequate amount usually produced convulsions in normal rabbits when the blood sugar fell to 46 mg. per 100 cc. The seventh was the discovery by Doisy, Somogyi and Shaffer of St. Louis, that insulin was precipitated at the iso-electric point....The eighth and final step was the development of methods of large-scale production by the chemical engineers of the Eli Lilly Co. of Indianapolis in 1922-23.

Chapter Nine: Honouring the Prophets (Pp. 212-33)

1. Rochester *Democrat-Chronicle,* June 4, 1923. See also N.Y. *Tribune,* July 8, 1923, "Wounded hero gave Insulin to humanity"; and BP, 47, p. 81, undated clipping. Paul De Kruif, perhaps the most famous popularizer of the triumphs of medical research, contributed his share to the myth-making in a long article syndicated by Hearst International in November 1923, and republished in his 1932 anthology, *Men Against Death.*
2. Toronto *Star,* August 24, 1923; Ibid., Feb 24; Toronto *Evening Telegram,* Jan. 18, 1923; New York *Tribune,* July 8, 1923; BP, 49, p. 22, undated clipping.
3. BP, 47, p. 44, undated clipping. Of Banting's speech, the reporter wrote that he was "painfully embarrassed, broke brisquely into his exposition without remark and literally fled from the hall after two minutes of nerve-cramped, graceless, almost inaudible explanation of his discovery." See also Banting 1940, p. 66, for his own recollection of how badly he spoke compared to Voronoff.
4. MP, folder 342, Macleod to Collip, Feb. 28, 1923.
5. BP, 26, desk calendar, May 8, 1923.
6. Mackenzie King Papers, *Diary,* Feb. 13, 1923; personal communication.
7. Toronto *Evening Telegram,* Oct. 13, 1933; New York *Tribune,* Oct. 14; McGill University Archives, Principal's files, 641, A.L. Lockwood to Sir Arthur Currie, Jan. 2, 1923 (misdated 1922); reply Jan. 23. I am grateful to Marlene Shore for bringing this exchange to my attention.
8. House of Commons, *Debates,* Feb. 27, 1923; pp. 729-30; Toronto *Star,* March 1, 1923.

9 BP, 26, desk calendar, Feb. 27, 1923. Billy Ross later told Lloyd Stevenson he had proposed the research professorship to the provincial government in mid-January. Stevenson 1946, p. 178.

10 BP, 26, desk calendar, March 13, 1923; the correspondence and the testimonials are in case 1, c. March 1923; see especially Banting to Mrs. Hughes, March 14, which may also have been drafted for Banting by Ross. In it Banting writes of Ross's letter: "I am not completely familiar with its contents, but know it is concerned with certain proposed actions of the Dominion Government with reference to myself."

11 A politically active medico on the Conservative side who was also helping Banting was Dr. Herbert Bruce. Both Ross and Bruce were simultaneously leading a struggle of old-guard Toronto practitioners against reforms in the university's faculty of medicine which were giving increased power to full-time faculty members like Duncan Graham. There is some evidence that Bruce and Ross found Banting a useful stick with which to batter further the university's establishment; he would have been more useful still had not his hero, C.L. Starr, been one of the chief establishment figures that they wanted to batter.

12 Interview with Edward Banting, Alliston, 1981.

13 King Papers, JI, 77025-30, Mulock to King, March 21, 23, 1923.

14 BP, 1, Lillian Hallam to Banting, March 6, 1923; reply March 13.

15 Drury Papers, Banting to Drury, April 4, 1923.

16 CP, UWO, Medicine, signed but undated second page of a letter from Collip to a friend, probably written in the summer or autumn of 1923, discovered in files stored in the biochemistry department. For Collip's other accounts see chapter 4, note 42.

17 Calgary *Herald*, May 30, 1923; Edmonton *Journal*, May 25.

18 King Papers, JI, 74205-09, Hughes to King, March 16, 1923, reply April 6; 77829-30, Ross to King, and reply.

19 King Papers, JI, 72674-6, King to Falconer, June 1, 1923.

20 BP, 47, p. 118, clipping from the *Star*, Nov. 7, 1923.

21 King Papers, JI, 72677-8, Falconer to King, June 2, 1923. The only other statement of Falconer's is in a letter to Dr. Lewellys F. Barker, May 11, 1923 (Falconer Papers), in which he says that the discovery "could not have been made unless Macleod and his associates in the Physiological Laboratory had not been ready to help Banting in his work."

22 King Papers, JI, 78836, Ross to King May 8, 1923; BP, 47, pp. 43, 69, clippings re Academy and CMA; Montreal *Gazette*, June 14, 1923.

23 House of Commons, *Debates*, June 27, 1923, pp. 4591-4.

24 BP, 1, Mulock to Ross, undated, in July 1923 file.

25 King Papers, JI, 80795-6, C.A. Stuart to King, June 27, 1923.

26 IC, unsorted, Fitzgerald to King, June 27, 1923.

27 BP, 1, Best to Banting, June 28, 1923.

28 BP, 1, Banting to Hipwell, undated in July 1923 file; Hipwell to Banting, July 28, 1923.

29 Best Papers, private, MMB scrapbooks, Banting to Best, July 15, 1923.

30 IC, unsorted, King to Fitzgerald, July 5, 1923; King Papers, JI, 80797-8, King to C.A. Stuart, July 3.

31 Banting 1940, p. 66; MacFarlane, *Florey*, p. 77.

32 MP, Macleod to Prof. A.R. Cushny, March 6, 1923.

33 Macleod 1923; BP, 26, desk calendar, July 26, 1923. In 1940, p. 66, Banting wrote of Edinburgh: "There could be no doubt in the minds of the

listeners that Macleod was the discoverer of the physiological principles of insulin."

34 The following account has been compiled almost exclusively from documents in the Nobel archive at the Karolinska Institute in Stockholm. In most cases, the identification of the documents in the text - Nobel Committee minutes, nominations, investigations, etc. - is sufficient identification to locate the original in the archive. The original documents are in Swedish or the native language of their authors. The standard history of the Nobel prizes is *Nobel, The Man and His Prizes,* by the Nobel Foundation (third revised edition, New York 1972).

35 The other scientists on the short list that year were Gustav Embden, Johannes Fibigers, Solomon Henschens, F.G. Hopkins, Arthur Loos, Thomas Morgan, and Fernand Widal. Fibigers, Hopkins, and Morgan were later awarded Nobel prizes.

36 Banting 1940, p. 72a; personal communication.

37 BP, 1, Banting to Best, Oct. 26, 1923.

38 *Star,* Nov. 2, 1923.

39 *Star,* Nov. 7, 1923.

40 *Star,* Nov. 8, 1923.

41 MP, folder 342, Macleod to B.P. Watson, Jan. 3, 1924.

42 IC, Krogh to Macleod, Oct. 25, 1923 (postscript Oct. 26). Macleod also received an interesting letter from a Western Reserve colleague, T. Wingate Todd, who had earlier written to British journals claiming that the greatest credit should go to Macleod. Todd was writing without certain knowledge of the award, having heard a rumour that it might go to Banting alone, or to the two of them:

> To give Banting the prize at all shows a deplorable lack of scientific judgment on the part of the Committee, but worse than that it shows also a complete lack of scientific common sense. It encourages the all-too-common lay belief in inspiration and indeed sets the stamp of approval of what the layman considers the high court of appeal. Apart altogether from the indignity which it offers to yourself it is a very severe and calamitous setback for medical science.... There is no doubt that time will put everything straight but it is heart-breaking to see how even friends and those whom we have hither-to regarded as not lacking in reason, have fallen under the spell of a malicious press-campaign.

MP, folder 342, Todd to Macleod, Oct. 29, 1923.

43 BP, 1, Geyelin to Banting, Nov. 7, 1923; reply Nov. 10.

44 Nobel archive, Miscellaneous correspondence, Zuelzer to G. Liljestrand, Dec. 22, 1923, 18 Aug. 1924; Paulesco to the president of the Nobel Committee, Nov. 5, 24, 1923.

45 *Star,* Nov. 27, 1923; also *Telegram, Globe,* and other clippings.

Chapter Ten: A Continuing Epilogue (Pp. 233-48)

1 Anecdote related by Dr. Keith MacDonald, Oct. 23, 1981.

2 Joan Didion, *The White Album* (New York, 1979), p. 53.

3 *Nobel,* pp. 224-5.

4 Robert Noble interview with Dr. Harold Ettinger, c. 1967.

5 Collip 1941; interview with Dr. Hugh Lawford, Oct. 16, 1981.
6 Interview with Dr. Rolf Luft, June 1981; Lusk 1928, p. 651.
7 BP, 1, Scott to Banting, Nov. 23, 1923. The fact that a copy of this letter is also in the E.L. Scott Papers, and was seen by his widow, makes A.H. Scott 1972 a nearly dishonest account. Scott also wrote H.H. Dale, at the time of Dale's attack on Roberts, Jan. 20, 1923, saying "I am sure that it is a matter of indifference to me whether or not Banting and Best are using my method or some modification of it and so long as they and most if not all others who have tried their preparation get the results which they do it can only give me a feeling of elation that perhaps I have played some part in the net result.

"I have never felt that I have not received all the credit that was coming to me in any of the publications from the Toronto Laboratory."
8 Murlin and Kramer, 1956.
9 IC, Myrtillin file; Allen 1927, 1928.
10 Henderson 1970; Allen papers, manuscript autobiography.
11 IC, unsorted, account of royalties compiled by Albert Fisher.
12 Falconer Papers, case 81, memorandum, Jan. 23, 1923.
13 Modest grants from the Connaught Fund supported the research necessary for this book.
14 BP, 47, pp. 94-9, clippings re Banting Research Foundation.
15 Lusk 1928, p. 655.
16 BP, 1, B.A. Bradley to Banting, Jan. 12, 1926.
17 Leyton 1923.

Sources

I. Manuscript Collections

Aberdeen University, Physiology Department Records

Frederick M. Allen Papers, in the collection of Alfred R. Henderson, M.D., Washington, D.C.

F.G. Banting Papers
1. Fisher Library, University of Toronto (61 cases)
2. Notebook, Academy of Medicine Toronto
3. National Research Council, Ottawa (4 cases)

C.H. Best and M.M. Best Papers
1. Privately held; to be transferred to Fisher Library
2. Best Institute, University of Toronto; transferred to Fisher Library

Canadian Diabetes Association Archives, Toronto

H.J. Cody Papers, University of Toronto Archives

J.B. Collip Papers
1. University of Western Ontario, Faculty of Medicine Library
2. University of Western Ontario, Library, Special Collections Division
3. Privately held

Connaught Laboratories Archives, Toronto

Sir Henry Dale Papers, Royal Society, London

E.C. Drury Papers, Public Archives of Ontario, Toronto

Eli Lilly and Company Archives, Indianapolis

Sir Robert Falconer Papers, University of Toronto Archives

W.R. Feasby Papers, Canadian Diabetes Association Archives

G. Howard Ferguson Papers, Public Archives of Ontario

Elizabeth Hughes Gossett Correspondence, Fisher Library, University of Toronto

Hannah Institute, Toronto, Oral History Collection

Havens Family Papers, privately held

Insulin Committee (University of Toronto) Records, University of Toronto Archives

W.L. Mackenzie King Papers, Public Archives of Canada, Ottawa

McGill University Archives, Montreal

J.J.R. Macleod Papers, Best Institute, University of Toronto; transferred to Fisher Library

Jonathan Meakins Papers, privately held

Medical Research Council Records, London, England

Nobel Archive, Karolinska Institute, Stockholm

E.C. Noble Papers, Fisher Library, University of Toronto

E.L. Scott Papers, U.S. National Library of Medicine, Bethesda, Md.

H.M. Tory Papers, University of Alberta, Edmonton

University of Toronto Archives, Oral History Collection

Sir Edmund Walker Papers, Fisher Library, University of Toronto

Wellcome Institute, London, England, Documents Relating to the Discovery of Insulin

II. Interviews

Dr. Ian Anderson
Edward Banting
William Banting
Dr. Murray Barr
Dr. Edward Bensley
Margaret Mahon Best
Dr. Henry Best
Dr. Walter Campbell
(by Robert L. Noble)
Dr. Kenneth Carroll
Spencer Clark
Dr. Robert Cleghorn
Stella Clutton
Dr. T.A. Crowthers
Dr. Harry Ebbs
Dr. Harold Ettinger
Dr. Harold Ettinger
(by Robert L. Noble)
Dr. Albert Fisher
Dr. W.R. Franks
Sadie Gairns
Dr. Robert Garry
Dr. Martin Goldner
Elizabeth Hughes Gossett
Maynard Grange
Mr. & Mrs. T.S.H. Graham
Dr. Reginald E. Haist
Sir Harold Himsworth
Mr. & Mrs. Henry Janes
Dr. Robert Kerr
Dr. Hans Kosterlitz
Dr. Hugh Lawford
Fannie Lawrence
Dr. Rolf Luft
Dr. Ian MacDonald
Dean C.J. Mackenzie
Dr. Neil MacKinnon
Linda Mahon
Dr. George Manning
Clara Mills
Dr. Peter Moloney
Dr. A.H. Neufeld
Dr. E. Clark Noble
(by Robert L. Noble)
Dr. Robert L. Noble
Jean Orr
Professor J.M. Peterson
Dr. Burns Plewes
Professor & Mrs. F.E.L. Priestley
Dr. I.M. Rabinowitch
Dr. Jessie Ridout
Dr. John W. Scott (Toronto)
Dr. John W. Scott (Edmonton)
Wallace Seccombe
Dr. Edward Sellers
Dr. O.M. Solandt
Dr. Randall Sprague
Dr. C.B. Stewart
Professor Keith Taylor
Lady Todd
Dr. Alan Walters
Dr. Harold Warwick
Geoffrey Wasteneys
Drs. Barbara and Jackson Wyatt
Professor Sir Frank Young

III. Articles, Books, Unpublished Accounts

Allan, Frank N. (1955). "J.J.R. Macleod." *Diabetes,* 4, 6 (Nov.-Dec. 1955):491-92.

______ (1972). "Diabetes before and after Insulin." *Medical History,* XVI, 3 (July 1972).

______ (1977). Letter to the Editor, *New England Journal of Medicine* (Aug. 4, 1977): 283-84.

Allen, Frederick M. (1913). *Studies Concerning Glycosuria and Diabetes.* Cambridge, Mass., 1913.

______, Stillman, E., and Fitz, R. (1919). *Total Dietary Regulation in the Treatment of Diabetes.* New York, 1919.

______, and Sherrill, James W. (1922A). "Clinical Observations on Treatment and Progress in Diabetes." *Journal of Metabolic Research,* II (1922): 377-455.

______, and Sherrill, James W. (1922B). "Clinical Observations with Insulin. 1. The Use of Insulin in Diabetic Treatment." *Journal of Metabolic Research,* II. (November 1922): 804-985.

——— (1923). "Clinical Observations With Insulin. 3. The Influence of fat and total calories on diabetes and the Insulin requirement." *Journal of Metabolic Research,* 3 (1923): 61-176.

———(1927). "Blueberry Leaf Extract." *Journal of American Medical Association,* 89 (Nov. 5, 1927): 1577-80.

——— (1928). "Present Results and Outlook of Diabetic Treatment." *Annals of Internal Medicine,* II,2 (Aug. 1928): 203-15.

——— (1949). Remarks, *Proceedings of the American Diabetes Association,* 9 (1949): 33-36.

Banting, Frederick G. (1922). Manuscript account of the discovery of insulin, September 1922. Banting Papers, University of Toronto.

——— (1923A). "Insulin." *Journal of the Michigan State Medical Society,* March 1923.

———(1923B). "The Value of Insulin in the Treatment of Diabetes." *Proceedings of the Institute of Medicine of Chicago,* 4 (1923): 144ff.

——— (1925), *Diabetes and Insulin.* Nobel Lecture, delivered at Stockholm, Sept. 15, 1925 (Stockholm, 1925).

——— (1929). "The History of Insulin." *Edinburgh Medical Journal* (Jan. 1929): 1-18.

——— (1937). "Early Work on Insulin," *Science,* 85, 2217 (June 25, 1937): 594-96.

———(1940). "The Story of Insulin." Unpublished manuscript, Banting Papers, University of Toronto.

———, and Best, C.H. (1922A). "The Internal Secretion of the Pancreas." *Journal of Laboratory and Clinical Medicine,* VII, 5 (Feb. 1922): 256-71.

———, and Best, C.H. (1922B). "The Internal Secretion of the Pancreas." A paper delivered to the Academy of Medicine, Feb. 7, 1922 (printed by the Academy, 1922).

———, and Best, C.H. (1922C). "Pancreatic Extracts." *Journal of Laboratory and Clinical Medicine,* VII, 8 (May 1922): 3-11.

———, and Best, C.H. (1923). "The Discovery and Preparation of Insulin," *University of Toronto Medical Journal,* I (1923): 94-98.

———, Best, C.H., Collip, J.B., Campbell, W.R., and Fletcher, A.A. (1922). "Pancreatic Extracts in the Treatment of Diabetes Mellitus. Preliminary Report." *Canadian Medical Association Journal,* 2, 141 (March 1922): 141-46.

———, Best, C.H., Collip, J.B., Campbell, W.R., Fletcher, A.A., Macleod, J.J.R., and Noble, E.C. (1922). "The Effect Produced on Diabetes by Extracts of Pancreas." *Transactions of the Association of American Physicians* (1922): 1-11.

———, Best, C.H., Collip, J.B., Hepburn, J., and Macleod, J.J.R. (1922). "The Effect Produced on the Respiratory Quotient by Injections of Insulin." *Transactions of the Royal Society of Canada,* Section V, 1922.

———, Best, C.H., Collip, J.B., and Macleod, J.J.R. (1922A). "The Preparation of Pancreatic Extracts Containing Insulin." *Transactions of the Royal Society of Canada,* Section V, 1922.

———, Best, C.H., Collip, J.B., and Macleod, J.J.R. (1922B). "The Effect of Insulin on the Excretion of Ketone Bodies by the Diabetic Dog." *Transactions of the Royal Society of Canada,* Section V, 1922.

———, Best, C.H., Collip, J.B., Macleod, J.J.R., and Noble, E.C. (1922A), "The Effect of Insulin on Normal Rabbits and on Rabbits rendered Hyperglycaemic in Various Ways." *Transactions of the Royal Society of Canada,* Section V, 1922.

———, Best, C.H., Collip, J.B., Macleod, J.J.R., and Noble, E.C. (1922B). "The Effect of Insulin on the Percentage Amounts of Fat and Glycogen in the Liver and Other Organs of Diabetic Animals." *Transactions of the Royal Society of Canada,* Section V, 1922.

———, Best, C.H., Collip, J.B., Macleod, J.J.R., and Noble, E.C. (1922C). "The Effect of Pancreatic Extract (Insulin) on Normal Rabbits." *American Journal of Physiology*, 62, 1 (Sept. 1922): 162-76.
———, Best, C.H., Collip, J.B., Macleod, J.J.R., and Noble, E.C. (1922D). "The Effects of Insulin on Experimental Hyperglycaemia in Rabbits." *American Journal of Physiology*, 62, 3 (Nov. 1922): 559-80.
———, Best, C.H., Doffin, G.M., and Gilchrist, J.A., (1923). "Quantitative parallelism of effect of insulin in man, dog and rabbit." *American Journal of Physiology*, 63, 3 (Feb. 1923): 391.
———, Best, C.H., and Macleod, J.J.R. (1922). "The Internal Secretion of the Pancreas." *American Journal of Physiology*, 59 (Feb. 1922): 479 (Proceedings of the American Physiological Society, 34th Annual Meeting).
———, Campbell, W.R., and Fletcher, A.A., (1922). "Insulin in the Treatment of Diabetes Mellitus." *Journal of Metabolic Research* (Nov. 1922): 547-604.
———, Campbell, W.R., and Fletcher, A.A. (1923). "Further Clinical Experience with Insulin (Pancreatic Extracts) in the Treatment of Diabetes Mellitus." *British Medical Journal* (Jan. 6, 1923): 8-12.
———, and Gairns, S. (1924). "Factors Influencing the Production of Insulin." *American Journal of Physiology*, 68, 1 (March 1924): 24-30.
Barach, Joseph H. (1928). "Historical Facts in Diabetes." *Annals of Medical History*, 10 (1928): 387-401.
Barker, Lewellys F. (1942). *Time and the Physician*. New York, 1942.
Barr, Murray L. (1977). *A Century of Medicine at Western*. London, Ont., 1977.
——— (1978), "James Bertram Collip (1892-1965): A Canadian Pioneer in Endocrinology." *Publications of the Hannah Institute for the History of Medicine*, I, 1, 1978, 6-15.
———, and Rossiter, R.J. (1973), "James Bertram Collip 1892-1965." *Biographical Memoirs of Fellows of the Royal Society*, 19 (Dec. 1973): 235-57.
Barron, Moses (1920). "The Relation of the Islets of Langerhans to Diabetes With Special Reference to Cases of Pancreatic Lithiasis." *Surgery, Gynecology, and Obstetrics*, xxxi, 5 (Nov. 1920): 437-48.
———, (1966). "The Discovery of Insulin." *Minnesota Medicine* (April, May, 1966): 689-90, 861-62.
Bart, Constantin, (1976). "Paulesco Redivivus." *Arch. Int. Claude Bernard*, 9 (1976): 31-55.
Battle, Constance U. (1967). "The Beginning of the Insulin Era in Historical Context." *Journal of the American Medical Woman's Association* (May 1967): 327-32.
Bayliss, W.M. (1923). "Insulin, Diabetes and Rewards for Discoveries." *Nature*, III, 2780 (Feb. 10, 1923): 188-91.
Best, Charles H. (1922). "A Report of the Discovery and the Development of the Knowledge of the Properties of Insulin." September 1922, copies in Feasby Papers, Best Papers, and Wellcome Collection.
——— (1924). "Recent Work on Insulin." *Endocrinology*, VIII, 5 (Sept. 1924): 617-29.
——— (1942). "Reminiscences of the Researches Which Led to the Discovery of Insulin." *Canadian Medical Association Journal* (Nov. 1942): 398-400.
——— (1947). "The Discovery of Insulin." *Physician's Bulletin*, XII, 5 (Sept. 1947): 16-18.
———. "The History, the Discovery and the Present Position of Insulin." *Commentarii*, Pontificia Academia Scientiarum. II, n. 57: 1-48.

——— (1956). "The First Clinical Use of Insulin." *Diabetes,* 5, 1 (Jan.-Feb. 1956): 65-67.
——— (1959). "A Canadian Trail of Medical Research." *Journal of Endocrinology,* XIX (1959): 1-17; reprinted in *Selected Papers.*
——— (1962A). "The History of Insulin." *Diabetes,* II, 6 (Nov.-Dec. 1962): 495-503.
——— (1962B). "The Internal Secretion of the Pancreas." *Canadian Medical Association Journal,* 87 (Nov. 17, 1962): 1046-51.
——— (1963). *Selected Papers of Charles H. Best.* Toronto, 1963.
——— (1971). "Fiftieth Anniversary of Insulin." *Modern Medicine of Canada,* 26, 12 (December 1971): 7-10.
——— (1972). "Recollections of 1921," in E. Shafrin, ed., *Impact of Insulin on Metabolic Pathways.* New York, 1972: 7-11.
——— (1974). "A Short Essay on the Importance of Dogs in Medical Research." *The Physiologist,* 17, 4 (Nov. 1974): 437-40.
———, and Macleod, J.J.R., (1923). "Some Chemical Reactions of Insulin." *Journal of Biological Chemistry,* LV, 2 (Feb. 1923).
———, and Scott, D.A. (1923A). "Possible Sources of Insulin." *Journal of Metabolic Research,* 3, 1 (Jan. 1923): 177-79.
———, and Scott, D.A. (1923B). "An Insulin-like Substance in Plants." *Transactions of the Royal Society of Canada,* Section V (1923): 87-88.
———, and Scott, D.A. (1923C). "Insulin in Tissues Other than the Pancreas." *Journal of the American Medical Association,* 81 (Aug. 4, 1923): 382-83.
———, and Scott, D.A. (1923D). "The Preparation of Insulin." *Journal of Biological Chemistry,* LVII, 3 (Oct. 1923): 709-23.
———, and Scott, D.A. (1924). "The Sugar of the Blood." *American Journal of Physiology,* LXVIII, 1 (March 1924): 124.
———, Scott, D.A., and Banting, F.G. (1923). "Insulin in Blood," *Transactions of the Royal Society of Canada,* Section V (1923): 81-88.
———, Smith, R.G., and Scott, D.A. (1924). "An Insulin-like Material in Various Tissues of the Normal and Diabetic Animal." *American Journal of Physiology,* LXVIII, 2 (April 1924): 142-65.
Best, Margaret Mahon. "The Discovery of Insulin," undated, unpublished manuscript, Best Papers.
Biedl, Artur (1913). *The Internal Secretory Organs: Their Physiology and Pathology.* London, 1913.
Bigwood, E.-J. (1923). "Le diabète et son traitement." *Bruxelles Médical* (March 22, 1923): 530-36.
Bliss, Michael (1978). *A Canadian Millionaire: The Life and Business Times of Sir Joseph Flavelle, Bart., 1958-1939.* Toronto, 1978.
Bliss, Sidney W. (1922). "Effects of Insulin on Diabetic Dogs." *Journal of Metabolic Research,* II (1922): 385-400.
British Medical Association. "Discussion on Diabetes and Insulin," Annual Meeting, 1923. *British Medical Journal,* 3272 (Sept. 13, 1923): 445-51.
Bruce, Herbert A. (1958). *Varied Operations.* Toronto, 1958.
Burgess, Norman, Campbell, J.M.H., Osman, A.A., Payne, W.W., and Poulton, E.P. (1923). "Early Experiences With Insulin in the Treatment of Diabetes Mellitus." *Lancet* (Oct. 6, 1923): 777ff.
Burrow, G.N., Hazlett, Barbara E., and Phillips, M. James. "A Case of Diabetes Mellitus." *New England Journal of Medicine,* 306 (Feb. 11, 1982): 340-43.
Burtness, H.I. and Cain, E.F. "A Thirty-Fifth Anniversary of Insulin Therapy." *Diabetes,* 7 (Jan.-Feb. 1958): 59-61.

Cammidge, P.J. (1913). *Glycosuria and Allied Conditions.* London, 1913.
______(1922A). "Insulin and Diabetes." *British Medical Journal* (Nov. 18, 1922): 997-98.
______(1922B). "Insulin and Diabetes." *British Medical Journal* (May 5, 1923): 787.
Campbell, Walter R. (1922). "Ketosis, Acidosis and Coma Treated by Insulin." *Journal of Metabolic Research,* II (Nov.-Dec. 1922): 605-635.
______(1946). "The First Clinical Trials of Insulin." *Proceedings of the American Diabetes Association,* 6 (1946): 97-106.
______ (1962). "Anabasis." *Canadian Medical Association Journal,* 87 (1962): 1055-61.
Camplin, John M. (1860). *On Diabetes and Its Successful Treatment.* London, 1860.
Carrasco-Formiguera, R. (1922). "Insulin and Diabetes." *British Medical Journal* (Dec. 9, 1922): 1143-44.
______(1972). "From the Preinsulin Age to the Banting and Best Era. Reminiscences of a Witness and Participant." *Israel Journal of Medical Science,* 8, 3 (March 1972): 484-87.
Cathcart, E.P. (1935). "John James Rickard Macleod, 1876-1935." *Obituary Notices of Fellows of the Royal Society,* I. London, 1932-5: 585-89.
Cheymol, Jean (1972). "Il ya a cinquante ans Banting et Best 'découvraient l'insuline'." *Hist. Sci. med.* 6 (1972): 133-51.
Clough, H.D., and Murlin, J.R. (1923). "Relative amounts of insulin obtained by extraction and by perfusion of the pancreas." *Proceedings of the Society for Experimental Biology and Medicine,* 20 (1922-23): 417-18.
______, Stokes, Arthur M., Gibbs, C.B.F., Stone, Neil C., and Murlin, John R. (1922). "Influence of pancreatic perfusates upon the carbohydrate metabolism of depancreatized animals." *Proceedings of the Society for Experimental Biology and Medicine,* 20 (1922-23): 66-67.
______, Stokes, Arthur M., Gibbs, C.B.F., Stone, Neil C., and Murlin, John R. (1923), "The Influence of Pancreatic Perfusates on the Blood Sugar, D:N Ratio, and Respiratory Quotient of Depancreatized Animals." *Journal of Biological Chemistry,* LV, 23: xxx-xxxi.
Clowes, George H.A. (1947). "In Retrospect." *Physician's Bulletin,* XII, 5 (Sept. 1947): 45-48.
______, (1948), "The Banting Memorial Address", *Proceedings of the American Diabetes Association,* 7 (1948): 49-60.
Clowes, G.H.A. Jr. "George Henry Alexander Clowes (1877-1958) A Man of Science for All Seasons," *Journal of Surgical Oncology* 18 (1981): 197-217.
Collip, J. Bertram (1916). "Internal Secretions." *Canadian Medical Association Journal* (Dec. 1916): 1-7.
______, (1920). "Antagonism of Depressor Action of Small Doses of Adrenalin by Tissue Extracts." *American Journal of Physiology,* 53 (1920): 477-82.
______ (1921). "Reversal of Depressor Action of Small Doses of Adrenalin." *American Journal of Physiology,* 55, 3 (1921): 450-54.
______ (1923), "Delayed manifestation of the physiological effects of insulin following the administration of certain pancreatic extracts", *American Journal of Physiology,* 63, 3 (Feb. 1923): 391-2.
______ (1923A). "The Occurrence of Ketone Bodies in the Urine of Normal Rabbits in a Condition of Hypoglycaemia Following the Administration of Insulin...." *Journal of Biological Chemistry,* LV, 2 (Feb. 1923).
______(1923B). "The Demonstration of an Insulin-Like Substance in the Tissues of the Clam (Myce Arenaria)." *Journal of Biological Chemistry,* LV, 2 (Feb. 1923).

——— (1923C). "The demonstration of a hormone in plant tissue to be known as 'glucokinin'." *Proceedings of the Society for Experimental Biology and Medicine,* 20 (1923): 321-23.
——— (1923D). "Glucokinin; a new hormone present in plant tissue." *Journal of Biological Chemistry,* 56 (1923): 513-43.
——— (1923E). "Effects of Plant extracts on blood sugar." *Nature,* III (1923): 571.
——— (1923F). "Glucokinin. Second Paper." *Journal of Biological Chemistry,* 57 (1923): 65-78.
——— (1923G). "Glucokinin." *Proceedings and Transactions of the Royal Society of Canada,* Section V (1923): 39-43.
——— (1923H). "Glucokinin; report of work in progress." *Transactions Sect. Pharmac Ther. American Medical Association* (1923): 41-42.
——— (1923I). "A Demonstration of a hypoglycaemic principle in the crystalline style of the clam." *Proceedings and Transactions of the Royal Society of Canada,* Section 5 (1923): 45.
——— (1923J). "Glucokinin; an apparent synthesis in the normal animal of the hypoglycaemic producing principle; animal passage of the principle." *Journal of Biological Chemistry,* 58 (1923): 163-208.
——— (1923K). "History of the discovery of insulin." *Northwest Medicine,* 22: 267-73.
——— (1923L). "The Original method as used for the isolation of insulin in semipure form for the treatment of the first clinical case," *Journal of Biological Chemistry,* 55: xl-xli (Proceedings of the American Society for Biological Chemistry).
——— (1924). "Some recent advances in endocrinology." *Canadian Medical Association Journal,* 14: 812-20.
——— (1935). "John James Rickard Macleod." *Biochemical Journal,* 29 (1935): 1253-56.
——— (1941). "Frederick Grant Banting, Discoverer of Insulin." *Scientific Monthly,* 52 (1941): 473-74.
——— (1942). "Recollections of Sir Frederick Banting." *Canadian Medical Association Journal,* 47: 401-403.
Colwell, Arthur R. (1968). "Fifty Years of Diabetes in Perspective." *Diabetes,* 17 (10 Oct. 1968): 599-610.
Connaught Laboratories (1923). *Insulin (Insulin-Toronto),* pamphlet. Toronto, 1923.
Corner, George W. (1964). *A History of the Rockefeller Institute.* New York, 1964.
Crofton, W.M. (1922). "Pancreatic Extract in Diabetes." *Lancet* (Dec. 2, 1922): 1195.
Dale, Sir Henry H. (1922). "Insulin," letter, *British Medical Journal* (Dec. 23, 1922): 1241.
——— (1926). *Lectures on Certain Aspects of Biochemistry.* London, 1926.
——— (1941). "Sir Frederick Banting." *British Medical Journal* (March 8, 1941): 383-84.
——— (1946). "The Silver Jubilee of Insulin." *The Diabetic Journal* (Sept. 1946): 392-99; reprinted in Dale, *An Autumn Gleaning.* London, 1954.
——— (1949). "Thomas Addison: Pioneer of Endocrinology." *British Medical Journal* (Aug. 13, 1949): 347-52.
——— (1954). "Address at the dedication of the Charles H. Best Institute." *Diabetes,* 3, 1 (Jan.-Feb. 1954): 30-35.
——— (1959). "Note by H.H. Dale, on the additional documents submitted by C.H. Best...for deposition in the Wellcome Historical Medical Library, and bearing, like others earlier received, on the 'Insulin Controversy'." Unpublished

manuscript, Wellcome Library, October 1959.
De Kruif, Paul H. (1932). "Banting: Who Found Insulin," in *Men Against Death.* New York, 1932: 59-87.
De Meyer, J. (1909A). "Contribution A L'Étude De La Pathogénie Du Diabète Pancréatique." *Archive Internationale de Physiologie* (1909): 121-180.
——— (1909B), "Action De La Sécrétion Interne Du Pancréas Sur Différents Organes Et En Particulier Sur La Sécrétion Rénale," *Archivio di Physiologia,* 7 (1909): 96-99.
Decourt, Philippe (1976). "La Véritable Histoire de la Découverte de l'Insuline. Introduction." *Arch. Int. Claude Bernard,* 9 (1976).
Defries, Robert D. (1968). *The First Forty Years. 1914-55. Connaught Medical Research Laboratories.* Toronto, 1968.
Dewitt, Lydia M. (1906). "Morphology and Physiology of Areas of Langerhans in Some Vertebrates." *Journal of Experimental Medicine,* 8 (1906): 193-239.
Doisy, E.A., Somogyi, Michael, and Shaffer, P.A. (1923). "Some Properties of An Active Constituent of Pancreas (Insulin)." *Journal of Biological Chemistry,* LV (1923): xxxi-xxxii.
Duncan, Garfield (1964). "Frederick Madison Allen 1879-1964." *Diabetes,* 13, 3 (May-June 1964): 318-19.
Feasby, W.R. (1958). "The Discovery of Insulin." *Journal of the History of Medicine* (1958): 68-84.
——— (1971). Excerpts from an unfinished biography of C.H. Best, *Medical Post* (Toronto), Sept. 7, 21, Oct. 5, 1971.
Feldberg, Wilhelm (1970). "Henry Hallett Dale." *Biographical Memoirs of Fellows of the Royal Society,* 16 (1970): 77-174.
Fitz, Reginald, Murphy, William B., and Grant, Samuel B. (1922). "The Effect of Insulin on the Metabolism of Diabetes," *Journal of Metabolic Research* (Nov. 1922): 753-66.
Fletcher, A.A. (1962). "Early Clinical Experiences With Insulin." *Canadian Medical Association Journal,* 87: 1052-55.
———, and Campbell, W.R. (1922). "The Blood Sugar Following Insulin Administration and the Symptom Complex - Hypoglycaemia." *Journal of Metabolic Research* (Nov. 1922): 637-49.
Flexner, Abraham (1910). *Medical Education in the United States and Canada.* New York, 1910.
Forrest, W.D., Smith, W., and Winter, L.B. (1923). "On the Change in the Nature of the Blood Sugar of Diabetics Caused by Insulin." *Journal of Physiology,* 57 (1923): 223-33.
Forschbach, J. (1909). "Versuche Zur Behandlung Des Diabetes Mellitus Mit Dem Zuelzerschen Pankreashormon." *Deutsche Medizinische Wochenschrift,* 35 (1909): 2035-55.
Foster, Nellis B. (1915). *Diabetes Mellitus.* Philadelphia and London, 1915.
Fulton, John F. (1956). "Sir Frederick Banting, 1891-1941." *Diabetes,* 5, 1 (Jan.-Feb. 1956): 64-65.
Funk, Casimir, and Harrow, Benjamin (1923). "Insulin: A Cure for Diabetes." *Strength* (Sept. 1923).
Gaebler, O.H. (1965). Letter to the Editor, *Bulletin of the Canadian Biochemical Society,* II, 3 (1965): 1.
Gastineau, C.F. (1971). "The Prelude to Insulin - The Years Before Banting and Best." Unpublished address to the Mayo Foundation History of Medicine Society, 1971.

Geyelin, H. Rawle, Harrop, George, Murray, Marjorie F., and Corwin, Eugenia, (1922). "The Use of Insulin in Juvenile Diabetes." *Journal of Metabolic Research* (Nov. 1922): 767-92.

Gibbs, C.B.F., Clough, Harry D., Stone, Neil C., and Murlin, John R., (1922). "The influence of pancreatic extracts upon the carbohydrate metabolism of depancreatized dogs." *Proceedings of the Society for Experimental Biology and Medicine,* 20 (1922-23): 67-68.

———, and Murlin, John R. (1922). "Influence on the respiratory metabolism of pancreatic extract administered by mouth to depancreatized dogs." *Proceedings of the Society for Experimental Biology and Medicine,* 20 (1922-23): 198.

———, and Sutter, C.C. (1923). "Clinical observations on the use of the anti-diabetic substance." *Proceedings of the Society for Experimental Biology and Medicine,* 20 (1922-23): 419.

Gilchrist, Joseph A., Best, C.H., and Banting, F.G. (1923). "Observations with Insulin on Department of Soldiers' Civil Re-Establishment Diabetics." *Canadian Medical Association Journal* (Aug. 1923).

Gley, Eugène (1922). "Action des Extraits de Pancréas Sclérosé Sur des Chiens Diabétiques." *Société de Biologie, Comptes Rendus,* 87 (Dec. 1922).

Goldner, Martin G. (1957). "Historical Review of Oral Substitutes for Insulin." *Diabetes* (May-June 1957): 259-61.

——— (1972A). General Discussion, "Insulin in Retrospect." *Israel Journal of Medical Science,* 8, 3 (March 1972): 492-93.

——— (1972B). "Reflections on the Early History of Insulin", unpublished address at the spring meeting of the Clinical Society of NYDA, May 18, 1972.

——— (1972C). Letter, *Annals of Internal Medicine,* 76, 2 (Feb. 1972): 329.

Graham, Duncan (1923). "Diabetes Mellitus and Insulin." n.p., nd.

Greenaway, Roy (1966). *The News Game.* Toronto, 1966.

Groen, J.J. (1972). "Discovery of Insulin Told as a Human Story." *Israel Journal of Medical Science,* 8, 3 (March 1972): 476-83.

Hall, G. Edward (1965). "James Bertram Collip - An Appreciation." *Canadian Medical Association Journal,* 93 (Aug. 28, 1965): 673.

Hall, W.E.B. (1941). "Sir Frederick Grant Banting, M.D., 1891-1941." *Arch. of Path.,* 31: 657-62.

Hare, Ronald (1970). *The Birth of Penicillin.* London, 1970.

Harris, Seale (1946A). *Banting's Miracle.* Philadelphia, London, Montreal, 1946.

——— (1946B). "An Appreciation of Banting & Best." *Proceedings of the American Diabetes Association,* 6 (1946): 118-34.

Hédon, E. (1909). *Société de Biologie, Comptes Rendus,* 66 (1909): 621-24.

Henderson, Alfred R. (1970). "Frederick M. Allen, M.D., and the Physiatric Institute at Morristown, N.J." *Academy of Medicine of New Jersey Bulletin,* 16, 4 (Dec. 1970): 40-49.

Henderson, J.R. (1971). "Who Discovered Insulin?" *Guy's Hospital Gazette* (June 19, 1971): 315ff.

Hill, L.W., and Eckman, R.S. (1921). *The Allen (Starvation) Treatment of Diabetes.* Boston, 1921.

Hipwell, Fred W. (1943). "Memories" (The Banting Memorial Address). *Proceedings of the American Diabetes Association,* 3 (1943): 111-27.

Hipwell, Lillian (1970). "Frederick Grant Banting as I remember Him." Unpublished memoir, 1970.

Hodgson, J.A. (1911). "Treatment of Diabetes Mellitus." *Journal of the American Medical Association,* LVII, 15 (Oct. 7, 1911): 1187-92.

Holt, Anna C. (1969). *Elliott Proctor Joslin. A Memoir, 1869-1962.* Joslin Foundation, 1969.
Horowitz, P.H. (1920). *Diabetes: A Handbook for Physicians and Their Patients.* New York, 1920.
Hoskins, R.G. (1921). "Some Current Trends in Endocrinology." *Journal of the American Medical Association,* 77, 19 (Nov. 5, 1921): 1459-62.
Houssay, B.A. (1952). "The Discovery of Pancreatic Diabetes: The Role of Oscar Minkowski." *Diabetes,* I, 2 (March-April 1952): 112-16.
Hudson, Robert P. (1979). "New Light on the Insulin Controversy." *Annals of Internal Medicine,* 91, 2 (Aug. 1979): 311.
Hunter, Andrew (1941). "Sir Frederick Grant Banting." *Transactions of the Royal Society of Canada* (1941): 87-93.
Inge, William R. (1924). *Personal Religion and the Life of Devotion.* London, 1924).
______ (1949). *Diary of a Dean.* London, 1949.
"Insulin: Fifty Years Ago." *Annals of Internal Medicine.* 75, 5 (Nov. 1971): 797-800.
Insulin Committee of the University of Toronto (1923). "Insulin: Its Action; Its Therapeutic Value in Diabetes, and Its Manufacture." *Journal of the American Medical Association,* 80, 25 (June 23, 1923): 1847-51; *Canadian Medical Association Journal* (July 1923).
"Insulin, the New Treatment for Diabetes." *The Prescriber* (Edinburgh) (June 1923): 240-47.
International Diabetes Federation (1971). "Report of the Special Committee set up to present a written summary of work leading up to the discovery of insulin." *News Bulletin of the International Diabetes Federation,* 16, 2 (1971): 29-40.
Jack, Donald (1981). *Rogues, Rebels, and Geniuses. The Story of Canadian Medicine.* Toronto, 1981.
Jackson, William P.U. (1970). "Diabetes Mellitus in Different Countries and Different Races. Prevalence and Major Features." *Acta Diabet. Lat.,* 7, 361, (1970).
Janes, Joseph M. (1967). "Sir Frederick Grant Banting, Orthopedic Surgeon." *Mayo Clinic Proceedings,* 42 (Aug. 1967): 501-10.
Joslin, Elliott P. (1917). *The Treatment of Diabetes Mellitus.* 2nd ed., Philadelphia and New York, 1917.
______ (1919). *A Diabetic Manual.* 2nd ed., Philadelphia and New York, 1919.
______ (1922). "Pancreatic Extract in the Treatment of Diabetes." Letter, *Boston Medical and Surgical Journal* (May 11, 1922).
______ (1923). "The Routine Treatment of Diabetes With Insulin." *Journal of the American Medical Association,* 80, 22 (June 2, 1923): 1581-83.
______ (1946). "Diabetes: Past, Present and Future." *Proceedings of the American Diabetes Association,* 6 (1946): 161-69.
______ (1956). "A Personal Impression." *Diabetes,* 5, 1 (Jan.-Feb. 1946): 67-68.
______, Gray, Horace, and Root, Howard F. (1922). "Insulin in Hospital and Home." *Journal of Metabolic Research* (Nov. 1922): 651-99.
Kahn, E.J., Jr. (1976). *All In a Century: The First 100 Years of Eli Lilly and Company.* Pub. by the company, 1976.
Keys, David A. (1965). "James Bertram Collip - An Appreciation." *Canadian Medical Association Journal* (Oct. 2, 1965): 774-75.
Kienast, Margate (1938). "I Saw a Resurrection." *Saturday Evening Post* (July 2, 1938).
King, E.J. (1941). "The Late Sir Frederick Banting." *Lancet* (April 26, 1941): 551.

Kleeberg, J. (1972). "A Personal Experience in the Early Days of Insulin." *Koroth,* 5 (1972): cxxx-cxxxv.
Kleiner, Israel S. (1916). "The Disappearance of Dextrose from the Blood After Intravenous Injection." *Journal of Experimental Medicine,* 23 (1919): 507-33.
______ (1919). "The Action of Intravenous Injections of Pancreas Emulsions in Experimental Diabetes." *Journal of Biological Chemistry,* 40 (1919): 153-70.
______ (1959). "Hypoglycaemic Agents - Past and Present." *Clinical Chemistry,* 5, 2 (April 1959): 79-99.
______, and Meltzer, S.J. (1914). "The influence of depancreatization upon the state of glycemia following the intravenous injections of dextrose in dogs." *Proceedings of the Society for Experimental Biology and Medicine,* 12 (1914-15): 58-59.
______ (1915). "Retention in the Circulation of Dextrose in Normal and Depancreatized Animals, and the Effect of an Intravenous Injection of an Emulsion of Pancreas Upon this Retention." *Proceedings of the National Academy of Science* (1915): i, 338-41.
______ (1959). "Hypoglycaemic Agents - Past and Present." *Clinical Chemistry,* 5, 2 (April 1959): 79-99
Leibel, Bernard S. (1970). "Banting, the Man I Knew." *Canadian Family Physician* (July 1970).
______, and Wrenshall, G.A. (1971). *Insulin.* Canadian Diabetic Association, Toronto, 1971.
Leibowitz, J.O. (1972). "The Concept of Diabetes in Historical Perspective." *Israel Journal of Medical Science,* 8, 3 (March 1972): 469-75.
Levine, Rachmiel (1970). "The Influence of Insulin on Scientific Thought of the Twentieth Century." *Acta diabet. lat.* 7 (suppl. 1) (1970): 429-33.
______ (1971). "The Endocrine Pancreas, Past and Present," in D.M. Klatchko, *et al.,* eds. *The Endocrine Pancreas and Juvenile Diabetes.* New York, 1977: 1-11.
Leyton, Otto (1923). "Insulin and Diabetes Mellitus." Letter, *British Medical Journal* (May 19, 1923): 882.
Luft, Rolf (1971). "Vem upptäckte insulinet?" *Läkartidninger,* 68, 44 (1971) (portions translated and printed in Pavel, 1976: 130-37).
Lusk, Graham (1928). *The Elements of the Science of Nutrition,* 4th ed., Philadelphia, 1928.
McCormick, Gene E. (1979). *Insulin: A Hope for Life.* Pamphlet, Eli Lilly and Company (rev. 1979).
McCormick, N.A. (1924). "Insulin From Fish." *Bulletin of the Biological Board of Canada,* VII (Dec. 1924).
______, Macleod, J.J.R., Noble, E.C. and O'Brien, K. (1923). "The Influence of the Nutritional Condition of the Animal on the Hypoglycaemia Produced by Insulin." *Journal of Physiology,* 57 (1923): 234-52.
______, Macleod, J.J.R., O'Brien, K., and Noble, E.C. "Experiments on the mechanism of action of insulin." *American Journal of Physiology,* 63, 3 (Feb. 1923): 389-90.
______, and Noble, E.C. (1924). "The Yield of Insulin From Fish." *Contributions to Canadian Biology,* II, part 1 (1924): 117-27.
MacFarlane, Gwyn (1979). *Howard Florey, The Making of a Great Scientist.* Oxford, 1979.
McGuigan, Hugh (1918). "Sugar Metabolism and Diabetes." *Journal of Laboratory and Clinical Medicine,* III, 6 (March 1918): 319-37.
Maclean, Hugh (1922). *Modern Methods in the Diagnosis and Treatment of*

Glycosuria and Diabetes. London, 1922.

Mackenzie, Wallis, R.L. (1922). "The Internal Secretion of the Pancreas and its Application to the Treatment of Diabetes Mellitus." *Lancet* (Dec. 2, 1922): 1158-61.

Macleod, John James Rickard (1913). *Diabetes: Its Pathological Physiology.* London & New York, 1913.

______(1922/78). "History of the Researches Leading to the Discovery of Insulin," September 1922; published in the *Bulletin of the History of Medicine,* 52, 3 (Fall 1978): 295-312.

______(1922). "Methods of Study of Early Diabetes." *Canadian Medical Association Journal* (Jan. 1922): 4-6.

______(1922A). "Pancreatic Extract and Diabetes." *Canadian Medical Association Journal* (June 1922): 423-25.

______(1922B). "The Source of Insulin." *Journal of Metabolic Research,* II (1922): 149-72.

______(1922C). "Insulin and Diabetes: A General Statement of the Physiological and Therapeutic Effect of Insulin." *British Medical Journal* (Nov. 4, 1922): 833-35.

______(1923). "Insulin." Lecture to the XIth International Physiological Congress, July 24, 1923. *British Medical Journal* (Aug. 4, 1923): 165-72.

______(1924A). "The Nature of Control of the Metabolism of Carbohydrates in the Animal Body." (Cameron Prize Lecture, Oct. 16-17, 1923), *British Medical Journal* (Jan. 12, 1924): 45-49.

______(1924B). "Insulin." *Physiological Review,* IV, 1 (Jan. 1924).

______(1926). *Carbohydrate Metabolism and Insulin.* London, New York, Toronto, 1926.

______, and Campbell, W.R. (1925). *Insulin. Its Use in the Treatment of Diabetes.* Baltimore, 1925.

______, Pearce, R.G., Redfield, A.C., and Taylor, N.B. (1920). *Physiology and Biochemistry in Modern Medicine.* 3rd ed., St. Louis, 1920.

"John James Rickard Macleod." Obituary, *Quarterly Journal of Experimental Medicine,* 25 (1936): 105-108.

Magner, Lois N. (1977). "Ernest Lyman Scott's work with Insulin: A Reappraisal." *Pharmacy in History,* 19 (1977): 103-108.

Major, Ralph H. (1923). "The Treatment of Diabetes Mellitus With Insulin." *Journal of the American Medical Association,* 80, 22 (June 2, 1923): 1597-1600.

Mann, F.C., and Magath, T.B. (1921). "The Liver as a Regulator of the Glucose Concentration of the Blood." *American Journal of Physiology,* 55, 2 (March 1921): 285-86 (Proceedings of the American Physiological Society).

______, and Magath, T.B. (1922). "Studies on the Physiology of the Liver. II. The Effect of the Removal of the Liver on the Blood Sugar Level." *Archives of Internal Medicine,* 30, 1 (July 1922): 73-84.

Matthews, Leslie G., *et al.* (1963). "Forty Years of Insulin Therapy." *The Chemist and Druggist* (May 4, 1963): 487-503.

Medical Research Council (1923). "Some Clinical Results of the Use of Insulin." *British Medical Journal* (April 28, 1923): 737-40.

Mellinghoff, K.H. (1971). *Georg Ludwig Zuelzers Beitrag zur Insulinforschung.* Düsseldorf, 1971.

Mosenthal, Herman D. (1921). *Diabetes Mellitus: A System of Diets.* New York, 1921.

Murlin, John R. (1922). "Properties and methods of preparation of the anti-diabetic substance (glucopyron) generated by the pancreas," *Proceedings of the Society*

for Experimental Biology and Medicine, 20 (1922-23): 70.
———, Clough, Harry D., Gibbs, C.B.F., and Stokes, Arthur M. (1923). "Aqueous Extracts of Pancreas: I. Influence of the Carbohydrate Metabolism of Depancreatized Animals." *Journal of Biological Chemistry*, LVI (1923): 253-96.
———, and Kramer, B. (1913). "The Influence of pancreatic and duodenal extracts on the glycosuria and the respiratory metabolism of depancreatized dogs." *Proceedings of the Society for Experimental Biology and Medicine*, 10 (1912-13): 171-73.
———, and Kramer, B. (1916). "Pancreatic Diabetes in the Dog." *Journal of Biological Chemistry*, 27 (1916): 481-538.
———, and Kramer, Benjamin (1956). "A Quest for the Anti-Diabetic Hormone 1913-1916." *Journal of the History of Medicine* (July 1956): 288-98.
———, Kramer, B., and Sweet, J.E. (1922). "Pancreatic Diabetes in the Dog. VI. The Influence of Pancreatic Extracts Without the Aid of Alkali Upon the Metabolism of the Depancreatized Animals." *Journal of Metabolic Research*, II (1922): 19-27.
Murray, Ian (1969). "The Search for Insulin," *Scottish Medical Journal*, 14 (1969): 286-95.
——— (1971). "Paulesco and the Isolation of Insulin." *Journal of the History of Medicine*, XXVI, 2 (April 1971): 150-57.
Myers, Victor C., and Bailey, Cameron B. (1916). "The Lewis and Benedict Method for the Estimation of Blood Sugar, With Some Observations Obtained in Disease." *Journal of Biological Chemistry*, 24 (1916): 147-61.
Nelken, Ludwig (1972). "Remarks," Insulin in Retrospect Symposium, *Israel Journal of Medical Science*, 8, 3 (March 1972): 467-68.
Nobel Foundation, *Nobel, The Man and His Prizes*. 1st ed., Stockholm, 1950; 3rd ed., New York, 1972.
Noble, E.C. "Who Discovered Insulin?" *Guy's Hospital Gazette* (Sept. 11, 1971): 452-53.
——— (1971). Manuscript account of the discovery of insulin, October 1971, Noble Papers, Fisher Library.
Noble, Robert L. (1965). "Memories of James Bertram Collip." *Canadian Medical Association Journal* (Dec. 1965): 1356-64.
Von Noorden, Carl H. (1912). *New Aspects of Diabetes: Pathology and Treatment*. New York, 1912.
"On Changing Terminological Horses." *Journal of the American Medical Association*, 198, 6 (Nov. 7, 1966): 190-91.
Opie, Eugene L. (1910). *Disease of the Pancreas*. Philadelphia, 1910.
Osler, William (1915). *The Principles and Practice of Medicine*. 7th ed., New York and London, 1915.
Palmer, Archie M. (1948). *Survey of University Patent Policies*. NRC, Washington, 1948.
Papaspyros, N.S. (1964). *The History of Diabetes Mellitus*. 2nd ed., Stuttgart, 1964.
Paulesco, N.C. (1920). *Traité de Physiologie Médicale*, II. Bucharest, 1920.
——— (1921A). "Action de l'extrait pancréatique...." *Comptes rendus des séances de la Société de Biologie*, LXXXV, 27 (23 juillet, 1921): 555-59. (Also in Pavel 1976).
——— (1921B). "Recherche sur le rôle du pancréas dans l'assimilation nutritive." *Archives internationales de Physiologie*, 17 (1921): 85-103. (Also in Pavel 1976).
——— (1923A). "Quelques Réactions Chimiques et Physiques Appliquées à l'extrait aqueux du pancréas pour le débarasser des substances protéiques en excès." *Archives internationales de Physiologie*, 21 (1923): 71-85.

______ (1923B). "Divers Procédes Pour Introduire l'Extrait Pancréatique dans l'Organisme d'un Animal Diabetique." *Archives internationales de Physiologie,* 21 (1923): 215-38.

______ (1924). "Traitement du Diabète." *La Presse Medicale,* V, 19 (5 mars, 1924): 202-204.

Pavel, I. (1976). *The Priority of N.C. Paulesco in the Discovery of Insulin.* Bucharest, 1976.

______, Bonaparte, H., and Sdrobici, D. (1972). "The Role of Paulesco in the Discovery of Insulin." *Israel Journal of Medical Science,* 8, 3 (March 1972): 488-500.

Piper, H.A., Mattill, H.A., and Murlin, John R. (1923). "Further observations on the chemical and physical properties of insulin." *Proceedings of the Society for Experimental Biology and Medicine,* 20 (1922-23): 413-14.

Poulson, Jacob E. (1975). "The Impact of August Krogh on the Insulin Treatment of Diabetes and Our Present Status." *Acta Medical Scandinavica,* 578 (1975): 7-14.

Pratt, Joseph H. (1954). "A Reappraisal of Researches leading to the Discovery of Insulin." *Journal of the History of Medicine,* 9 (1954): 281-89.

______ (1954ms). Manuscript, unrevised, of "A Reappraisal of Researches...." Faculty of Medicine Papers, University of Toronto Archives.

Pratt, T. Dennie (1979). *Joseph Hershey Pratt: A Family Memoir.* Privately published, 1979.

Purdey, Charles W. (1890). *Diabetes: Its Causes, Symptoms, and Treatment.* Philadelphia and London, 1890.

Rehberg, P. Brandt (1951). "August Krogh." *Yale Journal of Biology and Medicine,* 24, 2 (Nov. 1951): xv-xxxiv.

Rennie, John, and Fraser, Thomas (1907). "The Islets of Langerhans in Relation to Diabetes." *Biochemical Journal,* II, 1 (1907): 7-19.

Roberts, Ffrangcon (1922). "Insulin." Letter, *British Medical Journal* (December 16, 1922): 1193-94.

Ross, G.W. "Summary of History of Discovery of Insulin." Undated, unpublished, manuscript, Cody Papers, University of Toronto Archives.

Rossiter, R.J. (1966). "James Bertram Collip, 1892-1965." *Proceedings and Transactions of the Royal Society of Canada,* 4 (1966): 73-82.

Salter, J.M., Sirek, O.V., Abbott, M.M., and Leibel, B.S. (1961). "A Clinical Assessment of Foetal Calf Insulin." *Diabetes,* 10, 2 (March-April 1961): 119-21.

Sansum, W.D., Blatherwick, N.R., Smith, Florence H., Long, M. Louisa, Maxwell, L.C., Hill, Elsie, McCarty, Ray, and Cryst, J.H. (1923). "The Treatment of Diabetes with Insulin." *Journal of Metabolic Research,* III (1923): 641-78.

Scheller, James C., and Galloway, John A. (1975). "The Development of the Insulin Unit." *American Journal of Pharmacy* (Jan.-Feb. 1975): 29-32.

Scott, Aleita Hopping (1972). *Great Scott. Ernest Lyman Scott's Work with Insulin in 1911.* Bogota, N.J., 1972.

Scott, Ernest Lyman (1912). "On the Influence of Intravenous Injections of an Extract of the Pancreas on Experimental Pancreatic Diabetes." *American Journal of Physiology,* 29 (1912): 306-10.

______ (1913). "The relation of pancreatic extract to the sugar of the blood." *Proceedings of the Society for Experimental Biology and Medicine,* 10 (1912-13): 101-103.

______ (1923). "Priority in Discovery of a Substance Derived from the Pancreas, Active in Carbohydrate Metabolism." *Journal of the American Medical Association,* 81 (1923): 1303-4.

Sharpey-Schafer, E.A. (1916). *The Endocrine Organs.* London, 1916.
Shoule, Horace A. (1926). "The Preparation and Chemistry of Insulin." *Journal of Chemical Education,* 3, 2 (Feb. 1926): 134-47.
Shryock, Richard H. (1947). *The Development of Modern Medicine.* New York, 1947.
Slosson, Edwin E. (1923). "The Story of Insulin." *World's Work* (Nov. 1923).
Smith, Andrew H. (1889). *Diabetes Mellitus and Insipidus.* Detroit, 1889.
Smith, L. Muir (1956). "The Advent of Insulin." *Guy's Hospital Gazette* (1956): 620-22.
Somogyi, Michael (1951). "The Story of Insulin - Past and Current." *Bulletin of the St. Louis Jewish Hospital Medical Staff,* 7 (May 1951): 1-12.
Sönksen, P.H. (1977). "The Evolution of Insulin Treatment." *Clinics in Endocrinology and Metabolism,* 6, 2 (July 1977): 481-97.
Sprague, Randall G. (1965). "Convocation in Toronto. An Address." *Diabetes* 14, 1 (Jan. 1965): 37-43.
——— (1967). "Development of Endocrinology at the Mayo Clinic, 1913-1951." Unpublished address to the Mayo Foundation History of Medicine Society, Jan. 11, 1967.
——— (1969). "The Application of Investigative Knowledge to the Clinical Management of Diabetes." The Thirteenth Woodyatt Memorial Lecture, Center for Continuing Education, University of Chicago, Nov. 21, 1969.
——— (1971). "Fifty Years of Insulin." Unpublished lecture to ADA postgraduate course, Rochester, Minnesota, Jan. 21, 1971.
Stalvey, Richard M. (1971). "A Chat with Dr. Charles Best." *Nutrition Today* (Nov/Dec. 1971).
Stewart, G.A. (1974). "Historical Review of the Analytical Control of Insulin." *Analyst,* 99 (Dec. 1974): 913-28.
Stevenson, Lloyd (1946). *Sir Frederick Banting.* Toronto, 1946; 2nd ed., 1947.
——— (1979). "J.J.R. Macleod and the Discovery of Insulin." *Trends in Biochemical Sciences* (July 1979): N158-N160.
Striker, Cecil (1961). *Famous Faces in Diabetes.* Boston, 1961.
Sutter, C. Clyde, and Murlin, John R. (1922). "Three months study of the influence of the anti-diabetic extract on a case of severe diabetes." *Proceedings of the Society for Experimental Biology and Medicine,* 20 (1922-23): 68-69.
Telfer, S.V. (1923). "The Administration of Insulin by Inunction." *British Medical Journal* (April 8, 1923): 715-16.
Thomson, A.P. (1922). "Insulin and Diabetes." Letter, *British Medical Journal* (Nov. 11, 1922): 948.
Thomson, D.L. (1957). "Dr. James Bertram Collip." *Canadian Journal of Biochemistry and Physiology.* 35 (1957).
Turpin, R. (1923). "A Propos de l'Insuline." *Le Progrès Medical* (Paris) (May 26, 1923): 255-60.
Wallace, George B. (1910). "Recent Advances in the Treatment of Diabetes Mellitus." *Journal of the American Medical Association,* LV, 25 (Dec. 17, 1910): 2107-9.
Walls, Eric (1971). "The Tarnishing of Our Noble Prize." *Weekend Magazine* (Oct. 23, 1971).
Warwick, O.H. (1965). "James Bertram Collip - 1892-1965, An Appreciation," *Canadian Medical Association Journal,* 93, 331 (Aug. 14, 1965): 425-26.
Wilder, Russell M. (1946). "Twenty-Five Years of the Insulin Era." *Proceedings of the American Diabetes Association,* 6 (1946): 109-16.
——— (1963). "Recollections and Reflections from the Mayo Clinic, 1919-1950,"

in Dwight J. Ingle, ed. *A Dozen Doctors. Autobiographic Sketches.* Chicago, 1963.
______, Boothby, Walter M., Barborka, Clifford J., Kitchen, Hubert D., Adams, Samuel F. (1922). "Clinical Observations on Insulin." *Journal of Metabolic Research,* II (Nov. 1922): 701-27.
Williams, John R. (1921). "An Evaluation of the Allen Method of Treatment of Diabetes Mellitus." *American Journal of Medical Science,* 162 (1921): 62-72.
______ (1922). "A Clinical Study of the Effects of Insulin in Severe Diabetes." *Journal of Metabolic Research,* II (Nov. 1922): 729-51.
______ (1947). "A Note on Sir Frederick Banting." *Physician's Bulletin,* XII, 5 (Sept. 1947): 12-15.
Winter, L.B., and Smith, W. (1923A). "On a Possible Mode of Causation of Diabetes Mellitus." *British Medical Journal* (Jan. 6, 1923): 12-13.
______(1923B). "On the Nature of the Sugar in Blood," *Journal of Physiology,* 57 (1923): 100-12.
______(1923C). "Use of Yeast Extracts in Diabetes." *Nature,* No. 2784 (March 10, 1923): 327.
______(1923D). "Some Problems of Diabetes Mellitus," *British Medical Journal* (April 28, 1923): 711-15.
Wood, F.C., and Bierman, E.L. (1972). "New Concepts in Diabetic Dietetics," *Nutrition Today,* 7, 13 (May/June 1972): 4-13.
Woodbury, David (1962). "How Insulin First Triumphed over Diabetes," *Liberty* (June 22, 1962).
Woodyatt, Rollin T. (1921). "Objects and Method of Diet Adjustment in Diabetes." *Archives of Internal Medicine,* 28, 2 (Aug. 1921): 125-41.
______ (1922). "The Clinical Use of Insulin." *Journal of Metabolic Research* (November 1922): 793-801.
______ (1948). "Foundations of the Concept of Acidosis." *Proceedings of The American Diabetes Association,* 8 (1948): 17-31.
Wrenshall, G.A., Hetenyi, G., and Feasby, W.R. (1962). *The Story of Insulin.* Toronto, 1962.
Young, E. Gordon. *The Development of Biochemistry in Canada.* Toronto, 1976.
Young, F.G. (1970). "The Evolution of Ideas About Animal Hormones," in J. Needham, ed. *The Chemistry of Life.* Cambridge, 1970: 125-55.
Zuelzer, G.L. (1907). "Experimentell Untersuchungen Über den Diabetes." *Berliner Klinische Wochenschrift,* 44 (1907): 475-75.
______(1908). "Ueber Versuche Einer Specifischen Fermenttherapie Des Diabetes." *Zeitschrift für experimentelle pathologie und therapie,* 5 (1908): 307-18.
______, Dohrn, M., Marxer, A. (1908). "Neuere Untersuchungen Über den experimentellen Diabetes." *Deutsche Medizinische Wochenschrift,* 34 (1908): 1380-85.
______(1923). "The Overcoming of Diabetes. New Facts about 'Insulin'." Unpublished article, October 1923, copy in Banting Papers, University of Toronto.

Index